Praise for *Connecting Alaskans*

It wasn't long ago that Alaskans watched the evening news the next day, and personal messages were delivered to people in remote communities by radio shows called "Tundra Telegraph" or "Caribou Clatter." Alaska's leadership in adopting new telecommunications technologies has helped link the whole world. Heather Hudson's history comes from both a scholar and an expert who helped make giant leaps happen in a single generation.

—Mead Treadwell, Lieutenant Governor of Alaska, 2010-2014

Connecting Alaskans is a riveting story of how the people of a distant, isolated, and inhospitable land overcame isolation to emerge as leaders in adopting new digital ways of doing things, from telemedicine to distance education and egovernment.

It is also a book whose importance goes far beyond one state and one technology. As Heather Hudson shows in masterly fashion, Alaska has been a major success story for public and private sector collaboration in infrastructure, and a model that applies to other regions and platforms around the world. For all concerned with economic development, this book is an essential guide.

—Eli Noam, Professor of Finance and Economics and
Director of the Columbia Institute of Tele-Information,
Columbia University Business School

Heather Hudson has devoted her long and distinguished career to bring telecommunications to the rural areas of the world. She now brings her years of working in Alaska into a powerful policy and historical account. She remains a strong voice of advocacy for all communications, for all people everywhere. A must read!

—Emile McAnany, Emeritus Professor of
Communication, Santa Clara University

Connecting
Alaskans

We're a state of small towns and villages scattered over an entire subcontinent. The expanses involved are awesome, and because of our different economic bases and our different cultures, there's a strong tendency toward fierce regionalism. . . . So Alaska has some unique problems, and telecommunications is the key to bringing Alaskans closer together.

Alaska Lieutenant Governor Terry Miller, 1981

Connecting Alaskans

Telecommunications in Alaska from
Telegraph to Broadband

Heather E. Hudson

UNIVERSITY OF ALASKA PRESS
Fairbanks

University of Alaska Press
PO Box 756240
Fairbanks, AK 99775-6240

Library of Congress Cataloging-in-Publication Data

Hudson, Heather E.
Connecting Alaskans : telecommunications in Alaska from telegraph to broadband /
by Heather E. Hudson.
    pages   cm
    Includes index.
    ISBN 978-1-60223-268-6 (cloth : alk. paper)—ISBN 978-1-60223-269-3
(electronic)
    1. Telecommunication—Alaska—History.   I. Title.
    TK5102.3.U6H83 2015
    384.09798—dc23
    2015002921

Cover and interior design by Mark Bergeron, Publishers' Design and Production
Services, Inc.

Cover image by Heather Hudson. Microwave, mobile, and satellite facilities on the
North Slope.

This book is dedicated to the many people who have helped to connect Alaskans, and to the memory of Walter B. Parker, a true Alaska pioneer and patriot.

Contents

Acknowledgments

Many people helped to make this book possible, only a few of whom are mentioned here. Robert Walp frequently urged that the story of Alaska communications be told. He hoped for a book, a film, and a museum exhibit; I can fulfill only the first wish. About 15 years ago, Hilary Hilscher conducted interviews with Alaska pioneers and leaders in communications, and later donated the transcripts as well as documents from her work with Senator Ted Stevens to the University of Alaska Anchorage Consortium Library's archives. These materials were very valuable in adding personal reflections and insights from several of the key participants. Edwin Parker contributed documents, research, and analyses concerning the NASA biomedical satellite experiments and the state's efforts to obtain commercial satellite communications for all Alaska villages in which he played a major role. William Melody furnished his research, reports, and reflections on the policy and regulatory challenges that Alaska officials faced to acquire affordable and eventually competitive telecommunications services.

Theda Pittman, Karen Michel and Walter Parker contributed memories and reports from the early NASA educational and broadcasting experiments; Jennifer Wilke added insights about LearnAlaska. Martin Cary supplied information about the North Slope Borough's early innovations in distance education. Stewart Ferguson provided research on Alaska telemedicine. Fran Ulmer and Mead Treadwell offered sources and reminiscences. Lee Wareham described pioneering efforts to link Alaska with Kamchatka. Alex Hills and Douglas Goldschmidt provided studies and publications, as well as helpful comments.

The University of Alaska Anchorage Consortium Library archivists were very helpful in locating materials about Alaska communications; the staff of the Elmer E. Rasmuson Library at the University of Alaska Fairbanks also helped with historical documents and photographs. The Institute of Social and Economic Research at the University of Alaska Anchorage was a supportive "home" for me to complete the research and writing.

Preface

In 1993, in a keynote address at the "Visions of Alaska's Future" conference in Juneau, I stated: "What I learned from Alaskans . . . profoundly influenced the direction of my career." What drew me to this field was what I learned from rural people, including many in rural Alaska, about the importance of information. Access to information and the ability to share information are critical to the development process—to get help when you need it, to keep in touch with family and friends, to upgrade the quality of education and health care, to deliver government services, and to run businesses and nonprofit organizations. Throughout my career, I have tried to share what I learned in Alaska with people in other parts of the world. I hope this book will help to spread the word about what Alaskans have learned—and achieved—not only among Alaskans, but among others concerned with the importance of accessing and sharing information across the North and in other rural and isolated regions.

Heather E. Hudson
Anchorage, 2014

Chapter One

Introduction

How close they sound!

—Woman in an Alaska village participating in a statewide
audio conference, 1980[1]

From Telegraph to Broadband

Alaskans have pioneered in the use of telecommunications for rural development from the first telegraph lines across the tundra to satellite links for telephony and broadcasting to the Internet and broadband era. In the early years of the first army forts and trading outposts, the telegraph was a vital link for the U.S. military responsible for governing the new territory, and for traders to order supplies. The telegraph and later two-way radios brought news of the outside world. As wireless communications spread, doctors at regional hospitals and village health aides and teachers used two-way radios to get help during emergencies. In the 1970s, Alaskans experimented with NASA satellites to introduce basic telemedicine and distance education. State officials, researchers, and broadcasters then successfully advocated for commercial satellite service to extend reliable telephone service and television throughout the state, including the most remote villages. Today, distance learning over the Internet extends educational opportunities to isolated communities, village entrepreneurs market crafts and ecotourism over the Internet, and rural businesses from commercial fishing to mining to retail stores manage their logistics, banking, and payroll online.

Overcoming the challenges of connecting Alaskans scattered in remote communities to each other and the rest of the world has required both technological ingenuity and a commitment to provide service where networks are costly to build and maintain, and customers are few. But the story of connecting Alaskans involves much more than technological innovation and geographical challenges of vast distances and extreme climate. It includes advocacy by government agencies and the private sector, innovative strategies to attract investment, persistence by Alaska politicians and entrepreneurs, and creative techniques of putting telecommunications to use for Alaska's development.

In the mid-1800s, the telegraph introduced the era of electronic communications, as wires were strung along roads and railroad tracks in Europe and North America. In 1861, the transcontinental telegraph reached California, replacing the Pony Express, which took ten days to carry messages to the West Coast. Meanwhile, explorers and traders in Alaska could wait a year or more for news from the outside world or directives from their headquarters in St. Petersburg or London. However, it was not the communication needs of the northern frontier that caught the imagination of entrepreneur and adventurer Perry McDonough Collins, but an opportunity to link the United States with Europe. Early attempts to lay a submarine cable between Newfoundland and Ireland had failed, leaving a window of opportunity for an alternative solution—a terrestrial telegraph network from the U.S. Northwest through British Columbia and the Yukon Territory, across what was then Russian America, with a short submarine link across the Bering Sea to Siberia and then traversing Russia to Europe. Collins managed to convince Western Union to put up the venture capital to attempt to achieve his breathtaking vision of stringing telegraph wire through vast expanses of unexplored northern wilderness. Survey and construction crews were soon working their way north through British Columbia, while a team of surveyors and a naturalist explored routes up the Yukon River from the Bering Sea, and two teams pushed through the wilds of eastern Siberia. But the northern venture was soon abandoned. The transatlantic cable was finally laid successfully in August 1866, although the crews in Alaska did not receive orders to abandon their work and equipment until 11 months later.[2]

Soon thereafter, in 1867, Russia sold Alaska to the United States, but it was 35 years before a telegraph line would actually be built across Alaska. The U.S. Army, responsible for maintaining law and order and providing many public services in the territory, needed to link its posts and forts within Alaska and to connect them with the rest of the United States. Alaska governors repeatedly asked for funds to build a telegraph line in their annual reports to the U.S. Department of the Interior.[3] The first Alaska network connected at the border with a Canadian telegraph line through the Yukon Territory to Skagway, where messages had to be sent by ship to Seattle.[4] Eventually, submarine cables were laid from southeast Alaska to Washington State to complete an all-U.S. route. First known as WAMCATS (the Washington-Alaska Military Cable and Telegraph System) and later ACS (the Alaska Communication System), the network carried both civilian and military traffic but was owned and operated by the military until privatized by an Act of Congress in 1969 and sold to RCA (the Radio Corporation of America).[5]

For much of the twentieth century, Alaska's communications system resembled government-owned networks in Europe rather than the U.S. commercially owned and operated networks. Like European PTTs (post, telegraph, and telephone systems), the ACS received federal government allocations for operations and maintenance and could not reinvest its own revenues, but instead had to turn them over to the U.S. Treasury. Despite growth in Alaska's population and economy, there was little incentive for the military to upgrade and expand facilities for civilian services. However, World War II and the Cold War did provide the rationale and funding to improve military communications, with new technologies such as the White Alice troposcatter system and the U.S. portion of the DEW (Distant Early Warning) Line built to enhance national security on the northern Pacific and Arctic frontiers.

High-frequency (HF) radios had been the only link between most villages and doctors, police, and government agencies. These two-way radios were often both unreliable and inaccessible; the aurora borealis interfered with the signals, and the radios were typically kept in a teacher's residence or other private location and were not available for public use except in emergencies. The 1970s brought much-needed investment in Alaska's networks, but also a new wave of technological innovation, as satellite technology provided much greater bandwidth between Alaska and the rest of the United States. Comsat's Bartlett earth station near Talkeetna initially connected Alaska with the outside world on Intelsat's Pacific Ocean satellites designed for international connectivity among countries of the Asia-Pacific region. Tests of transportable Comsat earth stations and experiments on NASA satellites demonstrated that satellites could also bring reliable voice communications and broadcasting services to Alaska villages.

Experiments using NASA's ATS-1 satellite to link village clinics to regional hospitals demonstrated that reliable voice communications could make a difference in rural health care, not only to get help in emergencies but to enable doctors at regional hospitals to advise village health aides on diagnosis and treatment of their patients through daily "doctor calls."[6] Educators and broadcasters also experimented with the NASA ATS-1 and ATS-6 satellites to transmit community radio programs and educational videos to schools and community centers. Participants in these experiments became advocates for permanent satellite facilities for all of Alaska's remote communities. Yet demonstrations and experiments were short term; turning them into stepping stones to operational service required the commitment and ingenuity of federal and state government officials, Alaska business leaders, academics, and researchers. Their efforts culminated in an appropriation by the Alaska legislature in 1975 of $5 million for the purchase of satellite earth stations for more than 100 villages; the satellite facilities were installed and operated by RCA Alascom.

With the villages connected, Alaskans began to harness communications technologies to serve the needs of rural residents, businesses, and public services. Telemedicine equipment was installed in all village clinics, so that today, Alaska has one of the world's largest telemedicine networks, linking more than 240 sites.[7] As the result of a court case settled in 1976, village schools were required to offer kindergarten through 12th grade,[8] but teachers had little instructional material about Alaska, and no science labs. To help fill the void, in the 1980s, the LearnAlaska project produced video programs for village schools and licensed hundreds of hours of educational programming transmitted by satellite that teachers could download and record for later use. The Rural Alaska Television Network (RATNET—perhaps not the most fortunate choice of acronyms) was established so that villages could receive network television. Native representatives selected a mix of news, sports, and entertainment that was transmitted on a single satellite channel and rebroadcast in the villages. Alaska educational and commercial broadcasters solved the problem of how to offer television programs from all of the networks on one satellite channel without violating network distribution agreements by affiliating each receiving site with all of the networks. State agencies began to hold hearings using audio conferencing facilities in communities around the state so that residents could testify without having to take long and expensive flights to Juneau. Marveled one village participant: "How close they sound!"[9]

With the advent of the Internet era in the late 1990s, Alaska was once more a communications pioneer in offering online access to state government services ranging from hunting and fishing licenses to applications for annual Permanent Fund disbursements. Alaskans began to sell products including qiviut (muskox wool) scarves and hats, smoked salmon, and wild berry products online, and to promote winter activities such as viewing the Iditarod sled dog race and the northern lights, as well as Alaska summer vacations and adventures. Alaska's major commercial enterprises in aviation and shipping, fisheries, oil and gas, mining, retail merchandise, banking, and tourism now use communications networks for logistics, back office support, data analysis, reservation systems, financial transactions, and other services.

A new program mandated by the Telecommunications Act of 1996 introduced subsidies for Internet access to schools and libraries. Alaska soon had the highest percentage of participating schools in the country, most of which were in isolated villages where they connected to the Internet via satellite.[10] Today, Alaska remains one of the highest per capita beneficiaries of the Schools and Libraries Program, also known as the "E-rate." Another federal universal service fund subsidizing connectivity for rural health facilities supports Alaska's telemedicine and telehealth networks; Alaska now receives the largest absolute amount of funding as well as highest allocation per capita of any state from the Rural Health Care program.[11] And the High Cost Fund has provided critical subsidies to companies providing local telephone service where costs passed on to customers would otherwise have made the price of local service exorbitant. Today, the focus of federal subsidies to providers is on extending affordable access to broadband, with new subsidy regimes being introduced as part of the implementation of the National Broadband Plan.

Rural Alaskans continue to adopt new technologies and services as they become available in their communities. Most villages now have cellular service, although coverage remains limited offshore and on the land and rivers where emergency communications in harsh weather could save lives. And as rural Alaskans seek to participate in an increasingly information-driven global economy, they have realized that they need access to broadband. The goal of providing universal access to broadband that is both affordable for users and sustainable for providers is the latest Alaska communications challenge.[12]

Overcoming Challenges

Alaska's vast expanses, terrain ranging from tundra to mountains to dense rain forest, and unforgiving climate have posed challenges to the planners and builders of its telecommunications networks since the earliest days. Equipment shipped to the Alaska coast had to be hauled inland by mules or dog teams. Anchoring telegraph poles meant melting holes in the permafrost or building wooden tripods. A submarine cable across Norton Sound that kept breaking from the pressures of waves and sea ice was replaced with one of the world's first commercial wireless circuits to connect Nome with St. Michael. Further innovations in wireless technology brought the means to extend the networks, with White Alice troposcatter antennas and microwave relay towers. But it was satellite technology that made reliable communications for all of Alaska's settlements not only possible but achievable when innovative engineers designed small earth stations that could be assembled from components flown to villages in bush planes. Today, planners attempting to extend broadband to every community face similar challenges to power mountaintop repeaters and lay optical fiber across tundra and under grinding coastal ice.

The federal government has played many significant roles in Alaska telecommunications. The military both funded and operated the facilities that provided commercial communications services until 1969. Threats to national security prompted construction of the White Alice troposcatter network and the DEW Line. NASA satellites demonstrated the benefits of this new technology for reaching remote communities and provided the evidence state officials needed to make the case for investment in commercial satellite facilities. Federal loans for rural phone companies helped cooperatives and small "mom and pop" companies to install local networks and later to upgrade their facilities; federal subsidies for high-cost services have helped Alaska carriers to survive while keeping services affordable for their customers. Alaskans have also relied on federal programs to subsidize Internet access in schools and libraries and connectivity for rural health centers, and recent federal grants and loans have helped to extend broadband.[13]

Early governors of the Alaska Territory emphasized the need to connect Alaska with the rest of the world in their annual reports to Washington, DC. The chiefs of the Army Signal Corps reported on their progress in installing the telegraph networks across frozen tundra and mosquito-filled swamps. Later, they pressed the U.S. Department of Defense for communications facilities to protect Alaska and the Arctic. After statehood, Senators Bartlett, Stevens, and Gravel became strong advocates for satellite communications to reach all of Alaska's settlements with telephone and broadcasting services. Governors Miller, Egan, and Hammond recognized that communication technology could help to advance the economic development of the state. In the 1980s, the legislature created the Telecommunications Information Council (TIC); in the 1990s, the lieutenant governor headed a revitalized TIC that produced a technology plan for the state.[14]

The private sector, of course, was also critical to the expansion of facilities and operation of services. From the earliest days, the military contracted with private suppliers to build its networks, including the early submarine cables from southeast Alaska to Seattle. Local telephone companies founded by Alaska entrepreneurs sprang up to connect households and businesses to the military-owned Alaska Communication System (ACS). When the privatization of ACS ended the military's role in civilian communications, RCA Alascom became the state's long-distance carrier. GCI, formed by Alaska entrepreneurs who believed there were opportunities for new entrants even in Alaska's small market, became the first long-distance competitor in 1982. Today, several companies provide mobile communications services. Local radio and TV stations affiliated with national networks at first received news and sports by teletype, with the first video programs on tape delivered by plane. Satellite communications brought live programs from the outside and the formation of Alaska communications organizations to share content within the state. Business proprietors led by broadcaster A. G. ("Augie") Hiebert played key roles in advocating for satellite facilities for Alaska.[15]

Recurring Themes

The following chapters trace the development of telecommunications in Alaska from the earliest telegraph service to the present day. In addition to a historical overview of the introduction of telecommunications and broadcasting, the analysis includes a review of early satellite projects in the 1970s that demonstrated how telemedicine, distance education, and public broadcasting could serve the needs of Alaskans in remote communities. The next section reviews the very innovative strategies that involved the state government, federal government, private sector, and academia in the transition from these short-term experiments to operational services for all of Alaska's permanent communities. The book also examines the evolving roles of key players in the communications industry in the state, ranging from tiny "mom and pop" companies serving a few villages to rural cooperatives to large corporate enterprises. The impacts of federal policies concerning competition, rate integration, universal service funds, and rural broadband are analyzed. Chapters on distance learning and telemedicine discuss innovative applications, and the importance of federal subsidies for Internet access for schools and libraries, and connectivity for rural health facilities. A concluding section examines recent efforts to provide broadband throughout the state and discusses lessons learned from the Alaska experience for both Alaska and other rural and isolated regions.

An overarching theme throughout the book is the importance of *advocacy* by Alaskans including governors, state and federal representatives, industry officials, academics, and concerned citizens to extend communication facilities and services throughout Alaska. Many issues that emerge from this analysis are relevant not only for Alaska but also for the larger contexts of Arctic and global rural communications planning and policy. For example:

- Technological innovations that have been adapted for rural communications ranging from open wire telegraph lines to early wireless links to very small aperture satellite terminals (VSATs)
- Privatization of a government-owned monopoly operator
- The use of communications demonstrations and experiments as strategies to build a case for investment in rural infrastructure and services
- Advocacy by state agencies, the state legislature, and some private sector communications leaders on behalf of Alaska at the federal level
- Public-private partnerships such as state government funding of village earth stations that were then commercially operated, and a shared satellite transponder for public and commercial broadcasting

- The impact of external events on investments in Alaska telecommunications, including the transatlantic cable crossing, the second World War, the Cold War, and the growing importance of the Arctic
- The role of communications entrepreneurs, including pioneers in broadcasting and owners of local telephone companies
- Introduction of competition in an environment where a subsidized monopoly was assumed to be the only viable means of providing service
- Consolidation and remonopolization of some competitive communication services
- Government initiatives to fund investment in rural infrastructure, such as the state's investment in satellite terminals and federal programs for rural telephone companies and broadband networks
- The importance of federal operating subsidy programs, such as those for high-cost regions, low-income households, access to the Internet for schools and libraries, and connectivity for rural health facilities
- The roles of the rural indigenous population—Alaska Natives—as users of communications to stay in touch with scattered relatives and friends as well as to seek and share information, and as participants in providing services through telephone cooperatives and in producing content for broadcasting networks and online services.

While the specifics are about Alaska, many of these topics are relevant far beyond—across the North, for other indigenous populations, and for rural regions of the developing world.

Chapter 2

Alaska's First Information Highway

*The telegraph unleashed the greatest revolution in
communications since the development of the printing press.*
                           Tom Standage, *The Victorian Internet*[1]

A Communications Vision

The telecommunications era in Alaska traces its origins to Samuel Morse's
invention of the telegraph, and his famous message "WHAT HATH GOD
WROUGHT" transmitted over copper wire from Baltimore to Washington in
1844. Soon telegraph lines were being constructed between major American
cities. In 1852, *Scientific American* proclaimed: "No invention of modern times
has extended its influence so rapidly as that of the electric telegraph."[2] For the
first time, messages could be transmitted almost instantaneously rather than at
the speed humans could carry them by foot, horse, or ship. The Pony Express,
begun during the California gold rush, had taken ten days to deliver messages
from Missouri to California. By October 1861, a telegraph line paralleling the
newly constructed transcontinental railroad was completed across the country
to California so that messages could be received the same day.

Telegraph lines also began to bridge national boundaries. A submarine
cable across the English Channel carried the first cable message from London
to Paris in 1852. American entrepreneur Cyrus W. Field wanted to create a
telegraph link from America to Europe by laying a submarine cable from
Newfoundland to Ireland. His Atlantic Telegraph Company's first effort to
lay the cable by ship failed when the cable snapped while being laid in July
1857. During the next year, the cable snapped three times, but finally landed
on August 5, 1858.[3] However, the link proved unreliable and stopped working
less than a month after completion. After much effort to determine why the

cables were breaking and failing, the company tried again in 1865 to lay cable across the Atlantic, but the cable broke during splicing, disappeared in water two miles deep, and could not be retrieved.[4]

But what did these achievements and failures in electronic communication have to do with Alaska, then a Russian colony far from the U.S. mainland and much farther from Europe?

Another American entrepreneur, Perry McDonough Collins, who had made a fortune in the California gold rush, thought he could beat Field and his floundering Atlantic Telegraph Company in the race to link America with Europe. In 1855, Collins had been appointed American commercial agent in Siberia; after traveling through eastern Siberia, he concluded that the region had potential for American investment. He proposed building a telegraph line from Oregon through British Columbia and Alaska, across the Bering Strait, and from Siberia to St. Petersburg, to connect with the European telegraph— a distance of 17,000 miles through some of the world's most difficult terrain, much of which was not mapped and was frozen for half the year.

For the next eight years, Collins negotiated with the British Colonial Office, Russian Department of Telegraphs, and the U.S. Senate to secure rights of way through their territories. Meanwhile, he was also seeking financing and eventually convinced Western Union, the major operator of U.S. telegraph services, to invest in the venture. To raise the funds, Western Union created a special "extension stock" and gave existing shareholders first rights to subscribe for $100 per share. The estimated total cost of the project was $12 million. In 1861, Hiram Sibley, president of Western Union, announced "we will complete the line in two years, probably in one. The whole thing is entirely practicable."[5] In retrospect, the scale of the project and the estimated timeline are breathtaking, given the challenges of unexplored wilderness, extremes of climate, enormous distances, and logistics of supporting the venture by ships to the Arctic and then overland. Clearly Sibley had never been to the North, and Collins must have been an outstanding salesman. However, Secretary of State Seward praised the project, saying the overland line would inevitably lead to communication between "the merchant, the manufacturer, the miller, the farmer, the miner or the fisherman" of small American communities with producers and consumers in Siberia, Asia, Western Europe, and beyond,[6] and President Lincoln announced it in a message to Congress on December 6, 1984.[7]

The Overland Telegraph Company was formed to carry out the construction of the network. Colonel Charles L. Buckley, who had been a telegraph specialist during the Civil War, was put in charge of engineering, and a naturalist, Robert Kennicott, of explorations for the Yukon region and Alaska. Why a naturalist for a telecommunications project? Kennicott was the only member of the party who had previously been in Alaska; he had already explored the

Yukon River region to collect specimens for the Smithsonian Institution in 1861–1862.[8] In May 1865, ships carrying work crews, wire, insulators, and other supplies set out for British Columbia, Alaska, and Siberia.

In Alaska, the work crews set up a base camp at the Russian American Company trading post of St. Michael on Norton Sound near the Yukon River delta, where they spent their first winter. In 1866, on their way to a rendezvous with the team in Siberia, William Dall and Frederick Whymper stopped at St. Michael, where they learned that Kennicott had died, apparently of a heart attack near Nulato. They decided to stay on in Alaska to continue his exploration. According to Whymper, during the winters of 1865–1866 and 1866–1867, the Overland Telegraph Company had stations at the head of the Okhotsk Sea and at the Anadyr River in Eastern Siberia, at Plover Bay and Port Clarence on either side of the Bering Straits, two in Norton Sound, and one on the Yukon River, besides "numerous parties in somewhat lower latitudes" (apparently in British Columbia).[9]

Those landing in Siberia separated into two groups: one was to establish the best route southward to the Amur River; the other was to head north to the Anadyr River surveying the final leg of the route to the Bering Strait. They traveled first by pack horse and then by dog team, facing intense cold and huge snow drifts, trading with local people for game and dried fish to feed the dogs. Summer was no easier, as Captain R. J. Bush wrote to the company: "As the summer season approaches the deer migrate north to escape the clouds of mosquitoes. We are compelled to wear nets day and night, and can not take off our skin clothing even while sleeping under our blankets. In spite of all precautions, our hands and faces were badly swollen. The mosquitoes torment the dogs so that in desperation they tear all their hair off their backs with their teeth."[10] But they managed to finish their surveys, and hired Native people to cut and set poles hauled on sleds from distant forests. Adventurer George Kennan, leader of the party that headed north from Kamchatka to Chukotka, described the Siberian expedition in his book *Tent Life in Siberia*, calling the entire venture "the greatest telegraphic enterprise that had ever engaged American capital."[11]

Conditions in Alaska were equally daunting. In his memoir of the project, Whymper describes the tribulations of working during the Alaska winter:

> Our expedition was largely Arctic in its character, and affords perhaps the latest confirmation of the possibility of men enduring extreme temperatures and working hard at the same time. . . . Our men were engaged both exploring and building telegraphs at temperatures frequently below the freezing point of mercury. Minus 58° was our lowest recorded temperature in Russian America. Now, in such a climate, this work was no joke. The simple process of digging a hole to receive the telegraph pole became a

difficult operation when the ground was a frozen rock with 5 feet of snow on top of it, and where the pick and crow-bar were of more use than the spade or shovel. The axe-man, too, getting out poles and logs, found his axe ever losing its edge or cracking into pieces. All this was in addition to transporting materiel and provisions . . .[12]

In October and November 1866, they made their way overland and up the frozen Yukon River by dog team from Unalakleet, stopping at Kaltag, and on to the Russian fort at Nulato, the eastern limit of the Russian American Company's trading area, where they spent the winter.[13] After the breakup of ice on the Yukon, the set off by canoe on May 26, stopped at what is now Tanana, and reached Fort Yukon on June 23, 1867, in time for the annual fur trade between the Athabaskan people and the Hudson's Bay Company. After a few hot and buggy days in Fort Yukon, they set off again down river, reaching Nulato in only four days. There they received orders to continue to St. Michael, carrying all movable property of the Overland Telegraph Company. According to Whymper, they arrived at St. Michael on July 25, "but 15 1/2 days from Fort Youkon [sic], a distance of 1260 miles."[14] Along the way, they bought 30 to 40 pounds of salmon on the lower Yukon for five needles or less and duck eggs at the mouth of the Yukon at ten for a needle.[15]

In St. Michael, they learned that the Atlantic cable had been successfully completed at last, and that Overland Telegraph venture was disbanded. In fact, the cable had been successfully laid from Ireland to Newfoundland in 1866,[16] but the Overland Telegraph parties in the north did not find out until ships arrived the next summer. Ever the optimist, Whymper states: "This enterprise, in which the Company is said to have spent 3,000,000 dollars (in gold), was, in 1867, abandoned, solely owing to the success of the Atlantic Cable, and not from any difficulties in the way of the undertaking itself."[17] He notes that despite the cold and isolation, the men had persevered and "succeeded in putting up at least one-fourth of the whole line, and I can sympathise with the feeling of some of them at Unalachleet [sic], Norton Sound, on hearing of the withdrawal of our forces and the abandonment of the work, to hang black cloth on the telegraph poles and put them into mourning."[18]

Apparently at St. Michael, mourning about the news went beyond hanging black cloth. While waiting for the ship, some drank alcohol used to preserve specimens of toads and fish, and then ate the specimens![19] Whymper concludes: "A month later we sailed for Plover Bay, E. Siberia, meeting those who had wintered there, with the parties both from the Anadyr and Port Clarence, numbering in all 120 men. Shortly afterwards on the arrival of the Nightingale, our largest vessel, we set sail for San Francisco."[20]

The legacy in the north was initially many miles of telegraph line constructed in British Columbia but never completed, and the line constructed on the Seward Peninsula by the Alaska crews. Western Union redeemed the shareholders' stock in the Overland Telegraph Company, taking an estimated loss of $3 million. In Siberia, "the natives used poles for firewood, glass insulators for cups, wire for rope."[21] In British Columbia, the Kispiox people found an ingenious use for the abandoned equipment. At Hagwilgaet Canyon on the Bulkley River, they built a suspension bridge 100 feet long and 6 feet wide made of telegraph wire. Eventually, in 1880, Western Union sold the completed line running to Quesnel, BC, to the Canadian government for $24,000.

The Russian American telegraph project may have contributed to U.S. Secretary of State William Seward's initiative to purchase Alaska in 1867. When Collins and Hiram Sibley of Western Union visited St. Petersburg in 1864, Russian Foreign Minister Gorchacov questioned whether the Hudson's Bay Company would allow Western Union to have a right of way through British Columbia. Sibley responded that if they did not, Western Union would consider buying the HBC; it shouldn't cost more than a few million dollars. "Gorchacov laughed. For not much more than that, he suggested, Russia would sell all of Alaska."[22] Another legacy was the information compiled by the Overland Telegraph Company expeditions. According to Lyman Woodman, Colonel Charles Bulkley's survey parties produced records of exploration "which became valuable reference material later, including the period of Congressional deliberation on Alaska's purchase."[23]

A Telegraph Network at Last

After the demise of the Overland Telegraph Company, soon followed by the U.S. purchase of the Alaska territory from Russia in 1867, no one attempted to build a telegraph network in Alaska for more than 35 years. The affairs of the Alaska territory were initially administered by the U.S. War Department with small army garrisons. The U.S. Army Signal Corps had stationed weather observers in Alaska but soon withdrew them because it took too long for the data to get to Washington, DC, to be of any use.[24]

Gold was discovered near Dawson City in 1896, but word did not reach the outside world until 1897, when ships carrying newly rich miners docked in Seattle and San Francisco. Word of gold in the Klondike and then on the beaches of Nome in 1898 brought thousands of prospectors and carpetbaggers to the North. With these discoveries of gold in the Yukon and then Alaska, the need for faster communications increased dramatically. Congress called on the U.S. Army to establish law and order in Alaska. The headquarters of the Military Department of Alaska was established at Fort St. Michael on Norton Sound, with other garrisons near Nome, near Tanana, at Eagle, near Valdez, and at Haines Mission. The lack of communications among them, let alone to Army Headquarters in Washington, DC, made them difficult to administer: "It was not unusual for correspondence to take a full year to pass through channels from Fort St. Michael to Washington, DC, and return."[25]

Meanwhile, a telegraph line from Skagway, the port that prospectors and traders used to reach the Klondike over the Chilkoot Pass, reached Dawson City in 1899. The *Dawson City Daily News* announced: "By telegraph, the people of Dawson are now in direct touch with the wide, wide world. No, not exactly direct, for there is an interval of, say four days, between Skagway and Victoria or Seattle. . . . by comparison with the interval with the long journey by dog team that for more than half the year intervenes between here and the big busy world the interval of the run down 'the inside passage' is nothing."[26] The telegraph brought news, carried messages outside, and "put people living in the cabins, camps, Indian villages, trading posts, and policy outposts along the Yukon River in touch with each other."[27] Communication was not cheap; a message from Dawson to Skagway cost $3.75 for the first ten words and 20 cents for each additional word, but the prosperous citizens of booming Dawson "thought nothing of sending a $100 telegram."[28]

On May 26, 1900, Congress appropriated $450,550 to authorize the army to construct military and cable lines connecting the widely separated garrisons with their headquarters at St. Michael. Four months later, on September 25, 1900, the first telegraph line of what eventually became the Alaska Communications System was completed and put in service, with 24 miles of wire from Fort Davis near Nome to Port Safety.[29] In November, more than 600 miles of wire and cable were installed to link Fort St. Michael to Fort Gibbon (Tanana). The Army's Chief Signal Officer, General Greely, noted: "The importance and value of this connection may be estimated by the fact that the ordinary mail time between these two points is twenty-nine days."[30]

But the crews faced the same Arctic challenges as their predecessors working for the Collins Overland Telegraph 35 years earlier. The signal officer at Fort St. Michael reported: "The seasons have seemed to conspire against telegraph construction, the ground being almost impossibly boggy in the fall, the cold intense (–72° F.) in the winter, the snow soft and deep in the spring, and now in the summer hordes of appallingly ferocious mosquitoes drive the men of the working parties to the verge of insanity."[31]

The telegraph link between Fort St. Michael and Nome was especially challenging because it had to cross Norton Sound. The cable link under Norton Sound was a calamity from the beginning. During the laying of the cable, the cable ship hit a rock, sustained three holes in its bottom, and was declared a total loss. The cable was saved by offloading to a lighter ship, which was then towed by paddle wheeler to lay the rest of the cable. However, the cable soon broke with the force of the shifting polar pack ice; although restored and operated during spring and summer of 1901, it was broken again when a 60-kilometer section was ripped out by the shifting ice pack on November 24, 1901.

At that point, the military decided to try wireless, or what the chief signal officer called "the Fessenden system."[32] It awarded contracts to two companies to provide the wireless circuit between the two forts, but both failed. The U.S. Army Signal Corps had itself established a wireless system over 20 kilometers of water separating Fire Island and the Fire Island light ship off Long Island in 1899. Using its previous experience, the Signal Corps took over and developed its own wireless system, which opened on August 17, 1903.[33] General Greely reported that the wireless section between Safety and St. Michael was the "longest wireless section of any commercial telegraph system in the world," and its two masts at each station 210 feet high were the highest ever erected on the Pacific Coast.[34]

But there was still no connection between this internal Alaska military network and the rest of the world. To gain Canadian commitment to building a link through the Yukon, the U.S. Army's chief signal officer met with Canadian authorities in Toronto in 1900 "with a view to the extension of the existing Canadian telegraph lines, and the establishment of cooperation in telegraphic work between the Alaskan and Canadian systems. The distinguished premier of Canada, Sir Wilfrid Laurier, showed marked interest in the views advanced by the Chief Signal Officer and expressed his desire to further any plan that would bring the two countries into closer and more cordial relations, especially in the Northwest. The matter was immediately taken up in the same spirit by the acting minister of public works . . . and an appropriation was made by the Canadian Parliament for extending the Canadian telegraph line down the banks of the Yukon to the Alaska boundary."[35] Supplies were shipped that summer, and the Canadian telegraph line from Dawson City to Fort Egbert (Eagle) was completed on May 5, 1901, bringing the American territory on the Upper Yukon in direct communication with Skagway, from which steamers could reach Washington state in four days. Today, it is hard to imagine direct involvement of a prime minister in a communications project, almost immediate allocation of funds by parliament, shipping of supplies the same summer, and completion of a project by the next spring.

A submarine cable from Skagway to Juneau (which had become the capital in 1900 of what was then called the District of Alaska) opened for business on August 23, 1901. According to the chief signal officer, "[This cable] not only connects the military posts at Skagway with Juneau . . . but enables the

Territorial authorities to reach via Skagway and Dawson the points in the Yukon Valley now so remote that, as a rule, not more than two or three letters can be exchanged in a year."[36]

In 1901, Captain William "Billy" Mitchell, then only 21 years old, was sent to Alaska to oversee construction of the telegraph network in the interior. The first task was to complete the line between Eagle and Valdez, "a trackless wilderness of 400 miles," to connect with the submarine cable. In a 1904 article in the *National Geographic*, Mitchell described the process:

> First the line was surveyed, next the right of way was chopped, then the wire was run over the snow, insulators, brackets and nails being tied to the wire every quarter of a mile in sufficient quantities for the intervening distance. This was done because during the summer the pack animals can not pack the wire and more camps in the same manner that they can in the winter. In summer an animal can pack about 200 pounds, and in winter the same animal can pull on a sled from 800 to 2,000 pounds. It is also almost impossible to dig post-holes through the deep snow in winter.[37]

Instead of single poles, the crews set up tripods in swamp and muskeg, three long spruce poles tied together at the top with wire and insulator placed at the top of one pole; in areas where there were no suitable trees, they used iron poles made from pipe.

By August 1902, 1121 miles of landlines and submarine cable had been built and put in working order within a period of 24 months, including surveying, construction, installation, manufacture and inspection of material and instruments, and transportation over distances ranging from 4000 to 7000 miles. The army's chief signal officer stated in his 1902 report: "The accomplishment of such results would be most creditable . . . if Alaska were an ordinary country . . . . When one considers, however, the exceedingly difficult physical conditions within the Territory, the work must be considered simply phenomenal."[38] Everything but timber had to be supplied—food, shelter and human society. "One officer forcibly writes that what appalled him was the enormous distances from one handful of humanity to another in the valley of the Yukon."[39]

Mitchell's next task, during 1902 and 1903, was to survey a route and construct the line from Eagle to meet crews working inland from the Bering Sea. Mitchell and his men and dogs faced life-threatening frostbite in the winter while exploring a route that would be shorter than following the Yukon River from Eagle. Mitchell describes a remedy known as Perry Davis Painkiller: "This is the greatest medicine ever invented for use in the North. I do not know the ingredients, other than alcohol and some laudanum, but I would hazard a guess at red pepper, turpentine, and tabasco juice. You can take it internally or rub it

A tripod base in permafrost instead of a telegraph pole in the WAMCATS network.
(University of Alaska Fairbanks, UAF-1974-130-108)

on as liniment. For man or dog, it is one of the best remedies I know for frost-bite."[40] And it never froze.

From November 1902 to June 1903, 220 tons of supplies or materials were packed into the interior on sleds or pack animals over "the roughest imaginable trails" where "not 20 miles of constructed wagon road exists in the country traversed."[41] The chief signal officer reported in 1903: "Very early springs, late autumns, enormous snowfalls, summer floods, impassable canyons, and, last of all, a gold fever which stripped one officer of every civilian employee save one, have alternately impeded progress, but with energy and resourcefulness these officers have met and surmounted difficulties which seemed insurmount-able."[42] In the summer, insects and forest fires threatened men and beasts. In June 1903, a forest fire crept up the Tanana River valley: "From that time until the completion of the line the men worked directly through the fire, in some places the wire being taken through the smoldering embers by a man riding at a gallop on a mule."[43]

Finally, Mitchell met the Signal Corps party coming from the coast and personally crimped the wire to make the final connection on June 27, 1903, completing a network "with all the lines in the territory, [of] nearly 2000 miles of wire."[44] Mitchell concluded: "Never have I seen greater physical strength or endurance displayed by a group of men. There were twenty in each working party, great bearded fellows in blue denim clothing, high horsehide boots and slouch hats, with remnants of mosquito netting around the edges. Their faces were running sores from the terrible assaults of the mosquitoes and black flies. As they attacked the spruce trees, the forest seemed to fall in front of them. Without such men, the lines in the North could never have been completed. . . ."[45]

Thus, by the summer of 1903, the Alaska garrisons were linked to each other within Alaska and through Canadian government lines to Skagway, from which ships carried messages to Seattle. But Congress wanted an all-American route to bring the military stations in Alaska into direct contact with regional headquarters in Washington State, and appropriated funds for submarine cables down the coast. The U.S. Army cable ship *Burnside* laid 291 miles of cable between Sitka and Juneau, 1070 miles from Sitka to Seattle, and finally 640 miles from Sitka to Valdez, completing the all-American route in 1904.[46] The network now consisted of 2128 miles of submarine cable, 1497 miles of land-lines and wireless covering the overwater gap on Norton Sound of 107 miles.

Alaskans were now linked in nearly real time with the United States, Canada, and much of the rest of the world. In 1904, the Army's Chief Signal Officer proclaimed "The United States has brought southeastern Alaska, the Yukon Valley, and the Bering Straits region into telegraphic communication with the rest of the civilized world."[47] Reflecting the period of military expan-sionism after Philippine–American War from 1898–1902, he added: "The President or the Secretary of War can now reach, over strictly American lines of telegraph and cable, every important military command from the icy waters of Bering Strait to the tropical seas of the Sulu Archipelago, with the exception

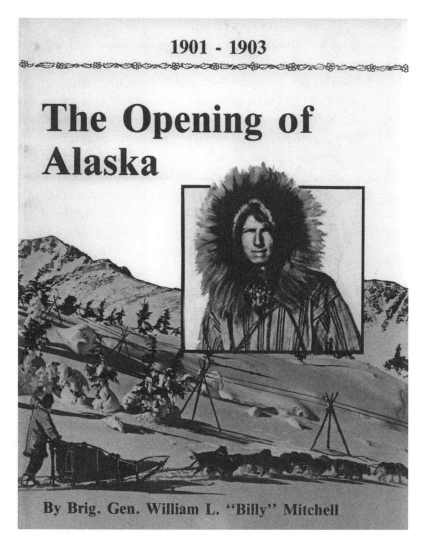

1901 - 1903

The Opening of Alaska

By Brig. Gen. William L. "Billy" Mitchell

Billy Mitchell's memoir about building the Alaska telegraph network. (Cook Inlet Historical Society)

of the legation Guard at Peking."[48] In fact, the signal officers selected to be in charge in Alaska had previous cable experience in the Philippines.

The chief signal officer reported: "the undertaking is unique in the annals of telegraphic engineering, whether one considers the immense extent of the territory, its remoteness from the United States, the winter inaccessibility of the regions, the severity of the climate, the uninhabited and trackless districts, or the adverse physical conditions. If plotted on a map of the United States this system would reach from Wyoming to the Bahamas, off the coast of Florida. The cables used would reach from Newfoundland to Ireland, and the landlines from Washington to Texas."[49] Military records indicate that the landlines in

Alaska were estimated to have cost the government $617 per mile, while submarine cables cost about $452 per mile, and annual maintenance of telegraph network cost $6.25 per mile.[50] But the dollars for maintenance did not reflect the personal cost to the men maintaining the line. The 1907 Signal Corps annual report noted: "A number of men were severely frozen and during the recent breaking up of the river, which washed away a hundred miles of lines, the men worked in ice cold water for days in order to restore communication."[51]

The chief signal officer proposed an even more expansive system in the North Pacific, with a wireless network to reach Unalaska/Dutch Harbor, and added that "the Signal Corps station at Safety Harbor could readily communicate with a wireless station established at a suitable point on the Asiatic shore of Bering Strait."[52] But direct telecommunications across the Bering Strait to Siberia did not commence for another 80 years (see Chapter 13).

The all-American network was also all-American built. Previously, the army had relied on British equipment, gutta-percha clad cables, and British-trained operators. Now they would use seamless American-made rubber-clad cables, laid by American ships and operated by American soldiers. But distance was still a challenge; the cable manufactured in New Jersey had to be shipped 12,000 miles around Cape Horn, although the second batch came across country to Seattle by train. Still, "The Alaska cable system involved not alone the telegraphic unity of American territory on this continent, but also American ability and resourcefulness in a new field."[53] And it demonstrated how quickly cable could be laid: "The celerity with which the Valdez-Sitka cable of over 500 miles in length was put under contract, manufactured, transported and laid, illustrates American possibilities."[54] In 1904, Congress appropriated the funds for the Valdez to Sitka cable on April 24, and the cable opened for commercial business in just 5 months and 12 days. Again, it is difficult today to imagine Congressional appropriation and project completion in such a short period.

WAMCATS serves Alaska

In the Appropriation Act of 1908, Congressed referred to the communications system as "the Washington-Alaska Military Cable and Telegraph System" It was to be known as WAMCATS for the next three decades. In the early days in remote villages, WAMCATS military personnel served as weather observers, local postmasters, U.S. commissioners, and even U.S. marshals.[55] To maintain the network, repair detachments of one Signal Corps repairman and two assistants were stationed at cabins about 40 miles apart. The logistics required to support these remote sites were formidable; the Army's Quartermaster's Department had to arrange shipping of 270 tons of rations and dog food to the telegraph stations, primarily by dogsled. The food and supplies had to be furnished a year in advance "as otherwise the men would be liable to perish of starvation."[56]

The Congressional Act of 1900 provided that ". . . commercial business may be done over these military lines under such conditions as may be deemed, by the Secretary of War, equitable and in the public interests and all receipts from such commercial business shall be accounted for and paid into the Treasury of the United States."[57] From the beginning of operations, demand for communications exceeded expectations, a theme that was to recur in Alaska communications until the present day. The army's chief signal officer reported that in 1904, 55,559 messages were sent, of which 31,020 were commercial and 26,539 were official messages, mostly for government business within the territory. Accounting was precise: the revenue collected for telegrams was $56,935.89, of which $12,208.93 for Alaska line tariffs was deposited in the U.S. Treasury and the rest turned over to other carriers for traffic originating or terminating outside the army network.[58] Thus in its first year of operation, 56 percent of the traffic was commercial, and more than 21 percent of the revenue was turned over to the Treasury.

In 1906, the governor of the District of Alaska praised the telegraph network, which was being extended to other towns, including Wrangell, Hadley, and Ketchikan: "Congress has conferred no greater benefit on Alaska than the construction of cables and telegraphs thruout [sic] the district."[59] In 1908, the governor reported that frequent interruptions of cable service between Seattle and Alaska during that year due to breaks in the cable and requested more funding from Congress: "The need of new cables is imperative and provision should be made by Congress to supply the need, if anything like a satisfactory service is to be maintained. The demands of the service have increased largely in recent years. . . ."[60] By 1910, the governor reported that the telegraph line had reached most settlements: "The military cable and telegraph system, which now extends to nearly every town and mining camp, continues to be of the greatest benefit to the people."[61]

The Alaska newspaper industry was a major commercial beneficiary of WAMCATS. Blanchard elaborates:

> In addition to running a news service, the System offered discounted rates
> to newspapers and transmitted news into and out of Alaska year round.
> As a result, news from the rest of the world appeared regularly in Alaskan
> newspapers, and news from Alaska appeared in papers world wide. This
> interchange of information connected Alaskans to the cultural life of the
> United States and was an important factor in the Americanization of
> Alaska. For example, after the 1906 San Francisco earthquake, Alaskans

raised money to aid the victims and rebuild. When Fairbanks burned down several months later, people from the outside offered to do the same.[62]

In 1914, the governor predicted that commerce on the West Coast would increase with the opening of the Panama Canal and asked Congress to fund an additional cable to Alaska to support expected increased economic activity: "With the development and growth of the Territory along industrial lines and the promise of a largely increased population the matter of an additional cable is worthy of consideration. This is emphasized by the increase in the commerce of the North Pacific Ocean that may be reasonably expected with the opening of the Panama Canal and the consequent expansion of local and over-seas traffic, which will bring Alaska into closer touch with the nations of the Orient."[63]

The governor also stated that message rates were too high and should be reduced. That request was apparently heeded, as in 1915, he noted that rates on commercial messages had been reduced 25 percent during the year, "the reduction being most acceptable to business men and others who make large use of the cables and land telegraph lines. A night service has also been established in some of the coastal cities."[64] However, he stated that the press rates between Seattle and Alaska's coastal towns were still too high, and he again requested funding for new and additional submarine cables and for wireless stations on Kotzebue Sound and in the Koyukuk region, where there were isolated permanent populations "without any telegraphic communication."

In 1918, the governor reported that there were 40 telegraph and cable stations and 11 radio (wireless) stations operated by the army, plus 11 other radio stations operated by the navy, 16 privately owned radio stations, and 1 operated by the Department of Education at Noorvik "with Eskimo operators." He pointed out the role of the telegraph in implementing the World War I draft, and its potential for increasing public safety: "The necessity of telegraph communication has been particularly noticeable in the execution of the selective draft. Additional service besides assisting in the development of the country would be valuable in performing rescue work and in apprehending the lawbreaker."[65] He again asked for more use of wireless to expand the network to remote settlements—at Point Hope, Point Barrow, and on the Kuskokwim and Koyukuk rivers. Possibly aware of the operation at Noorvik, he pointed out that local indigenous people could be trained as station operators: "A suggestion perhaps worthy of consideration is that Esquimo [sic] might be enlisted in the service and trained for permanent duty at the Arctic stations."[66]

After World War I, the governor pointed out that improvements in Alaska communications were needed to encourage economic development for the Territory, which had a total population of about 55,000—"less than one person to every 10 square miles".[67] "There are still considerable parts of Alaska which are

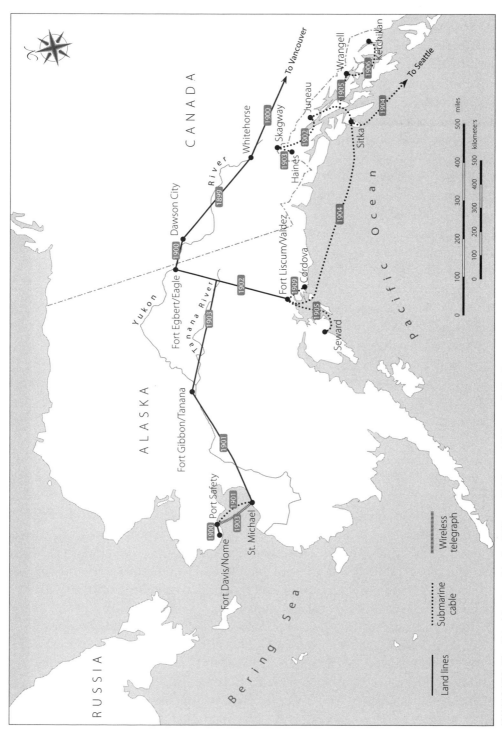

WAMCATS: The Washington–Alaska Military Cable and Telegraph System.

out of communication except by the old tortuous boat or sled travel. . . . There is no inducement to private enterprise to enter the field in completion or cooperation, so it logically devolves upon the Government to meet the growing needs of the population." He pointed out that the volume of commercial telegraph business over the military network was constantly increasing, with total revenue in 1915 of $294,745 and in 1918 of $570,199[68]—almost doubling in three years.

In 1918, there were 46 Army Signal Corps stations in Alaska plus nine radio stations operated by the Naval Communications Service, including both navy and army stations in Cordova, Seward, Sitka, Juneau, and Ketchikan, whereas there were naval radio stations only at Dutch Harbor, Kodiak, St. George, and St. Paul.[69] The naval wireless stations had been established for communication with coastal and North Pacific shipping, and were also to supplement and backup the army system to maintain communication between Alaska and the rest of the United States whenever the Signal Corps system was interrupted, but the governor noted that there was now duplication at some sites, with naval radio competing with army cable.

"The history of the WAMCATS was marked by constant technological development. Four distinct technologies were used on the WAMCATS between 1900 and 1936: the [wire] telegraph; undersea cable; early wireless; and vacuum tube wireless."[70] The arrival of the vacuum tube after World War I reduced the size and cost of wireless equipment. Although the WAMCATS cut back on the number of sites, businesses and communities could purchase their own wireless sets, which linked to the WAMCATS. Mines, canneries, and stores installed their own wireless stations so that they could connect with suppliers and customers and send funds by telegraph.

Individual Alaskans benefited from the telegraph network both directly as users and indirectly from the availability of news, goods in stores, and access to government services. Blanchard notes: "The most dramatic example of this indirect benefit is the WAMCATS' role during the 1925 Nome serum run. Faced with an outbreak of diphtheria after freeze-up, the WAMCATS transmitted the request for serum, coordinated the relief efforts and reported the story. In the process, they helped quell the epidemic and made the men and dogs that brought the serum to Nome by sled famous world wide."[71] Today, the serum run is reenacted annually in the Iditarod race from Anchorage to Nome, with online coverage available to a global audience.

Governors during the 1920s reported on the continued growth in telegraph traffic and upgrades in the network, such as the replacement of the cable between Seattle, Ketchikan, and Seward in 1924. In 1928, the governor noted that the rise and fall in commerce in Alaska was reflected in the rise and fall in the amount of traffic over the Signal Corps network. He also noted the importance of the network for transferring funds: ". . . the popularity of money transfer service . . . has been proven beyond any question," and had been extended to cover transfer between points in Alaska and the states. Western Union, the parent of the failed Collins Overland Telegraph venture, had reached Alaska at last!

The telegraph remained vital for the growth of commerce. In 1932, the governor's office successfully requested that the Signal Corps accept one message a month from local chambers of commerce to transmit to the Alaska Territorial Chamber of Commerce in Juneau: a report of industrial conditions of each region of Alaska for the previous month. The secretary of the central Chamber in Juneau was accorded free service for a consolidated wire to the editor of *Commerce Reports*, in Washington, DC. "Thus the Territory, through the courtesy of the Signal Corps, has been receiving valuable publicity each month through the publication of its wires."[72]

The growth in aviation also brought increased demand for communications. Weather stations were established in Fairbanks in 1929 and Nome in 1931. Working in cooperation with the Point Barrow station, they compiled meteorological data sent out twice daily over the Signal Corps system.[73] And in 1931, the army and navy finally sorted out the management of telegraph traffic that the governor had referred to as duplication in 1918. The handling of commercial and government traffic on the naval communication system was turned over by the navy to the Army Signal Corps. Messages from ships and communities served only by naval radio stations were to be accepted by the naval radio stations and delivered to the nearest Army Signal Corps station for further transmission.

The Alaska Communications System (ACS)

During the 1930s, voice communications came to Alaska, as radio telephone began to replace the wireless telegraph. The first radio telephone transmitters were installed in Marshall and St. Michael in 1931, with additions of Juneau, Ketchikan, Fairbanks, and Nome, and then Bethel in 1934.[74] The military began to close some radio telegraph stations and to transfer military radio equipment to local commercial entities such as Northern Commercial Company (NCC) stores. In 1936, in its appropriation of funds, Congress recognized the network as "The Alaska Communication system," (ACS) a name officially approved by the War Department and in general use instead of WAMCATS after that date.[75] By then, the ACS was handling traffic from 129 commercial radio stations as well as 11 stations operated by the territorial government.[76]

In the 1930s, ACS played several critical communications roles. During the Nome fire of 1934, the Nome station stayed on the air coordinating relief work and rescue agencies as fire destroyed two-thirds of the Nome business district.

When the plane carrying Will Rogers crashed in 1935, ACS at Point Barrow transmitted the news to the world that Eskimos had found the bodies of Rogers and his pilot, Wiley Post. ACS also provided communications for the round-the-world flight of Howard Hughes in July 1938. And in August 1939, ACS provided radio communications for a Japanese round-the-world goodwill flight while it was en route from Tokyo to Nome, Nome to Fairbanks, and Fairbanks to Whitehorse,[77] just weeks before the outbreak of World War II in Europe.

However, wireless reliability had become a problem. During 1937, there were numerous daily periods when high-frequency (HF) communications were interrupted by auroral disturbances. For much of the network, the army could no longer use cables as backup, as War Department policy was not to repair them as they failed. The Signal Corps expanded the use of medium frequency (MF), which was not affected by the aurora, although its range was much shorter. "By 1938, it was possible to handle traffic from Seattle to most of Western Alaska with minimum outage time due to auroral disturbances through the use of short relays."[78]

With the outbreak of World War II, ACS was expanded in 1939 to handle increased military traffic, including teletypewriter and TWX service and a direct high-speed radio channel to Washington, DC. The military again invested in its submarine cable, upgrading the Seattle-Ketchikan-Seward submarine cable in 1941 and converting it to a duplex circuit in 1942. During the war, ACS staff grew from 200 to more than 2,000 in 1944.

Alaska was the site of the only armed invasion of military troops on the North American continent during World War II. Just three years after their "good will flight" to Nome and Fairbanks, the Japanese attacked Dutch Harbor in 1942 and occupied the Aleutian islands of Attu and Kiska. The War Department realized that immediate upgrading of Alaska's communications was necessary. "Communications were so poor during the early days of the war that the Japanese were able to bomb Dutch Harbor . . . with unalerted American aircraft only forty miles away on Umnak Island."[79]

In 1943, ACS communication teams followed U.S. combat troops ashore at Attu, Kiska, Shemya, and Amchitka and established radio communications from these isolated posts back to the main headquarters at Adak, Kodiak, and Anchorage.[80] Air Force General William Butler commended the ACS for its role in the Aleutians: the ACS, integrated with other communications facilities, "was of material assistance to the Eleventh Air Force in the campaign which drove the Japanese from the Aleutian Islands."[81]

However, concerned that the Japanese could intercept radio communications along the Pacific coast, the U.S. military made planning and construction of a landline voice communications system inland through Canada a high priority. The solution was to build an open-wire pole line parallel to the Alaska Highway that had been constructed in 1942 from Dawson Creek, BC, through the Yukon Territory to Delta Junction in Alaska. The first leg of the landline to reach the highway from Edmonton, Alberta, to Dawson Creek, BC, was rushed to comply with a deadline of December 1, 1942, set by military headquarters in Washington, DC. "Men floundered through two meter snow drifts and strung wire by oil lamps, flashlights, and automobile headlights in temperatures of 30 degrees below zero. . . . In all, there were over 500 men who literally jammed the line through the teeth of the Canadian winter."[82] But they met the deadline, with a call from Dawson Creek received in Washington at 7:40 pm on December 1.

By May 22, 1943, the line was completed to Whitehorse, a distance of 2540 km (1580 miles) with almost all of the construction done during the Arctic winter. Most of the remaining 1000 km (620 miles) to Delta Junction and on to Fairbanks were built during the summer of 1943. By September, there were still about 50 miles of swamp to cross just east of the Canada/Alaska border. Once again, the Army faced the tribulations that the crews of the Collins Overland Telegraph confronted in 1865. And like the Signal Corps at the turn of the century, they loaded coils of wire on pack horses to haul into the bog. "Pack horses bellied themselves in the muskeg and the men were constantly wet for four weeks at a time. The mosquitoes and white socks can only be imagined." By October 1, the first line was through, with the permanent line completed on November 10, 1943. When completed, the pole line to Alaska was the longest open-wire communications line in the world, stretching from Edmonton to Fairbanks.[83] A telephone system that would have taken years to build in normal times was completed in 15 months. The 255th Signal Construction Company, which had been given the task of stringing wire through the muskeg, had lived up to their motto: "Through muck and mire we string our wire."[84]

Setting the Stage

Many of the themes that recur in Alaska telecommunications originated in the early days. Alaskans adopted new technologies and adapted them to local conditions, such as the wireless link across Norton Sound, tripod poles for telegraph lines, and American rubber-clad undersea cables. The role of the brutal climate recurs with each new expansion or upgrade of the communication networks, sometimes with disheartening results, like the repeated destruction of the cable under Norton Sound by shifting ice. Ice scouring remains a challenge for proposed installation of optical fiber along Alaska's coast.

The military assumed the key role as monopoly provider of civilian communications following the failed attempt by commercial entrepreneurs to build the first telegraph network across Alaska. Although the federal government invested in a network that might not have been built until much later, the military's priorities, the requirements to turn revenue over to the U.S. Treasury, and the many Department of Defense approvals needed to commit funds for upgrades and expansion left civilian Alaskans frustrated about the prices and quality of services and the lack of access in much of rural Alaska. However, the exigencies of war enabled the military to justify significant expenditures in infrastructure to link its installations and to connect with the chain of command outside Alaska.

Beginning with the delays in completing the first Atlantic cable, external events influenced investment in telecommunications for Alaskans. The era of military expansionism was reflected in the installation of American military telegraph cables around the Pacific and army signal officers with experience in the Philippines heading projects to expand Alaska's cable network. World War II brought wireless networks to reach isolated posts in the Aleutians and the construction of the open-wire landline network along the ALCAN highway from Dawson Creek, BC, to Fairbanks. The Cold War with the Soviet Union would justify investments in highly reliable (and costly) communications facilities.

From the days of the gold rush to the beginnings of aviation and the expansion of commerce, Alaska's leaders recognized that telecommunications would be critical for the economic future of the territory. Governors pleaded with Washington for more funds for communications. Businesses pleaded for lower rates to use the network. And as would be true in later years, traffic on the network exceeded expectations.

Chapter 3

Expansion of Telecommunications after World War II

The most exciting creature in Alaska today goes by the name of White Alice. . . . White Alice is no lady. Hers is the code name for Alaska's sensational new Government-financed long-distance phone and telegraph system. . . .

The Wall Street Journal, 1957[1]

Expansion after World War II

Commercial expansion in Alaska after Word War II increased the demand for telecommunications. In April 1947, ACS provided Alaskans with their first commercial landline telephone services to the United States and Canada over ACS and Canadian facilities. The inaugural call was a conference between the ACS commanding officer in Seattle, Alaska delegate Bob Bartlett in Washington, DC, and the mayors of Seattle, Portland, Anchorage, and Fairbanks at their offices. This conference call over an all-wire connection of more than 5500 miles was believed to be "the longest all-wire call on record."[2]

In 1948, Alaska Delegate Bob Bartlett told Congress: "Since 1900 the Signal Corps has been part and parcel of the Alaska scene. [ACS] ranks high in the esteem of Alaskans and has played an important part in the development of the Territory."[3] In 1949, marking its 49th anniversary in Alaska, ACS stated that it was "providing for Alaska an integrated network of telephone and telegraph service which already offers to many localities in Alaska a service similar to a combination of AT&T and Western Union service offered the public in the U.S."[4] It also provided a ship-to-shore service, press service, money transfer service, and radio broadcast service, transmitting live programs such as Jack

Benny, the World Series, and election results for rebroadcast on Alaska's local radio stations.

ACS estimated the value of its plant, equipment, and general facilities in 1949 at $12 million, including the wireless network, submarine cable, small cable ship, landline facilities along the Alaska Highway in conjunction with the Canadian government, and VHF radio facilities between Juneau, Haines, and Skagway.[5] By 1954, it operated 45 stations within Alaska, with several hundred government and privately owned wireless stations connected to the system. During 1955, it generated commercial revenue of $2.9 million[6] and added 340 miles of coaxial submarine cable in southwest Alaska as the first phase of a project designed to provide 36 additional communication channels between Alaska and the lower 48 states.

Meanwhile, the territorial government had established a Department of Communications governed by the Alaska Aeronautics and Communications Commission, consisting of five members, one from each of the four judicial jurisdictions and the state attorney general. Departmental funds were appropriated by the territorial legislature on a biennial basis. The governor elaborated: "The Department of Communications is authorized to provide for the establishment, operation, and maintenance of radio-telephone stations in isolated communities where no similar services are available and which may best serve the public interest by insuring maximum protection of life and property, and at the same time insure direct commercial communication to the Alaska Communications System operated by the United States Signal Corps."[7] He noted that "Radio telephone communications are a vital part of Alaska communities" and required investment by the Territory: "By the end of 1955 our 65 stations will be of the latest type, employing five or more pretuned frequency channels to provide for the three basic classes of service—aviation, fixed public, and coastal harbor."[8]

The search for oil on the North Slope also required communications support. Early communications on the National Petroleum Reserve were primitive: "Air-to-ground communications in the Arctic once consisted of dropping a message inside a roll of toilet paper from the airplane. The real advantage to using the toilet paper roll instead of the traditional rock was that the paper would unravel on its way down and mark the landing spot."[9] During its initial petroleum exploration program in what was known as NPR-4 or PET-4, the navy had no voice communications, but had to send telegraph messages through the ACS office in Barrow. From 1954 to 1959, messages could be sent only at scheduled times. In 1959, ACS extended its service throughout the working day: "It could only be hoped that emergencies would not happen after five o'clock or on weekends."[10] In the early years of operations, barges carrying supplies for the Reserve through the Bering Sea had no communication with their points of destination. "Barge captains had no idea what weather or ice conditions awaited them."[11]

In 1959, the Department of Defense determined that it needed to upgrade the World War II wireline network from Alaska through Canada. Canadian National Telecommunications (CNT), the crown corporation that had taken over operations of communications in the Canadian North from the Canadian military, was awarded the contract to build a microwave network from Grande Prairie, Alberta, to the Alaska border, where it would connect with ACS. At a cost of $25 million, CNT replaced the pole line with 42 microwave towers, installed diesel generators at 36 sites where no power was available, and built tramways to haul workers and materials to remote repeater sites. From Whitehorse, where he was participating in the official opening of the microwave network in July 1961, Canadian Prime Minister John Diefenbaker called President Kennedy.[12] The two leaders had never been close, but Diefenbaker must have felt he won this round: "It was the best communication that money could buy, and the American military had paid the shot."[13]

Ham Radio

In addition to the military-run ACS, amateur "ham" radio operators often played a critical role in Alaska communications. In emergencies, ham radio could be a lifeline. When a mother in Lazy Bay on Kodiak Island sought help for her five-year-old son suffering from pain and high fever, a local ham sent out a CQ URGENT message. A ham operator in Seattle picked up the message and described the symptoms to a doctor at the Marine Hospital in Seattle, who diagnosed a ruptured appendix. The hospital called counterparts in Anchorage, who arranged a medevac flight, saving the boy's life.[14] When the Kennecott copper mine, deep in the mountains north of Valdez, was struck by a massive snowslide, its wireline link was cut. But a local ham radio operator ". . . poked around the debris for his equipment, finally found all of it, put it back together, and within a short time notified the outside world of the disaster. Help was soon on the way."[15]

Officially, the U.S. Signal Corps had a monopoly on radio in the Territory, except on amateur radio. Other entities that wanted to set up their own two-way radio systems, such as canneries and airlines, had to get approval from the Signal Corps before they could start operating. The Signal Corps charged for messages, while ham operators passed messages for free. Alaska hams had heard that the radio inspectors would not allow them to handle messages of a commercial nature. This distinction could be very inconvenient for isolated Alaskans far from a Signal Corps station, such as a trapper on the Koyukuk

River who had broken the last mantle for his lamp and barely had time to radio an order for a batch of mantles to be sent on the last boat of the season.[16] A ham in Seattle heard about the supposed restriction, and contacted the local Federal Communications Commission (FCC) office, which referred the matter to Washington. Back came the official response: "As long as the amateurs do not accept remuneration for their services, the messages can be of any nature—barring, of course, indecent or profane language."[17]

Radio also helped overcome isolation in the bush. "Through radio, the party line of the Far North, whatever happened in almost any part of the land became common knowledge. We felt the cares and problems of distant strangers almost as if they were neighbors."[18] Women as well as men in isolated communities took up ham radio: "Through her station K7ANQ, on lonely Wosnesenski Island, Lily Osterback Evans handled hundreds of messages each month to distant parts of the Interior and Southwestern Alaska."[19] And radio operators helped to pass the news. "We used to get the news over our small radios, and relay it to distant communities by way of the phone. It was illegal to tune in on a commercial news broadcast and relay the news directly by putting our own transmitter in front of the broadcast. . . . But there was no law against memorizing the news as it came in, and then repeating the gist of it over the radio phone."[20]

Yet radio could also be unreliable, as Jay Ellis Ransom in Stevens Village remembered in the 1940s: "During the long, murky winter, radio communication developed troublesome phases. . . . Sometimes the radio waves traveled everywhere with amazing clarity. At other times we simply could get nowhere, either receiving or sending. Occasionally my transmitter, though operating perfectly, could not be heard beyond a range of ten miles. At other times it could be heard hundreds of miles away."[21] After being out of contact for several days, Ransom made contact with the Weather Bureau operator in Fairbanks, who gave a brief summary of the news. "It was in this way that we learned of the beginning of World War II. . . . It fairly raised us off our seats, and when the schedule was over, the trader and I tore through the village shouting the news."[22]

Alaskan hams in the Interior had many contacts in Canada; others on the Bering Sea coast had been communicating with hams in other countries as far away as Australia. But a few days after Canada followed England into World War II in 1939, the FCC notified the operators that no further contact between hams in Alaska and those in Canada was to be permitted. "America was to remain strictly neutral, and so must break off the friendly relations the radio hams had with all foreign countries." Apparently, ham communication was viewed as a threat to neutrality, and later as a threat to national security. However, when the United States entered the war, "we Alaska operators were permitted to continue service because we were needed in the hinterland to maintain commercial contact with the supply cities."[23]

Amateur radio played other significant roles in World War II. In 1940–1941, the FCC authorized an experimental license for a 1000-watt amateur radio facsimile transmitter developed by Augie Hiebert and Stan Bennett of KFAR to conduct tests with New York inventor Austin Cooley. They sent the first experimental facsimile transmissions over shortwave between New York City and Fairbanks. It took 8 to 10 minutes to transmit an 8- by 10-inch photograph, whereas mail delivery took a week.[24] The successful tests interested the military; the result was that facsimile equipment was manufactured during the war by *The New York Times* and used by the military for weather and other transmissions.[25] In 1942, the FCC's Radio Intelligence Division authorized the ham radio station K7XSB in Fairbanks to return to the air to transmit Siberian weather information intercepted by the FCC's secret receiving site at the University of Alaska to San Francisco, where it was decoded and sent to the Pentagon by radioteletype and used to plan bombing of the Kuril Islands off Kamchatka, then occupied by Japan.[26]

The Talking Lady of the North

In April 1957, a story in the *Wall Street Journal* exclaimed:

> The most exciting creature in Alaska today goes by the name of White Alice. She's expensive—a $100 million plus baby. And it happens you're paying part of her upkeep.
>
> She's dangerous. Fourteen men have died for her. . . . White Alice is no lady. Hers is the code name for Alaska's sensational new Government-financed long-distance phone and telegraph system. . . .[27]

In the 1950s, in addition to civilian demand, the threat of war again spurred the expansion of Alaska communications. With the outbreak of the Korean War in 1950, ACS started a construction program that lasted until 1956, upgrading the ALCAN highway network to the Canadian border, and in 1956, laying a new submarine coaxial cable from Skagway to Ketchikan, where it met a new AT&T deep sea twin submarine cable from Port Angeles, Washington. The new cable network was capable of handling 48 voice calls; each telephone channel could be used for 18 telegraph circuits.[28] ACS also installed new manual switchboards to handle calls at Tok, Delta Junction, Fairbanks, Anchorage, Juneau, and Ketchikan[29] and constructed new toll buildings at Anchorage, Fairbanks, and Ketchikan.

However, ACS was not the only communications system in Alaska. Other federal agencies constructed their own dedicated communications networks. The CAA (predecessor of the Federal Aviation Administration, or FAA) installed VHF repeater systems for aviation. The Alaska Railroad, then under the Department of the Interior, built an open wire system from Anchorage to Fairbanks along the tracks, and a mixture of open wire and microwave from Anchorage to Seward. The result was a lot of waste and duplication: "Between the Air Force, the Army, Navy, the CAA . . . and the Alaska Communications System, there were 14 different communications systems that you couldn't even tie together."[30] The incentive—and funding—to construct a single higher capacity network was the perceived threat of Soviet attack in what became known as the Cold War. In 1954, the military formed the Alaska Communications Study Group to develop a plan for a combined defense-government-civilian communications system and commissioned a study from AT&T to "recommend a suitable and economic way of creating such a network."[31]

AT&T proposed a new technique for over-the-horizon radio called "forward propagation tropospheric scatter," or troposcatter, that covered distances up to 200 miles. A feed horn on a tower sprayed signal against a curved movie-screen sized antenna, which then beamed the signal outward. Much of the signal was lost, but a small amount was deflected downward by the troposphere and received by another antenna over the horizon which could amplify and retransmit the signal.[32] The giant receiving antennas could pick one-ten-trillionth (1/10,000,000,000,000) of the signal sent out a couple of hundred miles away.[33] The entire network of 33 transmitting and receiving stations was designed to provide 3100 route miles of voice and telegraph channels using troposcatter plus some line-of-sight microwave links in high-density areas.

Troposcatter was also being used in the international Distant Early Warning (DEW) line, a communications defense network being built from Alaska across the Canadian Arctic to Greenland to protect the continent from aerial attack across the pole. Military strategists were concerned that enemy bombers might take the shortest flight path, the great circle route across the Arctic and central Canada, to reach major American industrial centers. Early detection of incoming bombers would provide an extra hour of warning to get interceptor planes in the air, alert anti-aircraft defenses, and allow civilians to take cover. The first phase approved in December 1952 was a trial installation of stations in Alaska. The U.S. Air Force awarded AT&T a contract to undertake full responsibility for engineering, construction, installation, and operation of a chain of radar and communications stations on Alaska's northern coast and have them functioning within a year.[34] The Air Force Research and Development Command justified its choice: "The Bell System with its integrated development, manufacture, supply and operating units appears to be uniquely qualified to assure

DEW Line site. *(www.beatrice.com/bti/porticus/bell/dewline.html)*

the success of a project which is of the highest urgency and importance to the defense of the Country."[35]

It appears no coincidence that in 1955, Western Electric (the equipment subsidiary of AT&T) got another military contract to build the section of the DEW Line across the Canadian North, as reported in *The New York Times*: "The Air Force announced today selection of the Western Electric Company, Inc., to build the radar warning system across the Canadian Arctic. . . . The United States will meet the full construction cost of the project. Estimates of the cost vary from $200,000,000 [$200 million] to $1,000,000,000 [$1 billion]."[36] In 1955, Western Electric also received a letter contract to begin construction of the Alaska troposcatter network that it had proposed. There is no evidence of any competitive bidding for either the initial study or the contract to build the system, both of which went to AT&T. Western Electric was already building the U.S. section of the DEW Line, had supplied the new submarine cable from Washington State, and had previously built radio-telephone facilities to link Anchorage and Seattle in 1941, as well as supplying equipment and crews to work with the army in constructing the telephone network along the Alaska Highway in World War II.[37]

White Alice was the code name the air force assigned to the project of building the new network, "Alice" apparently standing for "Alaska Integrated Communications Extension," the continuation of the toll network in the lower 48 states. The network would serve army and navy installations, the CAA, and the people of the Territory. It was also designed to be a vital part of the continent's air defense, enabling combat centers of the Alaskan Air Command to receive reports and relay information about aircraft detected by the DEW Line to NORAD Headquarters in Colorado. Thus, while White Alice became the backbone for much of Alaska's civilian communications, its primary purpose was "to provide reliable communications facilities vitally needed for the continent's defense against air attack from the north."[38]

At the peak of construction in 1956, more than 3500 workers including military personnel, civilian employees, and contractors were involved in building the network.[39] Bush planes, helicopters, snow vehicles, and dog teams hauled in the 14 tons of equipment needed to erect a 50-foot temporary tower at each site. Lee Staheli was one of the pilots who hauled gear: "In theory, Lee was a pilot, but when the rocks proved too tough for his little ski-plane, he took to a big tracked 'Cat.' When this kept breaking through the ice . . . he took to a smaller tracked 'Bug.' When this tumbled off a ridge and landed upside down on Lee . . . he continued by dog-team. Mission accomplished, he ended up in a hospital for 'a little rehabilitation.'"[40]

Again, the northern climate took its toll. The winter of 1955–1956 was the most extreme in Alaska in many years; ice and wind crumbled two test towers and delayed moving gear.[41] The huge antennas were protected using a de-icing system with three oil-fired heaters installed behind the antenna face generating "enough heat for 15 six-room houses."[42] By mid-1957, 14 men had died, mostly in airplane crashes.[43]

Near Anchorage in late 1956, an air force officer spoke across the first completed leg of the network: "White Alice had begun to talk."[44] The last link was placed in service between Bethel and Romanzof in February, 1958. "On March 15, 1958 . . . with a single telephone call, which symbolized the completing of the White Alice Network, Govern Michael A. Stepovich, Alaska's last appointed chief executive, enabled a quarter million square miles of wilderness to become articulate. . . . it was a just cause for celebration by thousands of bush-bound sourdoughs. Previously, they had been denied a reliable link with the vast 'outside.' Henceforth—literally—they would be 'as near as their telephone.'"[45]

The air force noted: "Supply and logistical problems have been enormously simplified with the advent of reliable communications. Manned stations along the DEW Line, for example, can make their wants known as readily in foul weather as they used to do in fair. Right down to the actual scheduling of cargo flights."[46] But the cost of adding the White Alice network had been much higher than anticipated. At the conclusion of the first phase in 1958, the initial construction cost, originally estimated at $39 million, had exceeded $113 million. With expansion to Shemya in the Aleutian chain and additions to the

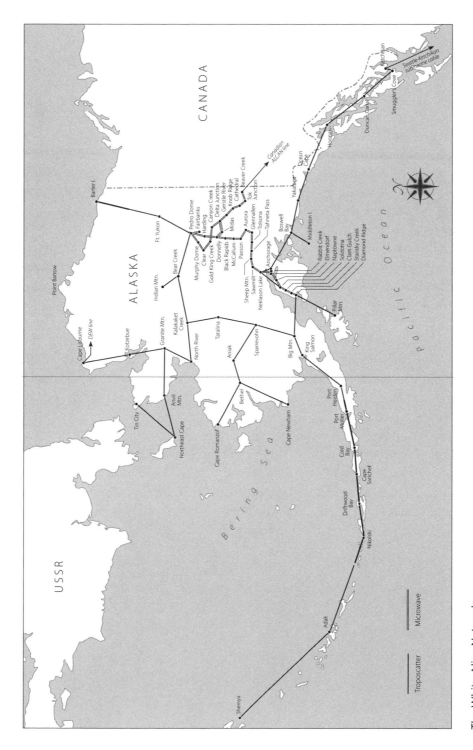

The White Alice Network.

microwave system in interior Alaska, the cost by 1962 reached almost $300 million.[47] It appeared that the Cold War allowed expediency to trump economy.

Earthquakes and Floods

In the 1960s, telecommunications again played critical roles in Alaska's history. After the Good Friday earthquake of March 27, 1964, at the Anchorage toll building, "women operators refused to leave their jobs despite the evident danger that another quake or tremor would bring down the damaged toll building. Just one and one-half hours after the quake, the emergency generator was turned on, and by 7:30 pm eleven toll circuits to Seattle were operational." Emergency centers were set up in Anchorage to let residents send telegrams outside. ACS reported that more than 12,100 calls were completed to other states in the first four days, and some 20,000 telegrams were delivered to Alaskans. A special CS press wire service transmitted 300,000 words of copy, equivalent to 20,000 telegrams.[48] More than 400 ham radio operators also sent an estimated 40,000 messages to other hams outside Alaska, who made collect calls to friends or relatives.[49] The day after the quake, the amateurs got help from the Military Affiliated Radio Station (MARS) at Elmendorf Air Force Base, which handled 7100 messages in the first three days, working through other MARS units around the country.[50] Local radio stations also broadcast thousands of emergency messages.

In August 1967, Fairbanks experienced the worst flood in Alaska history, as rampaging waters from the Chena and Tanana rivers destroyed much of the downtown and reached within five inches of the ACS toll and administration building. "The toll center became the only point in the city from which calls could be made or received. It became the command center for military and civilian officials . . ." including Governor Wally Hickel.[51]

Chapter 4

Early Broadcasting

Radio is the party line phone of the new Alaska.

Jay Ellis Ransom, *Alaska Sportsman*, 1946[1]

Early Radio Broadcasting

Wireless was first used in Alaska to extend the range of the open-wire telegraph network for point-to-point communications, and later for other services such as weather monitoring and aviation, and by amateur (ham) radio operators. In the 1920s, entrepreneurs began to install wireless transmitters for broadcasting (or point-to-multipoint communications). The earliest amateur license granted to a "special land radio station" was for station 7WQ, which operated in Kake for a short time in 1915. The first commercial radio station licensed by the Department of Commerce in Alaska began broadcasting in Fairbanks in 1922.[2] Licensed to the Northern Commercial Company (NCC), in addition to transmitting news and entertainment, WLAY was also a vehicle for selling radio sets. NCC initially set up a loudspeaker in front of the store for patrons who hadn't yet purchased sets.

The engineer from Seattle sent to install the equipment said that "on account of the absence of static, electrical or mineral disturbances and the general contour of the country," he anticipated that the 50 watts of power would cover a radius of 200 to 400 miles.[3] The military soon reported that the signal had reached Circle, a distance of about 160 miles. In winter, the signal traveled much farther, with reports in the *Daily News-Miner* that the WLAY signal had been received not only in Whitehorse but in Massachusetts, New York, Rhode Island, Pennsylvania, and several locations in the Midwest.[4] However, the station was apparently short-lived. Alaska broadcasting historian Tom Duncan could find no references in the *News-Miner* to WLAY after Jan 25, 1923.[5]

Before radio: Alaska Natives listen to a phonograph for the first time.
(Alaska Historical Society)

Juneau followed Fairbanks with commercial radio. Two months after
WLAY began broadcasting in Fairbanks, Alaska Electric Light and Power in
Juneau announced that its station KFIU was on the air in Juneau. KFIU was
initially assigned the same frequency as WLAY, but by 1927 it had changed to
1330 kc (KHz) with an authorized power of 10 watts. However, while Fairbanks
soon lost its station, KFIU continued in operation until KINY was established
in Juneau in 1935.[6] In May 1924, KFQD became the first radio station in
Anchorage, licensed to the McDonald Radio Shop, again apparently with the
intent of stimulating demand for radio sets. KFQD began operation at 870 kc
(KHz) with power of 100 watts[7] and continues to broadcast on AM and FM
in Anchorage today. Ketchikan's first radio station, KGBU, went on the air in
1927. Like KFIU in Juneau, it was established by the local power company.[8] But
the Ketchikan station had another clandestine purpose—to signal rumrunners
between Prince Rupert and Ketchikan: "A prearranged code consisting of certain words given at certain times by KGBU announcers informed boat operators
and others interested when the coast was clear."[9]

In the early years after the invention of telephony, some industry pioneers
thought that its most promising application would be transmission of concerts over telephone lines. In Alaska, this means of sharing entertainment was
adopted for a short time in Juneau before KFIU went on the air. In October

1922, the Juneau *Sunday-Capitol* newspaper reported: "At present, the government wireless station enjoys the advantage of nightly concerts, broadcast by the Seattle P-I and frequently they connect up with telephones in the homes or offices of city friends."[10]

Recognizing the importance of radio to reach remote residents, in 1931, the territorial legislature granted $10,000 to Alaska radio stations to continue a station message service to remote listeners "to disseminate news, executive proclamations and inquiries, information about the Territory, etc."[11] To ensure that listeners in town could hear local radio stations, municipal governments passed ordinances to limit electrical interference when the stations were on air. In April 1930, the Anchorage city council passed an ordinance prohibiting electrical interference with radio reception between 6 pm and midnight and empowered the city electrician or his assistant to enter premises at "[A]ny reasonable time for inspection purposes."[12] Violators found guilty were subject to a fine up to $100 and 5 to 50 days in jail. It appears that the city council was not concerned about constitutional rights to privacy.

Until the late 1930s, radio stations aired programs primarily produced in their own studios or transmitted from nearby locations over telephone lines. Sometimes local newspaper staff with access to wire services broadcast news reports. Some stations also bought programming on discs mailed from national distributors. The first interconnection between an Alaska station and national network occurred on Christmas Day 1937, when the U.S. Signal Corps received a program via shortwave from NBC's Red Network that was broadcast over KINY Juneau.[13]

Although other commercial radio stations had been licensed in Alaska, by 1939, only three stations survived—in Juneau (KINY), Anchorage (KFQD), and Ketchikan (KGBU). However, "after a sixteen-year absence from local civilian radio service, Fairbanksans welcomed KFAR in 1939. . . ."[14] The call letters KFAR, "Key for Alaska's Riches," were chosen by a community competition, with the slogan "From the Top of the World to You."[15] KFAR was funded as a gift to the pioneers of the Alaska interior from Austin E. Lathrop, known throughout Alaska as "Cap" Lathrop, a former river boat captain and mining magnate who later founded Midnight Sun Broadcasting.

One of the first employees was a young engineer, and later influential broadcaster, named Augie Hiebert, who recalled "before radio, Fairbanks was pretty much out of touch . . . the only time they'd get outside radio was when the weather was good and that was just in the winter time, maybe one or two or three weeks when it was dark."[16] Hiebert lived with colleague Stan Bennett at the KFAR transmitter, site of the tallest tower in the Alaska Interior, some 60 miles from the Fairbanks studios. Hiebert and Stan Bennett took turns with three- to four-hour shifts transcribing Morse code from Trans-Radio Press, received on shortwave using directional antennas pointed at New York and San Francisco. Hiebert remembered that the code was sent at 45 to 50 words per minute (two and a half times the speed required for commercial telegraph license for operating ships), a challenge to transcribe. They then took turns driving the 60 miles to town to deliver copy for the noon and evening news. However, the shortwave reception was unreliable, sometimes with fadeouts because of sunspots, "and when that happened, our announcers would read two- or three-week old *Time* magazines and the news would be pretty outdated."[17] They continued to copy bulletins in Morse code for the station until 1943, when KFAR was able to receive the AP wire service via radio teletype.

As in the early days, radio broadcasts were not only a source of national news and popular music; they provided important information for those traveling in the region, and a connection from Fairbanks to Interior villages as well as among the villages. "Weather Permitting" announced the air taxi schedules for the day so that people in villages and mining camps could plan for trips or cargo and could clear their runway before the flight arrived. If the mining companies didn't know when the goods were going to be delivered, "by the time they'd get the Cat train out there, bears could have broken into the supplies and just totally destroyed them. This way they would be out there and meet the plane."[18] "Tundra Topics" was an evening program of messages for the bush. People could call in or take messages to the radio station to be announced on the air about travel plans and delays, and news of relatives in the hospital or away at school. Sometimes the messages saved lives. Hiebert tells of a family that had bought potatoes in Fairbanks and taken them home to Fort Yukon. The Northern Commercial Company discovered that it had sold the family highly toxic treated seed potatoes and asked KFAR to announce that if anyone knew how to reach the family, to tell them not to eat the potatoes. "Somebody got into a boat and went upriver to their house, and they were boiling the potatoes . . . they were going to have them for a meal very soon."[19]

On December 7, 1941, Hiebert heard about the bombing of Pearl Harbor via short wave and notified the military in Alaska; they had not yet heard about the attack.[20] When war broke out, with support of the military, the Federal Communications Commission (FCC) authorized KFAR to upgrade its power from 1000 watts to 10,000 watts on clear channel 660 KHz in order to transmit news and entertainment to troops in the Pacific and to provide navigational service to military planes in the area. KFAR then was programmed as an Armed Forces Radio Station for the duration of the war.[21] In May 1942, *The New York Times* described KFAR's contribution to military morale: "The Japanese on Kiska Island call it 'the cesspool of information,' but to United States troops on remote, cold assignments, it is the last connecting link with civilization, being their primary source of news and entertainment. Such is the role of one of America's more interesting and important stations, KFAR, in Fairbanks, Alaska. . . ."[22]

The station was also an important source of sports news, both local and national. KFAR covered the sled dog race from Fairbanks to Livengood in 1940, the first dog race broadcast by radio.[23] Studio announcers would recreate play-by-play coverage of professional baseball games in the lower 48 from telegraph transmissions. KFAR also served as a community messenger in emergencies. On Christmas Eve 1946, fire destroyed the Fairbanks telephone exchange. Hiebert organized a network of ham radio operators throughout the city; twice a day KFAR broadcast a check-in by each of the amateur operators to keep the public informed.[24]

After the war, radio stations proliferated in Alaska, with more stations going on air in Anchorage, Fairbanks, and Juneau, and communities getting their first local stations in Seward, Sitka, Cordova, Glenallen, North Pole, and Soldotna. The advent of the transistor radio in the early 1960s meant more Alaskans bought radio sets, especially in the bush. In 1960, Nome's KICY, established by the Arctic Broadcasting Association owned by the Covenant Church, went on the air on Easter Sunday. KICY became Alaska's first "non-profit commercial radio station" with religious ownership that accepted ads and sponsorship. Transmissions from its 250-foot antenna were apparently popular in eastern Siberia. During the Cold War, Radio Moscow's Siberian AM transmitters covered Alaska's west coast with programming in English and Inuit languages.[25]

Alaska's first FM radio station went on the air in Anchorage in September 1960 as Northern Broadcasting's KTVA-FM, later changing its call letters to KNIK-FM. A commercial station, it was the first FM station to carry the Metropolitan Opera broadcast sponsored by Texaco. Marvin Weatherly, the station's engineer and first announcer, later became director of the Governor's Office of Telecommunications, executive director of the Alaska Educational Broadcasting Commission (AEBC), and a commissioner on the Alaska Public Utilities Commission (APUC).

Government-owned military radio stations operated in Alaska for several decades. In 1941, military radio stations were established in Sitka and Kodiak, the latter by army staff from Fort Greely. Also, the Nome Civil Defense Commission had started a carrier current radio station using existing power lines, telephone lines, and pipes as conductors to create a closed-circuit radio station.[26] Kodiak's KODK broadcast "local programs from its miniature 'radio city' studio, as well as big commercial programs intercepted on a 15-tube set and fed into KODK's transmitter. . . . Bing Crosby, Bob Hope, Fred Allen, Jack Benny—none is too big for KODK to pirate off the air and make its own. KODK sells no advertising, but does not object to relaying the toothpaste commercials right along with the highjacked music and comedy."[27]

However, the Armed Forces Radio Service (AFRS) was not officially established until May 1942 "to provide education, information and orientation for our Armed Forces overseas by means of entertainment and special events broadcasts."[28] In this case, "overseas" included Alaska. One of the initial activities of its first director was a trip to Alaska to see firsthand the soldiers' needs for information, education, and recreation in the field. Director Tom Lewis later said that an encounter with a soldier in Alaska helped him see the value of radio for troops in wartime. Lewis's driver complained that he didn't know why the military had sent him to Alaska. "All I do is drive VIP's back and forth on this road. I don't know anything else."[29] Lewis believed that if American forces were going to win the war, they needed to understand why they were fighting, and radio could help provide that information, as well as boost morale.

Thus, "American Forces Radio and Television Service literally began in Alaska during the first weeks of World War II. . . . It is the only state where such broadcasts could be regularly heard by anyone dwelling near a military facility with such service. . . ."[30] By the end of the war there were 23 military radio stations covering troop installations and nearby civilian communities;[31] the military also bought time on commercial stations and provided programming for them.

By late 1959, the White Alice system fed all Alaska military stations and others at U.S. installations in the Canadian North. The number of military radio stations increased from 9 to 32 by 1964; most were repeater stations retransmitting programs from Elmendorf Air Force Base near Anchorage over White Alice. The peak of military broadcasting in Alaska was in 1971 when 39 radio and 8 television stations were operating. Some 21 Armed Forces Radio Stations were still in operation in 1986. With the increase in local radio stations and network television affiliates, and the capability to receive TV over local cable systems and directly via satellite to small antennas, Alaska military installations ceased to resemble isolated overseas outposts as they gained access to domestic communications like their counterparts in the lower 48. The Armed Forces Network signed off the air in Alaska in June 2001.[32]

The first noncommercial and nonmilitary radio station in Alaska was KUAC, established in 1962 at the University of Alaska Fairbanks. Government-supported public broadcasting did not expand in Alaska until the 1970s (see Chapter 10).

Early Television

After World War II, broadcasters were eager to invest in the new medium of television. However, in September 1948, the FCC imposed a "freeze" on television licensing, determining that the new technology had outgrown the existing regulatory structure. The regulatory thaw did not occur until 1952, with the release of the FCC's Sixth Report and Order including new rules for television and making available 70 additional channels using UHF spectrum. Nineteen VHF channels were reserved for Alaska for Anchorage, Fairbanks, Juneau, Ketchikan, Seward, and Sitka.[33] But until the freeze was lifted, only stations with construction permits could go on air, and no new license applications were accepted.

In 1952, Hiebert, who had moved to Anchorage to establish Midnight Sun Broadcasting's radio station KENI, worked on a television license application for Anchorage. However, Midnight Sun's founder, "Cap" Lathrop, had died in 1950, and the executors of his estate were concerned about the cost of the project and its potential threat to the income from the estate-controlled radio and theatre properties.[34] Hiebert eventually resigned as manager of KENI, and with other KENI employees and Anchorage businessmen founded Northern Television Inc., which applied for the TV license. The FCC issued a construction permit for channel 11, with 3200 watts, and KTVA was the "first official commercial television station in Alaska to go on the air" on December 11, 1953, choosing a date the same as its channel number.[35] Vic Fischer, then Anchorage's first city planner and later an author of the state's constitution, remembered: "It was just wonderful. . . .We were always able to use radio, but here we were able to have visuals and show maps of existing conditions and the proposals and talk directly to people who could see what we were presenting."[36]

A California-based partnership filed at the FCC a week after Northern Television for channel 2, KFIA (First in Anchorage), which actually went on the air on October 15, but had initial transmission problems.[37] Midnight Sun, managed by Al Bramstedt, Sr., another KFAR pioneer, purchased KFIA in December 1954, and invested in a new transmitter tower in 1955 when the station became KENI-TV.[38]

Hiebert estimated that there were 5,000 to 6,000 TV sets in Anchorage when KTVA signed on, growing to about 15,000 sets for a population of 35,000 by the end of the first year.[39] However, unlike the early radio broadcasters, the television entrepreneurs did not base their business model on selling TV sets but on commercial advertising. KTVA combined commercialism with public service and patriotism in its opening proclamation:

> One hundred years from now the citizens of Anchorage will not remember what has happened here at KTVA on the 11th day of December 1953. That it was KTVA's first official telecast will be of little concern to them. . . . However, there will be something begun here today which will be important to them, something with which the people of 2053 will be concerned

and with which they will have to deal in their own time. It is the question of whether or not KTVA maintains its integrity as a public servant, its support of civic institutions, local business, its loyalty to Alaska and unfailing allegiance to the United States. This, the Anchorage citizens of 2053 will be vitally concerned with, for this will directly effect [sic] them and their way of life. . . .[40]

Hiebert soon got requests from Fairbanks: "The business people there asked if I would come up and start the first TV station there, too. So I went up and sold a little more stock, enough to start a station in Fairbanks."[41] Fourteen months after KTVA, Northern Television's KTVF was the first TV station to go on the air in Fairbanks, in February 1955. Midnight Sun Broadcasting, which had bought the assets of KFIA in Anchorage, soon followed to open KFIF in Fairbanks, which became KFAR-TV. A year later, KINY-TV Juneau began broadcasting in February 1956.[42]

Of course, Alaskans outside the major cities also wanted to watch television. The solution at the time was TV translators, which, like radio translators, would amplify and retransmit the TV signal to reach an isolated community. The first permanent TV translator in Alaska was installed in 1959 to reach Suntrana village in a valley about 125 miles from Fairbanks, and then extended to Usibelli Mine and Healy. Augie Hiebert and local miners climbed a 2500 foot ridge above the mine to install a test translator: "We didn't carry a pole for the antenna because we expected to cut down a tree—only to find that we are above tree level when we got there. We finally found a crooked stick, mounted the TV antennas on it, turned on the equipment and asked the Chief Engineer on the ground what he saw. . . . The picture from KTVF Fairbanks was excellent, and we decided to leave the translator on the air . . . so the kids could watch 'Lassie' and then we came down the mountain."[43] Later, translators were installed at Delta-Fort Greely and Nenana, and down the Kenai Peninsula from Anchorage to cover the Kenai/Soldotna area.

In 1966, Midnight Sun Broadcasting was the first to introduce color television in Alaska. Al Bramstedt, Jr. remembers his father watching test color transmissions aired in the middle of the night on KENI-TV when the station was normally off the air, to keep his color plans secret from competitor Augie Hiebert.[44]

Broadcasting Themes

As with telegraphy and telephony, Alaskans were pioneers in adapting new technologies for broadcasting to link residents to each other and provide information from the outside world. However, the driving force for broadcasting was commercial enterprise, and not the military or other government sources that had funded and operated telecommunications services. Augie Hiebert noted in his announcement of the plan for KTVF in Fairbanks: "It has been our opinion that the key to successful television operation in Alaska is when ownership is retained by the businessmen and investors of the city which the station is to serve."[45]

Yet markets were small, advertisers often few, and national content (if not pirated) expensive. The affiliate model which began with recorded content sent by mail from national networks, and later evolved to satellite feeds, became a lifeline for commercial broadcasters. Local ownership also led to market concentration, with dominance by key entities such as Midnight Sun Broadcasting and Northern Television. Cross-ownership with other media such as the *Fairbanks News Miner* (owned by "Cap" Lathrop from 1929 to 1950) also concentrated media ownership.

Tom Duncan points out that "One of the most significant developments in Alaska was government-funded broadcasting for American residents."[46] As with telecommunications, the earliest government entity involved with broadcasting was the military, through its operation of Armed Forces radio and later television stations. Although their broadcasts could be heard in the surrounding communities, the military's mandate was not to serve civilians, in contrast to its operation of the ACS. Later government involvement included state and federal support for public broadcasting stations established primarily in the 1970s.

From the late 1960s, the evolution of broadcasting in Alaska became increasingly linked to satellite communications and other new technologies. Expansion of radio and television geographically throughout the state and thematically into educational and other noncommercial programming is discussed in the following chapters. However, comprehensive coverage of broadcasting in Alaska is beyond the scope of this book. Much of this history through the 1970s can be found in research papers by Tom Duncan, formerly an employee at KUAC, at the University of Alaska Fairbanks, which were key sources for material in this chapter.[47]

Chapter 5

Privatizing the Alaska Communications System

The sale of ACS was unusual because it looked to the future: it took into account Alaska's telecommunications needs, rather than just going for the highest bid.

Institute of Social and Economic Research[1]

An Outmoded Network

In the early 1960s, ACS consisted of nearly 12,000 km (7,500 miles), with major components of the network being the White Alice System, the DEW line, its Alaska extension, and the BMEWS (Ballistic Missile Early Warning System) routes. The air force installed the Alaska Switching System to increase the capacity of the network, which previously relied on many dedicated channels. Its name was later changed to Alaska Telephone Switching System "to facilitate a more appropriate acronym."[2] However, by mid-decade, ACS was not keeping up with commercial demand and changing technology. Most of the White Alice routes available for commercial traffic were saturated, and quality of service was poor. In the late 1960s, it was increasingly difficult to get through to the long-distance operators at the manual switchboards, and busy signals on long-distance circuits were the rule rather than exception. For calls within Alaska, "Delays were never less than thirty minutes, some more than an hour." Articles in Alaska newspapers complained: "A Long Distance Call? Be prepared to wait . . ." or "Why it's a pain to call Long Distance."[3]

Most villages still relied on high-frequency radio if they had any service at all: "Villages . . . existed in a communications vacuum a great deal of the time."[4] In 1965, even Sitka was connected to the network only by HF radio via Hoonah and Biorka. The air force had no funds to upgrade the connection and said it would oppose any private applications to the FCC to install cable. There were long waits for circuits and bad connections, and the automatic typesetter purchased by the *Sitka Sentinel* to print wire service stories directly off the wire from Seattle wouldn't work.[5]

Yet long-distance prices kept going up and were much higher than for comparable distances in the lower 48 states. In 1965, the lowest interstate rate from Anchorage to Seattle was $5.25 for first three minutes, whereas AT&T's rate for a three-minute evening call from New York to Seattle was $1.50. In 1966, an eight-minute call (the average length of Alaska calls) from Juneau to Seattle cost $17, whereas a call in the lower 48 over a similar distance from New York to St. Louis cost $4.25. Teletype private line leased services for airlines serving Alaska were more than double Outside rates.[6] The *Anchorage Daily News* editorialized: "In this era of enlightened communication, these are fantastic penalties assessed the state of Alaska for the use of one of its most essential instruments of conducting day to day business operations. It curtails our activities and blights our attempts to bring costs to the proper levels."[7]

Alaska had become the 49th state in 1959, but still had what amounted to a colonial communications system, owned by the federal government and operated by the military. The Congressional legislation that had authorized the army to build a telegraph system to connect military posts in 1900 had authorized the army to accept commercial traffic: "Commercial business may be done over these military lines under such conditions as may be deemed by the Secretary of War, equitable and in the public interests."[8] The statute required that net revenues be paid into the U.S. Treasury, so these funds could not be used for operation or upgrades of the system. Each year's operational costs had to be provided by a congressional appropriation initiated by the military.

Thus the Alaska Communications System was much like European government-owned telecommunications systems of that era, which typically had to turn over revenues to their national treasuries and depended on annual government budget authorizations. Yet ACS had additional layers of bureaucratic disincentives, as it was a military system that had a secondary function of carrying commercial traffic and no mandate for expansion to meet its nonmilitary needs. Its annual requests had to go up the chain of command in the Pentagon, often frustrating officers operating the network, who realized the need for modernization and expansion.

Options for the Future of ACS

In mid-1962, responsibility for ACS was transferred from the Signal Corps to the air force. Beginning in the early 1960s, the army, air force, and the Bureau of

the Budget had indicated an interest in selling ACS.[9] In 1963, the air force commissioned a study by the Middle West Service Company on the financial status
of Alaska's long-lines communications and their potential salability.[10] In 1965,
the air force carried out a study for the Secretary of Defense that recommended
the sale of ACS, proposed guidelines for disposal of the system, and suggested
short-term alternative options if the sale were not authorized or completed.[11] It
also explored alternative strategies of financing ACS to apply a more "business-
like" approach to its operations. The Federal Field Committee for Development
Planning in Alaska, which had been established to plan for economic redevelop-
ment of the state after the 1964 earthquake, worked closely with the Secretary
of the Air Force's office in framing the Department of Defense (DOD) proposal
for sale and recommended to the parent Cabinet Committee on Alaska that
the sale be authorized.[12]

In 1965, Alaska's Senator Bartlett stated in the Congressional Record: "It
has become clear recently that the volume of business passing over the system
is approaching the point at which revenues will equal expenses. . . . Since 1959,
no capital improvements of any consequence have been added to the ACS plant
in Alaska."[13] ACS had actually programmed and requested funds for upgrades
but had been turned down since 1960 by the Defense Communications Agency
(DCA) in the Pentagon. Meanwhile, Senator Bartlett had asked several members of his staff to carry out a study of ACS and explore future options. The
report was completed in August 1965.[14] The staff found that starting in 1964,
ACS brought more revenue into the Treasury than the military had requested
in appropriations for the network, and that some ACS employees themselves
thought that ACS ought to be "in the black" in three to five years.[15] They also
sought opinions from industry officials familiar with ACS. The chairman of
BC Tel (then majority-owned by GTE) thought that ACS could be operated at
a profit if several changes were made, including replacing operators with direct
distance dialing (DDD), reducing military overstaffing, introducing low "after
8 (pm)" rates to generate additional revenue, introducing TWX instead of leased
circuits for teletype, and renegotiating connecting contracts with AT&T.[16]

Senator Bartlett's staff discovered that the Middle West Service Company
had made similar recommendations in the study it had carried out for ACS in
1963. This report had never been released outside the air force until Bartlett's
team made a special request for it and visited the Pentagon; the report was then
released to them but marked "For Official Use Only." The Middle West recommendations included:

1. Existing contracts with connecting telephone companies should be
   renegotiated to bring in more revenue per message to ACS and less to
   the connecting companies;

2. ACS should "enter upon a concerted promotional campaign through
   the medium of advertising";
3. ACS headquarters should be moved from Seattle to Anchorage;
4. ACS should develop a true cost accounting basis for its operations.[17]

Bartlett's staff identified several options for the future funding of ACS.
They included different operational structures such as an industrial fund like
that established for the Alaska Railroad, with operations funded from revenues
and loans from the U.S. Treasury for improvements, or a federal corporation
like the Panama Canal Company. Another option was to place ACS directly
under the secretary of defense rather than the air force so that it did not have
to go through so many bureaucratic levels for approval for funding. ACS could
also contract out its operations or lease its facilities. And finally, ACS could sell
some of its assets such as its toll centers to local telephone companies, or the
government could authorize an outright sale of the entire system.[18]

Senator Bartlett's staff concluded "while not regulated by the FCC, ACS
has used its influence as a Government agency to create and develop a total
monopoly of intra-Alaska and inter-Alaska long distance communication."[19]
While they were conducting their study, ACS exercised its monopoly power
and further antagonized Alaska residents and businesses by raising some of its
rates. ACS introduced rate classifications of person-to-person and station-to-
station, but initially increased the person-to-person rates above the previous
unclassified rates. "With no advance notice to the Governor, the congressional delegation, to Alaska local utilities or to chambers of commerce, the new
rates were announced on April 19, 1965. There was 'intense public reaction
to the new rates.'"[20] ACS officials responded that their decision was intended
to enable ACS to lower its station rates substantially below the previously
undifferentiated rates, as they had been instructed that they could make
whatever changes they wished to the rate structure as long as the system continued to return to the U.S. Treasury the amount it generated in 1964, a total
of $8.9 million.

Business customers also protested the high rates and antiquated services. A
business in Kodiak complained about teletype: "we, in Alaska, are not allowed
to share a line to Seattle. A one-party line is prohibitive in cost. The year rate,
quoted for a comparable distance in the South 48 is equivalent to what we would
pay for one month." The Bartlett staff, however, found that in Whitehorse, Telex
or TWX had been introduced, "and in very short time, 100 terminals were
installed."[21] The local telephone companies were also frustrated with lack of
air force investment to get direct distance dialing (DDD); they proposed to buy
most of the equipment themselves but never did so. ACS staff themselves had
made several requests for Telex or TWX on an experimental or pilot project

basis that were rejected at the Pentagon level, as were their requests to install switches for DDD.

Toward the Privatization of ACS

In 1966, Senator Bartlett introduced a bill to sell ACS, and Senate subcommittee hearings were held in the spring. Elmer Rasmuson, mayor of Anchorage and chairman of the board of the National Bank of Alaska, testified at the hearing that the bill to permit disposal of ACS was "one of the most important considerations by Congress since statehood."[22] He noted the growing importance of data communications for Alaska businesses: "High speed computers have been installed by the state, banks and industry for data processing, but we cannot get full utilization by tying in branch locations. We cannot get economical rates for data processing, even at nighttime when general usage is low. . . . Separate accounting and machine installations would be eliminated if we have economical message transfer. The key to reduced interest rates and the tapping of central investment funds is through cheap and constant communication with money centers."[23] Emphasizing that "the service is poor and charges excessive," he demanded prompt passage of a bill to sell ACS, and said he believed a privately owned company would be able to reduce rates: "I'd be surprised if it wouldn't be at least 20 percent. . . ."[24]

Joseph H. Fitzgerald, chairman of the Federal Field Service Committee for Development Planning for Alaska, told the Senate subcommittee that the "primary focus of the government in disposing of the commercial telecommunication system is to assure the creation of a public utility system adequate to serve the needs of the state."[25] Concerning the commercial viability of the system, he stated: "The government is, in no sense, seeking help from industry. Alaska is a market with real growth potential, and its location, spanning much of the distance between North America and Asia, as well as having polar position central to all continents of the Northern Hemisphere will place the successful bidder in a potentially strategic world position."[26]

The *Anchorage Daily News* pointed out that Alaska had the lowest number of lines per capita (or teledensity) of the 50 states, with 29 lines per 100 inhabitants, compared to the U.S. average of 46 lines per 100 population.[27] Of course, many villages still had no telephone service at all. Under the headline "The Sale of ACS: Let's Get on with It," the *Anchorage Times* editorialized, "Hopefully the men of the Senate won't stand still too long for the same busy signal Alaskans have been getting for years when they called to complain about high rates and poor service. The inequities which exist in our present service are simply too evident to be ignored."[28]

Local telephone companies also endorsed selling ACS. Harry Reimer of the Alaska Telephone Association (ATA) urged a negotiated sale of ACS long-lines facilities to the ATA member companies. USITA, the national association of independent (non-AT&T) telephone companies, also endorsed the bill, asserting that commercial operation of the toll network should result in increased settlements for local companies, allowing them to reduce rates for local service.[29]

ACS remained recalcitrant about rates. In June 1966, Lt. Col. William G. Giel, the commander in charge of ACS, told members of the Anchorage Chapter of the Alaska Society of Professional Engineers: "There's been a lot of yak-yaking about rates." He stated that the rates were established using an AT&T schedule based on the number of installed phones divided by revenues received; there were only 57,000 phones in Alaska compared to 90 million in the rest of nation. Giel said the City of Anchorage (owner of the Anchorage Telephone Utility) received 94 cents from ACS for all long-distance calls using city equipment and wanted $1.25 per call, but ACS thought $.77 was the proper figure. But he added: "Offer me $35 million bucks and I'll willingly sell you the entire system."[30]

Giel's rate analysis was contradicted the next week in a speech to the Anchorage Chamber of Commerce by Robert Ely, an Anchorage lawyer on Senator Bartlett's staff who had worked on the Bartlett staff study: "I believe the system can be operated in a manner in which significant rate reductions can be made. . . . And I'm happy to report that as a result of the recent hearings in Washington on the proposed sale, there is growing support for getting the system into private industry. . . . Senator Bartlett was not completely convinced that the transfer should be made until those hearings, but now he and many others in Congress will work for the sale."[31]

In October 1966, ACS introduced a 10 percent rate reduction on interstate calls. ACS had negotiated a reduced cable rate in a new contract with AT&T after threatening to get lower rates by using a BC Tel cable. The annual savings of $1.33 million per year were to be passed on to customers.[32] Congress moved slowly. Meanwhile, a year later, on October 1, 1967, ACS shifted to an "industrial funding operation" with telephone and telegraph services operating on a self-sustaining basis, paying their own way and paying for improvements. But on October 2, 1967, the House approved the Senate bill authorizing the sale of ACS.[33]

The Alaska Communications Disposal Act[34] was finally passed on November 14, 1967. In introducing the original Senate bill at the 1966 hearings, Senator Bartlett had said: "Bidding (for the system) should not be a tug of war to see how high a price the federal government can get." Thus the legislation included the stipulation that in addition to stating the price they proposed to pay for the system, "offerors must specify the rates they propose to charge for service and the improvements in service which they propose to initiate." And the previously unregulated ACS was to become a regulated carrier: "the rates and charges for such services applicable at the time of the transfer shall not be changed for a period of one year from the date of such transfer unless approved by a governmental body or commission having jurisdiction"[35]

Douglas Jones, then a Congressional staff member who helped draft the legislation to sell ACS, explained the strategy:

> Central to the legislation, we felt, was an innovative provision intended to maximize the chance for a longer-term public-interest outcome. We would require a bidding process that (1) would establish a single "upset price" for the assets, which all bidders had to meet, and (2) bidding would be based on how much new investment in the system would be made and how much the carrier would lower telephone rates to customers. The winning bid, then, would be determined not on price, but rather on the "best" combination of improved service and lower charges. As might be imagined, the Treasury Department was less enthusiastic about the concept (of not getting "top dollar"), but ultimately acquiesced. The beauty of doing it this way is that it allowed future Alaska telephone subscribers to pay prices *lower* than would be the case if ACS had been "auctioned off to the highest bidder: the winning company having to recoup its higher-cost investment in higher tariffs under traditional public utility regulation."[36]

It appeared that the large U.S. telephone companies would be the major potential bidders: AT&T, GTE, RCA, ITT, and Continental Telephone. Other possible purchasers were the local telephone companies in Alaska, Alaska investors, and perhaps even the state of Alaska. Given AT&T's many military contracts for Alaska communications equipment, it was considered the leading candidate. However, an AT&T reorganization had placed Alaska in the "foreign division," and its overall appraisal of the Alaska market was pessimistic.[37] It had been Western Electric, the equipment division of AT&T, rather than its long-lines operating division, that had benefited from the military business. RCA Service Company operated and maintained the White Alice network from 1960 to 1969, as well as other major government projects in Alaska under separate contracts including the Ballistic Missile Early Warning System (BMEWS) site at Clear Air Force Station and the NASA satellite detection and tracking facilities at Gilmore Creek. ITT Federal Electric (Arctic Services) operated and maintained the DEW Line system during the same time period. In 1968, the air force sent bid invitations to GTE, United Telephone, Continental Telephone, RCA, Western Union, Western Union International, and Comsat (the satellite company that planned to build an earth station near Talkeetna—see Chapter 6).

While the planning of procedures for the sale of ACS dragged on, service continued to deteriorate. Lack of modernization was used as justification not to reduce rates further. ACS in Seattle thought that any further station-to-station rate reductions would result in so much more traffic over ACS circuits that long-distance operators would not be able to handle them. Only DDD could handle such an increase in traffic, but the air force refused to finance the investment "because of the anticipated sale."[39]

In December 1967, the Anchorage Telephone Utility (ATU) was still handling handwritten toll tickets and estimated that DDD would save more than $281,000 per year.[40] ATU said it was ready to install DDD equipment with little cost to ACS, but that ACS refused to convert from manual operations pending the sale of the ACS system to be sure the equipment was compatible with that used by the new owner. ATU emphasized that compatibility was not a problem: "In diplomatic terms, the Anchorage telephone utility says this is poppycock." By August 1968, ACS had 54 manual switchboards in Anchorage, handling more than 10,000 calls per day. With traffic growing about 20 to 25 percent annually, ACS described the problem as a "shortage of operator stations."[41]

The Sale of ACS

By 1969, oil had been discovered on the North Slope, with the promise of greatly expanded business activities in Alaska and enormous tax revenues for the state. Senator Bartlett had died of a heart attack in December 1968, and Ted Stevens, an Alaska lawyer and state legislator, was appointed to fill his term (and subsequently elected and re-elected, serving until January 2009). On February 28, 1969, the day before bids were opened, Stevens made a speech on "Modern Communications in Alaska" in the Senate: "Alaska, experiencing what may be the greatest oil strike of the century, Alaska, one of the fastest growing States in the Union, is hogtied by inadequate communications. . . . It is reasonable to believe that a privately owned company could operate the system and do so, over the longer run, with every prospect of having a profitable operation. This belief is strengthened, if . . . Government business is carried on the system."[42] He reiterated the requirements Senator Bartlett had included: "Transfer of the Alaska Communications System must be carried out in such a way as to serve the interests of the Alaskan citizenry by obtaining firm commitments to improve service and lower rate schedules, and by obtaining a timetable for the implementation of both, as an integral part of any bid proposal action. . . . It will be to the benefit of the Federal Government, as a customer, to have improved service and lower rates. And assured Government business, as part of the rate base, would go far toward making these lower rates possible."[43] Stevens would champion efforts to extend modern telecommunications throughout Alaska during the following decade.

On June 25, 1969, President Nixon approved the sale of ACS to RCA Global Communications, Inc. (RCA Globcom). The letter from the Office of the Secretary of the Air Force to Howard Hawkins, president of RCA Globcom, listed the terms of the sale:

> to purchase all the property the Federal Government offered for sale (except the Healy exchange) at a total estimated price of $28,431,132; to expend $27,683,000 within three years of the Transfer Date for system improvement and modernization, including new service to 142 remote communities and the commencement of Direct Distance Dialing service by late 1970 for the Anchorage area, early 1971 for the Fairbanks area, and mid-1971 for the Juneau and Ketchikan areas, and including purchase of a 50% ownership interest in the COMSAT Earth Station currently planned to be constructed at Talkeetna; and to continue present service at the rate schedules proposed in your Company's final offer, providing as well a new lower rate 'after midnight' service and a bush and marine rate reduction, so as to realize the total rate reductions stipulated in your final offer.[44]

RCA then had to seek approval from the Federal Communications Commission (FCC) to operate in Alaska. In its application, RCA summarized its commitments for the network:

> In addition to paying $28.4 million for the existing ACS, RCA Alascom proposes to reduce interstate and intrastate rates by nearly one-third overall. . . . During the three years following the acquisition of the ACS, RCA Alascom also proposes to make service improvements and extensions at a cost estimated to be in the order of approximately $27.6 million. . . . It will provide the people of Alaska with new types of communications services that are not now available to them, including Telex, Data Phone, data transmission, full-time telephone service to 142 small villages, and others. It will play an integral part in the establishment of satellite communications for the Alaskan people involving exciting new concepts. It is anticipated that the system will be capable of providing all types of services, including furnishing live educational and commercial television to all parts of the state at a cost RCA Alascom believes will be well within the means of the State of Alaska.[45]

RCA informed the FCC that it expected to generate only $25 million of long-lines revenue in the first year and to retain only about $13 million after payouts to local telephone companies, the military, and others for leases of long lines. The company forecast that it would operate at a loss for at least the first two years, and that several years would pass before it could earn "a reasonable return on its investment."[46]

Western Union International (WUI) raised antitrust issues: "It is difficult to imagine an arrangement less compatible or consistent with the antitrust laws than one grounded upon absolute monopoly, which appears to be RCA's view of the ACS transaction."[47] However, the FCC included antitrust provisions concerning RCA sales of equipment and services and using its facilities to favor any affiliated entity or disadvantage competitors. It prohibited "RCAA (Alascom) or an affiliated company from providing a computer utility service or operating a CATV system or commercial broadcast station in the State of Alaska for 10 years after the sale without the consent of the Department of Justice . . ." and prohibited "RCAA from restricting the attachment of customer provided equipment with its communications facilities, except as necessary to protect the technical operation of the facilities. . . ."[48] It further ordered that RCAA would be regulated by both the FCC and a State Commission, stating "we believe that rendition of common carrier service in Alaska by a nongovernmental entity subject to appropriate government regulation will have a salutary effect on the development of communications in the State."[49]

Concerning the establishment and funding of RCA Alascom (RCAA), the FCC ordered the parent company to fund its new subsidiary: "Prior to July 1, 1970, RCA Globcom will purchase 1,050,000 shares of RCAA stock for $21 million and RCA will advance RCAA $14 million, so RCAA will have initial cash assets of $35 million."[50] The air force finalized the sale of ACS to RCA Alascom on Jan 10, 1971. The company had three years to invest $28 million in modernization and expansion, reduce rates by $40 million, and provide telephone service to 142 bush communities.

State Certification

Before statehood, a territorial statute imposed a duty on public utilities to serve the public without discrimination, but in general, regulation of utilities was left to municipalities; if outside incorporated cities, there was provision for formation of public utilities districts with power to construct, maintain, and operate public utilities, issue bonds, and levy taxes to pay for the bonds. In 1959, after statehood, the Alaska Public Service Commission Act was enacted, creating a regulatory commission within the Department of Commerce; the legislation became effective in 1961.[51] In 1970, the Alaska Legislature established the Alaska Public Utilities Commission (APUC) to replace the APSC.[52]

RCA required a certification from the APUC to operate in Alaska. Jones and Tuck note: "the APUC was itself a young body, with limited experience in the regulation of long line communications, but with a fair amount of experience in the regulation of local exchanges. Turf battles over who would be allowed to offer which services erupted before RCA Alascom got to carry its first long distance call."[53] RCA's application was extremely broad and viewed by some of the existing carriers as restricting services that local exchanges could provide in the future, including direct distance dialing and service to the bush and areas not yet served by local exchange carriers. The Alaska carriers also raised objections about the lack of specific toll-sharing agreements and other operating agreements.

The APUC granted RCA Alascom its operating certificate in August 1970. It did not, however, grant exclusive authority to Alascom, leaving the door open to future competition: ". . . the Commission does not want to foreclose the possibility of applications by other for authority to provide some form of long lines service that future circumstances may justify."[54] Noting that RCA had committed to providing telephone service to 142 villages within three years, the APUC also stated: "The public interest requires that the applicant's plans for expansion and improvement be subject to continuing review by the Commission. . . ."[55]

The APUC recognized the complexity of the issues raised, but also that resolving them could take years. It therefore decided to exclude from consideration what it identified as "peripheral" issues that would be considered in future proceedings and to focus only on the provision of long-distance telephone and telegraph services.By deferring the "peripheral issues," the APUC was setting the stage for many of the regulatory and policy debates that followed. As Jones and Tuck point out: "What was really at issue was the future shape of the Alaska telecommunications market."[56]

The Alaska Model

As noted earlier, the structure of ACS was similar to many European telecommunications providers, which were initially set up as government departments, typically under the post office. Although the post office was not involved in U.S. telecommunications after its early sponsorship of Samuel Morse's telegraph link from Washington, DC, to Baltimore, the federal government through the military remained responsible for Alaska telecommunications from 1900 to 1971. However, unlike the European systems that were intended to provide civilian communications, ACS was run and funded primarily as a military network, with the commercial traffic as an added, and subsidiary, function. It remained "the only military operated telephone and telegraph system under the American flag."[57] What started out as a temporary solution, to provide limited commercial service over the Army telegraph, became a permanent anachronism.

Like the European PTTs (Post, Telegraph and Telephone networks) and their colonial counterparts in the developing world, ACS was required to turn over its commercial revenue to the U.S. Treasury and to apply each year to Congress through the Department of Defense for operational funding. It thus had no control over its own revenues and no ability to reinvest them in modernizing the network. And there was little institutional incentive to do so. Even ACS operations staff who understood the needs for upgrades such as automatic switching were unable to get requests approved for congressional submission at the higher levels of the Department of Defense.

Senator Bartlett's staff showed more ingenuity and foresight than many officials of other national governments confronted with the requirement to corporatize and then privatize their PTTs. They considered the corporatization option, which would have transformed ACS into a government-funded corporation like the Alaska Railroad. This was the model initially adopted in the Canadian North, with Canadian National Telecommunications (CNT) set up as a crown corporation. Instead, they recommended immediate privatization, but with terms designed to contribute to Alaska's economic development and consumer needs rather than simply selecting the highest bidder, by requiring specific investment amounts for upgrading the network, specific reductions in rates, and telephone service to be extended to 142 then unserved villages.

The model they proposed, and the FCC required, was a regulated monopoly utility, which was a major advancement over the military long-distance de facto monopoly. At the time, there were few challengers to the view that telecommunications was a "natural monopoly" best provided by a single unified utility, although cracks in that theory were beginning to appear, with the Carterfone case in 1968 that allowed "foreign attachments" (equipment not made by AT&T's Western Electric subsidiary) to be attached to AT&T's network, and MCI beginning to offer private line business services via microwave. But the public long-distance network was still operated as a monopoly, and it seemed unlikely that there would ever be competition in rural and remote areas. In fact, in its submission to the FCC, RCA specifically sought to forestall any future competition or "fragmentation" based on their obligations to serve rural communities, pointing out that its proposals in the bid were made on the basis that "there would be no fragmentation of the service that the long-lines carrier would provide, at least so long as it operates at a loss or at less than a reasonable profit. Any erosion of this position would undermine the economic viability of the system and would be inconsistent with the responsibilities of ownership and operation of the ACS in the manner proposed to the Defense Department and approved by the President of the United States."[58] It reemphasized arguments against "fragmentation," referring to the possibility of competition, because of the high cost and low revenues expected from the isolated communities it would be required to serve: "Moreover, unique to Alaska is the large proportion of communities that are both tiny and remote. The minimum essential service to such communities will be extremely costly per capita. Such communities could not even begin to pay the full cost of the necessary communication facilities to meet their basic requirements of health, safety and education. Rates they can pay can be provided only by a State-wide carrier."[59]

However, the advent of satellite communications would raise the possibility of competition in Alaska during the next decade.

Chapter 6

The Beginning of the Satellite Era

Satellite technology was invented for Alaska.

Alaska broadcaster Augie Hiebert[1]

NASA and Comsat: Opportunities for Alaska

While Alaskans were complaining in the 1960s about the lack of investment in modernizing their telecommunications system, a technology that was to have a profound impact on their future communications was being developed. In 1957, the Russians launched Sputnik, the first communications satellite, spurring the space race and the establishment of NASA (the National Aeronautics and Space Administration) in 1958. Although Sputnik merely beeped as it passed overhead, it demonstrated the potential of satellite communications. In 1945, Arthur Clarke had foreseen the possibilities of placing a transmitter in space above the equator at an altitude of 22,300 miles (36,000 km) where it would revolve around the earth once in 24 hours and thus appear stationary as the earth rotated. By the 1960s, the invention of the transistor plus improvements in rocketry developed in World War II made it possible to launch such a geostationary satellite.

Congress passed the Communications Satellite Act of 1962 to exploit the potential of this new technology. The Communications Satellite Corporation (Comsat) established by the act was to be the U.S. vehicle for developing domestic and international satellite services. In 1964, Comsat played the leading role in establishing Intelsat, an international cooperative with the goal of providing global communications via satellite. In 1965, the first Intelsat satellite, known as Early Bird, was launched. In 1967, an Intelsat satellite was launched to provide coverage of the Pacific, and in 1969, a satellite over the Indian Ocean completed global coverage.[2]

Comsat built and operated the U.S. earth stations used to communicate with the international satellites at sites in California, Washington State, Hawaii, and Guam for the Pacific basin satellite, whose footprint also covered much of Alaska. Perhaps the earliest supporter of satellite communications for Alaska was broadcaster Augie Hiebert; in 1964, his Northern Television Company (NTV) bought shares in Comsat "as a demonstration of faith in satellite communications as an investment in Alaska's future."[3] That year, Hiebert also became president of the newly formed Alaska Broadcasting Association (ABA): "as Augie saw it, a structured group would better attract interest in satellite communications in Alaska."[4] To mark the centennial of the Alaska purchase in 1967, Hiebert invited Federal Communications Commission (FCC) Chairman Rosel Hyde and George Sampson, vice president for operations of Comsat, to speak at the ABA annual meeting in Fairbanks and to tour the state to see its communication problems firsthand.

In 1961, rocket scientist Wernher von Braun, then developing launch vehicles for NASA, had visited the University of Alaska and Barrow at the invitation of the Civil Air Patrol (CAP) to promote aerospace education.[5] Six years later, Hiebert and Senator Barrett revived Alaska's NASA connections to request that NASA demonstrate satellite television for the ABA conference. However, the Israeli-Egyptian Six-Day War intervened, and military priorities preempted the transport of the NASA earth station to Alaska.[6] Von Braun was to return to Alaska in 1974. (See Chapter 7.)

Although the broadcasters wanted satellite facilities to receive live television, Comsat said that live TV was unlikely because of the high cost and small Alaska market. At that time, Hawaii, with a much larger population, could not afford to receive live TV full time; Comsat charged $3600 for an hour-long telecast from Washington State to Hawaii, so that live transmission of a football game cost $16,000.[7]

The Talkeetna Earth Station

Both Governor Hickel and Senator Bartlett were also enthusiastic about the possibility of satellite communications for Alaska. Comsat's Sampson organized a task force to plan for an Alaska earth station after his return and sent a team to Alaska in March 1968. Senator Bartlett took credit for the Comsat visit: "At my urging, the Communications Satellite Corporation (COMSAT) has been studying the feasibility of bringing satellite communications to Alaska. As you know, a team of five COMSAT officials recently completed an inspection tour of your state in connection with this study. As so often happens when people visit Alaska for the first time, pessimism or guarded optimism turns into enthusiastic support. That is exactly what has happened at COMSAT. . . ."[8] Governor

Hickel also announced that the state would assist financially in any program that might develop Alaska's communications network: "I can't go into details, but it [a satellite] could be the solution to our communications problems."[9] In April, the Comsat board approved building an earth station between Anchorage and Fairbanks, and a site near Talkeetna was chosen.

The *Anchorage Daily News* recognized that the Talkeetna earth station could be a major advancement in telecommunications for Alaska:

> . . . there are few developments that will change the nature of life in Alaska as dramatically or permanently as the establishment and operation of a modern communications system. We are so far behind the rest of the nation—and much of the world—in our communications facilities that the gap is almost too far to see across into the future.
>
> The list starts with live television. An event that occurs in the United States can be viewed instantly in Canada, Mexico, Japan, and most of Europe. But not in Alaska. Enormous toll costs for long distance telephone calls is [sic] inhibiting to those who would like to talk to friends or family Outside, and those who must use the long distance system for business. All Alaska news media must content themselves with primitive news gathering facilities that cheat readers, viewers and listeners of most of the day's available national and world news copy and photos.[10]

The editors also recognized that pricing would be critical for the services available through the new earth station to benefit most Alaskans:

> . . . the only way that a major communications corporation can develop the largely fallow Alaska communications network is to provide improved service at a cost that will be immediately attractive to those who can use it. High rates won't do it. Limited service won't do it. Alaska is largely an untapped communications market that can be developed in the same manner that other markets are developed—with a good product at economically realistic price, and skilfully promoted.
>
> The decision to build an Alaska satellite ground station is an event of enormous significance to Alaska. How much of a milestone it represents will depend almost entirely on how the benefits are made available to those who live in Alaska.[11]

Comsat and ACS

The efforts to persuade Comsat to locate an earth station in Alaska were under way while ACS was up for sale. The satellite station was intended primarily for telephone service for ACS, but Comsat was determined to start construction in summer 1968 despite the sale, and briefed prospective buyers on its plans. Senator Bartlett was well aware of the potential benefits of the Talkeetna station for ACS:

> Obviously, this plan must be taken into consideration by firms interested in purchasing the Alaska Communications System. On the one hand, if the purchase of ACS did not intend to make full use of the COMSAT facility, it might not be economically feasible for COMSAT to construct the earth station. On the other hand, if the purchase of ACS did not take advantage of the COMSAT facility, the purchaser eventually would have to build duplicate facilities of his own, thereby passing an unnecessary cost along to the consumer in the form of higher rates. However, I think the purchaser of ACS would be anxious to make use of this facility.[12]

In addition to the Talkeetna earth station, Senator Bartlett thought that Alaska should have a satellite communications system instead of merely expanding the ACS network to connect with the lower 48, despite the terrestrial network being less expensive: "I do not believe Alaska should saddle itself with an outdated communication system just to secure some very small short-range reductions."[13]

In April 1968, Governor Hickel announced the appointment of a seven-member Satellite Communication Task Force with Hiebert as chairman, to work with Comsat in coordinating potential users including the military, state government, industries, and education in support of the Alaska earth station, which was to provide not only another connection between Alaska and the lower 48 states but to link Alaska with the Asia-Pacific and the world.[14] After the death of Senator Bartlett in December 1968, Hiebert launched a successful campaign to name the Talkeetna earth station for him.

In February 1969, the cities of Anchorage and Fairbanks issued a joint resolution on telecommunications that stated: "The City Councils of Anchorage and Fairbanks strongly endorse and support a thorough review of the ACS sale and Comsat license request for an earth station. . . . The Councils respectfully urge that the Alaska Delegation take all necessary steps to provide as soon as possible a modern communications system to serve the growing development in Alaska . . ."[15] Newly appointed Senator Ted Stevens responded to George Sullivan, the mayor of Anchorage: "The resolution approved by the two councils calls for a modern communications system for Alaska as soon as possible. I believe that this can best be achieved by the construction of the earth station at Talkeetna and by the sale of the ACS to a modern, experienced and highly capitalized telephone company."[16]

In May 1969, the FCC authorized Comsat to build a satellite earth station near Talkeetna, at a capital cost not to exceed $4.5 million, to provide communication services between Alaska and the rest of the United States, Japan, and other Pacific area locations. Comsat proposed that the earth station would initially provide service through the Intelsat II, III, and IV series satellites.[17]

Local Alaska telephone companies Glacier State and Continental Telephone opposed the sale, protesting that Comsat rates would be excessive and arguing that the existing landline system should be upgraded and maintained. They also stated that Comsat should not be permitted to construct the earth station until finalization of the ACS sale, since the buyer of ACS would be the common carrier for the earth station traffic.

The Department of Defense withdrew initial approval, citing the pending ACS sale and rates as issues that should be reviewed through a hearing process. However, realizing that time was of the essence because of the short Alaska construction season, Hiebert briefed Senator Ted Stevens, who met with Secretary of Defense Melvin Laird. As a result of the meeting, the Department of Defense withdrew its dissent, as did the telephone companies, and the Comsat construction permit was approved; construction began in the spring of 1969.[18]

Alaskans also benefited from Secretary Laird's largesse that summer, when he approved use of a military ground station and satellite to bring live lunar coverage to Alaska. On July 21, 1969, Alaskans watch Neal Armstrong walk on the moon through a transmission to Anchorage stations of a CBS broadcast anchored by Walter Cronkite. Alaska Airlines offered a special package of $39.90 for fares from Fairbanks to Anchorage for the week of the broadcast.[19]

Pointing to further monopolistic consolidation of the Alaska market, Western Union International filed a petition to deny RCA's application for a microwave link between Anchorage and the Comsat earth station under construction in Talkeetna. Senator Ted Stevens supported RCA:

> When RCA became the successful bidder for the ACS I reached the conclusion that I was in favor of its basic proposal to acquire the ACS and operate the communication system in Alaska on an essentially unfragmented basis. This position stemmed from no personal favoritism for RCA, but rather from the belief that this approach would best promote the providing of maximum service at the lowest possible rates to the entire State of Alaska, particularly to the many remote areas of the state which have low populations.[20]

Communications Planning for Alaska

In August 1969, Senator Ted Stevens convened a conference in Anchorage "to assess Alaska's future needs for a satellite communications system," "to explore a long range total systems approach of a satellite communications system" and "to bring together in Alaska those who are vitally interested in a rapid development of satellite communications in our state."[21] Delegates included representatives of government agencies including NASA, the FCC, the Department of Commerce, Department of Defense, the White House Office of Telecommunications Policy, The Department of Health, Education and Welfare (HEW), and the Economic Development Administration; organizations representing broadcasters including National Education Association (NEA) and Joint Council on Educational Telecommunications (JCET), and the Corporation for Public Broadcasting (CPB); the Ford Foundation; telecommunications industry representatives from RCA, Comsat, Western Union International, and Page Communications; space systems contractor Raytheon; and oil companies active in Alaska including Atlantic Richfield, Humble Oil, and Mobile Oil.

The agenda included presentations on satellite communications technology and costs, educational communications via satellite, and needs for satellite communications for the state government, broadcasting and publishing, and major industries including mining and logging. Alaska Senator Mike Gravel's staff presented a proposed experiment in satellite communications for Alaska. Comsat's Chairman of the Board made a presentation on "A Satellite Communications Network for Alaska."

Earlier in 1969, the Communications Working Group of the Federal Field Committee for Development Planning in Alaska produced a report on communications needs for Alaska. Based on this report, the advice of the Governor's Communications Task Force, and information provided at the August satellite conference, the Field Committee requested $250,000 for a planning study from the Department of Commerce to estimate the communication requirements for Alaska, estimate unmet needs in Native villages, estimate revenue requirements, determine administrative requirements and statutory authority required by State of Alaska, and recommend rate structuring and revenue sharing.[22]

The exploitation of oil reserves on the North Slope and the resulting economic growth in Alaska would stimulate demand for telecommunications. In late 1969, the Brookings Institution conducted a policy conference in Anchorage, where Alaskans drew up guidelines and recommendations to the state legislature on use of Alaska's new revenues from the North Slope oil discovery and the $900 million lease sale. Broadcaster Augie Hiebert presented a paper entitled "Anatomy of Alaska's Telecommunications Satellite in Alaska's Immediate Future," advocating the establishment of a state telecommunications policy: "There is an urgent need for adequate telecommunications services of all kinds. . . . Alaska is at the communications crossroads with an unparalleled opportunity to shape the future. . . . No other state or country of the developed world has had the opportunity to proceed to the crest of economic explosion using a telecommunications system whose great cost did not have to be amortized before a new stage of development could be initiated."[23]

While the Bartlett earth station was under construction, Comsat continued to propose steps toward providing a satellite network for communications within the state. In June 1969, Comsat offered to do satellite demonstrations using earth stations in Washington State and Hawaii, with two smaller 10-meter antennas and the NASA experimental ATS-1 satellite (see Chapter 7).[24] Comsat President Joseph Charyk wrote to Governor Miller: "We would hope, however, that such a demonstration would lead into an operational system providing regular communication services within Alaska."[25] Comsat Chairman James McCormack also wrote to Senator Ted Stevens, proposing to use Intelsat satellites for telecommunications and TV throughout Alaska, and then to transition to its own domestic satellite: "At such time as a domestic satellite program, for which an application is pending before the FCC, is approved and implemented, the domestic system can be utilized effectively and efficiently to fulfill the Alaska communications requirements."[26]

Continuing Comsat's efforts to sell the concept of an intra-Alaska satellite network, McCormack also wrote to RCA president Robert Sarnoff:

> The announcement . . . of RCA's successful offering for the Alaska Communications System gives me the reason for writing to you to emphasize the aspect of communications in our 49th state which seems to me to be of greatest interest. That is satellites for communications *within* Alaska.
>
> [The earth station at Talkeetna] can serve equally well as the keystone in a network of ground facilities for Alaskan state-wide services. . . . We in Comsat are convinced that with forward-looking joint planning RCA and Comsat can in one giant stride help move Alaska communications from the poorest in our nation to a place along with the best. . . .
>
> I want to give all the weight I can to the idea of a major joint endeavor by RCA and Comsat toward the wide-scale introduction of satellite communications in Alaska. We should definitely include the possibility of a satellite designed specifically for Alaska, as well as the prospects for adding Alaska to the proposed overall U.S. domestic satellite system. A specially tailored Alaskan satellite system could well be the pilot for the larger system, an idea with very interesting potentials.[27]

In November 1969, Comsat proposed a satellite system to RCA to meet the needs of the Trans Alaska Pipeline System (TAPS) that was to be built to deliver oil from the North Slope to tankers at the port of Valdez. The network was proposed to include three large antennas at Valdez, Fairbanks, and Prudhoe

Bay, and 25-foot antennas at each of the 11 pump stations. Comsat estimated the total revenue requirement of $13 million per year, to come from TAPS and commercial services. Comsat proposed to own the earth stations and to install, operate, and maintain them on a joint venture basis with RCA.[28]

A year later, in October 1970, Comsat actually filed an application with the FCC for a domestic satellite system consisting of two 24-transponder satellites "to provide common carrier services within the 48 contiguous states and . . . the capability of providing services between Alaska and the contiguous states."[29] However, Comsat's domestic satellite system was not launched until 1976. By that time, RCA had decided to build its own domestic satellite.

The Bartlett Earth Station Begins Operation

On June 30, 1970, Comsat's Talkeetna earth station was dedicated as the Bartlett earth station, in memory of Senator "Bob" Bartlett, who had been a very effective advocate for modernizing Alaska's telecommunications. Comsat's president, Joseph Charyk, Governor Keith H. Miller, and Senator Bartlett's widow were among the participants. Using the Intelsat III satellite over the Pacific Ocean, the huge 30-meter antenna, weighing more than 300 tons, would communicate with other U.S. gateway earth stations in Washington State, California, Hawaii, and Guam.[30] Via the Bartlett station, Alaskans received a message from President Nixon at the White House, a film from Expo 70 in Japan, and a transmission from Waikiki. Alaska broadcasters also transmitted a taped program about Alaska to Japan to be broadcast by Nippon Educational Television (NET).[31] The first live television transmission from Alaska was the Department of the Interior's Environmental Impact Hearings on the Trans-Alaska oil pipeline on February 24, 1971. Later that year, the Bartlett earth station transmitted coverage of Nixon's meeting with Japan's Emperor Hirohito in Anchorage.[32]

Broadcaster Augie Hiebert's KTVA station in Anchorage announced the first live satellite transmission from Alaska:

KTVA—CBS NEWS RELEASE
FEBRUARY 24, 1971—4:30 PM

SATELLITE TELEVISION COVERAGE OF THE DEPARTMENT OF
INTERIOR'S ENVIRONMENTAL HEARINGS ON THE TRANS-ALASKA
PIPELINE WAS AIRED TONIGHT IN THE LOWER 48 STATES. THE
BROADCAST WAS THE FIRST SUCCESSFUL COMMERCIAL SATELLITE
TRANSMISSION GOING SOUTH FROM ALASKA.

THE COVERAGE AIRED FROM SYDNEY LAWRENCE AUDITORIUM, WAS TRANSMITTED THROUGH THE BROADCAST CENTER-KTVA TO THE BARTLETT EARTH STATION AT TALKEENTNA, THEN TO THE SATELLITE, AND RECEIVED AT THE COLUMBIA BROADCASTING SYSTEM'S HEADQUARTERS IN NEW YORK.

"TODAY ALASKANS CAN BE PROUD THAT A MAJOR NEWS EVENT HAPPENING IN THEIR STATE WAS SEEN BY MILLIONS ELSEWHERE ON A SAME-DAY BASIS," ACCORDING TO A. G. HIEBERT, PRESIDENT OF NORTHERN TELEVISION, INCORPORATED. PREVIOUSLY, TELEVISION NETWORK FILM CREWS HAD TO SHOOT FOOTAGE HERE AND SHIP IT SOUTH FOR BROADCAST. THE DELAY WAS USUALLY 24 HOURS OR MORE. . . .

THE COVERAGE WAS AIRED AS A FIVE MINUTE SEGMENT OF THE CRONKITE NEWS. CBS NEWS CORRESPONDENT TERRY DRINKWATER REPORTED FROM HERE FOR THE NETWORK. . . .

"THE TRANS-ALASKA PIPELINE CONTROVERSY IS OF VITAL INTEREST IN THE LOWER 48 AS WELL AS HERE. THE DEBATE ON THE SUBJECT IS REALLY A NATIONAL ONE. BUT WHAT ALASKANS HAVE TO SAY IS DOUBLY IMPORTANT. THAT IS WHY THIS SATELLITE FEED FROM KTVA TO CBS WAS HIGHLY SIGNIFICANT FROM THE NEWS STANDPOINT," DRINKWATER SAID.[33]

A Satellite Television Demonstration for Rural Alaska

With the Bartlett earth station online, the Governor's Satellite Communications Task Force was disbanded. To help fill the void, the Alaska Educational Broadcasting Commission (AEBC) appointed a Telecommunications Advisory Committee, again with commercial broadcaster Augie Hiebert as chair. Hiebert used the Alaska Public Service Commission hearings on the RCA purchase of ACS (APSC approval was required in addition to approval from the FCC) to continue to promote the need both for a satellite to serve Alaska and for a state telecommunications policy:

Satellite communications offer the bright promise to accomplish the philosophy and infinite benefits to a total Alaskan Telecommunications System. A satellite . . . will illuminate every Alaskan city and village, offering equal opportunity for all Alaskans to communicate in whatever mode of transmission suits their needs and requirements. . . . The economic, social, technological factors are all present—the timing is perfect—the scenario can be, and must be written to provide for adequate communications wherever there are Alaskans who need to communicate within our state. Someone must speak for the isolated villages; someone must be concerned about opening the world educationally. . . . Someone with policy making powers and judicial wisdom must care about Alaskans outside the mainstream of advantages we take for granted. . . .[34]

Meanwhile, Alaska broadcasters wanted to move forward on using satellite communications for television throughout the state. Broadcaster Augie Hiebert drafted a letter for signature by Governor Egan requesting that Comsat provide a television demonstration. Egan's letter explained: "The project's goal . . . is to prove that high quality television and voice transmission can be achieved through use of satellites and small, low-cost earth stations. This must be achieved if Alaska's far-flung and remote communities are to enjoy the many kinds of information which Americans in other states have long taken for granted. . . ."[35]

Hiebert also wanted a trial of smaller satellite antennas for Alaska similar to the 4.5-meter (16-foot) antennas that he had seen at Comsat headquarters in Washington, DC. Senator Mike Gravel requested "an exceptional, but temporary variance" to permit the 16-foot antennas, which were smaller than the FCC had authorized, pointing out that "Existing terminals of this small size can provide satisfactory television reception for local viewing and reliable two-way telephony."[36]

Comsat agreed to the request, and the FCC granted authority for the demonstration, subject to Comsat filing a comprehensive public report of test results and permitting other parties to send observers. Once more concerned about the threat of competition, RCA opposed the demonstration and challenged Comsat's permit from the FCC for the test, but was overruled by the Commission.[37]

Comsat announced that the purpose was to provide TV and voice communications via satellite in cooperation with the state of Alaska "to demonstrate the advantage of satellite communications in meeting the unique communications requirements of small communities in remote areas, to determine the effects of extreme climatic conditions on satellite communications and analyze possible interference from the aurora borealis. . . . Data gathered during the program will help determine the feasibility of using small aperture earth station terminals for commercial satellite services to remote areas."[38]

The demonstration was delayed until the spring and summer of 1972 in order to use capacity on the Intelsat IV Pacific Ocean satellite, launched in

January 1972. This was the same satellite that had relayed coverage of President Nixon's trip to China in February 1972. CBS had sent correspondents Walter Cronkite, Eric Severeid, and Dan Rather to Anchorage to originate the *CBS Evening News* when Nixon flew back through Anchorage after meeting with Chinese Premier Chou En-Lai in Beijing.[39]

The two-month Alaska trial demonstrated the feasibility of using small aperture earth stations in remote areas and northern latitudes. The first site was Juneau. On May 4, 1972, the *Southeast Alaska Empire* reported: "A television signal traveling 45,000 miles at the speed of light brought Juneau its first taste of space-age communications last night. An hour's television broadcast from Anchorage to Juneau via satellite went off without a hitch. It was the opening segment of a demonstration designed to show that existing satellite technology can overcome the communications problems posed by Alaska's vast and rugged expanses."[40]

The 16-foot antenna and other equipment were transported by plane from Juneau to Kodiak, Bethel, Nome, Barrow, and Fort Yukon. The demonstrations lasted only about three days at each site, with at least ten days required to move the equipment between sites. Once more, the Alaska climate and local conditions challenged a telecommunications installation. Rain got into a connector during erection in Juneau; some antenna rib welds broke on the way to Nome; power voltage fluctuations in Nome damaged a low-noise amplifier (but the team carried a spare); and in Barrow: "an excessive number of mosquitoes were drawn into the ventilation system on the HPA (high power amplifier) causing over-heating." In each case, the problems were resolved and the unit then functioned properly.

During the trial, a two-way circuit for telephony was established between the remote station and the Bartlett station connected by RCA's microwave link to Anchorage so that calls could be routed over the now-available direct distance dialing (DDD) service anywhere in the state. The one-hour per day of television included remarks by Governor Egan, a "Space Puppet" sequence, and a film on the history of Alaska during the first day; news programs on the second day; and on the third day, a University of Alaska program[41] and the PBS children's program *Electric Company* from KUAC in Fairbanks.[42]

In Nome, live TV was shown in the rear of the Northern Commercial store and carried on Nome cable TV (which usually transmitted tapes from the lower 48). At Fort Yukon, the equipment was left in place so that a technical team from the University of Alaska's Geophysical Institute could collect data on satellite transmission and ionospheric phenomena.[43] However, although the transportable earth station remained in Fort Yukon for two months, residents were not able to use it during this extended period to make telephone calls or to watch television.

In its report on the demonstration to the FCC, Comsat stated: "The recently completed Alaskan Program served two major purposes. First, it successfully demonstrated the capability of a small aperture earth station for receiving television and providing voice communications in remote areas. Second, tests and investigations were conducted which provided information for assessing the operation of earth stations under conditions found at high northern latitudes."[44] The elevation angle of the antenna was only 8.9 degrees at Fort Yukon and a maximum of 18.2 degrees at Bethel. However, satellite acquisition proved to be no problem, and despite minor difficulties, there were no surprising performance variations as the equipment was moved from site to site. Further, there were no severe fadings due to ionospheric effects.[45]

Yet perhaps only Governor Keith Miller (who served in 1969 and 1970 after Governor Hickel resigned to become President Nixon's Secretary of the Interior) had the vision to regard the demonstration as part of a larger strategy to extend television reception by a variety of means to rural Alaska. He originally proposed that some communities receive TV on videotape, including Barrow, Bethel, Cordova, Wrangell, Petersburg, Haines, Skagway, Hooper Bay, Little Diomede, Juneau, Sitka, and Ketchikan.[46] But videotaped programming did not reach rural Alaska until several years later.

Policy Considerations

While Alaska was experimenting with satellite communications, the United States determined that domestic satellite service should not be a monopoly, so that any qualified company could launch a satellite to serve the country. The FCC's "Open Skies" decision in 1972 literally opened the new industry to competition, rejecting the monopoly model then in force for international satellite services. Also, Comsat was required to separate its international and domestic operations. As a result, Comsat contracted to sell the Bartlett earth station to RCA Alascom in 1973. Ironically, this decision appeared to strengthen the telecommunications monopoly in Alaska. FCC Chairman Dean Burch attempted to refute this view in a letter to Senator Ted Stevens in August 1973: "You state that (Comsat) has interpreted . . . the Commission's Memorandum Opinion and Order of December 1972 in its domestic satellite proceeding to mean that Comsat must remove its communication satellite operation from the State of Alaska, and express concern that RCA Alascom may be afforded a virtual monopoly."[47] Burch added that Comsat was not legally precluded from offering domestic services by means of its facilities although "we have recently been apprised that Comsat has voluntarily contracted to sell the Bartlett earth station to RCA Alascom."

Burch then elaborated on the policy issues concerning the provision of telecommunications to Alaska's small and scattered population:

The establishment and use of <u>domestic</u> satellite facilities to serve Alaska posed a difficult problem in reconciling State interests (which are themselves somewhat self-conflicting) with the overall public interest objectives being sought in the introduction of this new technological capability. The State requested, on the one hand, incorporation of Alaska rates and services into the nationwide, rate-averaged services now offered by monopoly suppliers of message telephone service; and on the other hand, the authorization of competing carriers. To accommodate the former request, the Commission authorized only RCA Alascom—the only existing inter- and intra-state long lines carrier in Alaska—to establish and use domestic satellite facilities for the provision of <u>interstate message toll</u> telephone service to Alaska; and ordered RCA and AT&T jointly to develop and present to the Commission a plan for incorporating this service into the nationwide, rate-averaged tariff structure. To accommodate the latter request, the Commission did not grant RCA Alascom any monopoly position with regard to the provision of other domestic communication services, whether intrastate or interstate. . . .

As a practical matter, of course, there is considerable doubt that the traffic volume either to or within Alaska is sufficient at the present time to justify on a commercial basis the establishment of duplicate satellite or earth station facilities. . . . If the market legitimately will support only a single supplier, government efforts to divide the market seem bound to increase the costs for all users. Since these costs are to be borne in part by other carriers and rate-payers, under the nationwide rate-averaging approach, it is even difficult to conceive of the rationale for or method of implementing any such market division. *Indeed the commission does not authorize competitive facilities for the provision of interstate MTS* (message toll) *service anywhere in the United States.*

. . . let me reiterate that neither Comsat nor any other domestic satellite supplier is precluded by Commission policy from going after the market for specialized (i.e. non-MTS) services to and within Alaska—to the extent this market will support such activities. Comsat and other carriers (excluding RCA Alascom) are precluded from offering MTS services for the simple reason that the requested subsidization of such services by other carriers and rate payers is inconsistent with both the concept and practicality of competitive suppliers.[48] (underlining in original)

Thus, it could be argued that Alaska's politicians and broadcasters were years ahead of the federal government in their assertions that there actually could be competition in Alaska telecommunications. In 1973, as Chairman Burch pointed out, the FCC's position was that there could be competition in delivery of value-added services, but that message toll (regular long-distance services) must remain a monopoly. The struggles to obtain operational satellite service for Alaska communities and to open Alaska telecommunications to competition would occupy much of the rest of the decade.

Chapter 7

The NASA Satellite Experiments

I use the satellite radio every, every, every day.

Health aide Rose Ambrose in Huslia, 1973[1]

Experimental Satellites in Context

Alaska's involvement with communication satellites began in 1957, very soon after the launch of the Soviet Sputnik: "Alaska has been tracking satellites and developing satellite communication since about one hour after [Sputnik] I was launched.... A radio-star tracking facility at the University of Alaska Fairbanks was converted to a satellite tracking station in a few minutes."[2] The University's Geophysical Institute was later awarded a contract by the newly established National Aeronautics and Space Administration (NASA) to design and install a mini-tracking station at Gilmore Creek, north of Fairbanks.

Between 1966 and 1974, NASA launched a series of Applied Technology Satellites (ATS). The ATS-1 and ATS-6 satellites played important roles in Alaska, demonstrating not only the technical viability of satellites to provide reliable communications to Alaska villages, but also applications including telemedicine, distance education, and radio and television broadcasting that became the foundation for operational services throughout rural Alaska. A planned seventh satellite (ATS-G) for which the state also proposed experiments was canceled in 1973 because of federal budget cuts.[3]

In 1966, NASA launched the first of its Applied Technology Satellites, known as ATS-1, which was put in orbit over the Pacific Ocean with a footprint covering Alaska and the Pacific Islands. In 1968, NASA requested the Geophysical Institute to evaluate use of the VHF transponder on ATS-1 for transmission into Alaska's remote areas. Researchers carried out tests using modified 100-watt taxi radios and a handmade antenna and found that, although the signal faded during electromagnetic storms, it came back in a few minutes. This performance was far superior to the high-frequency (HF) radios in the villages that were at best "practicable for only a small percentage of the time and may not be available at all for several consecutive days."[4]

The potential of satellites for domestic communications was already receiving national attention as one of the key issues addressed by the President's Task Force on Telecommunications Policy established by President Johnson in August 1967. Its final report authored by Eugene Rostow recommended pilot projects to investigate possible satellite applications: "Before we reach any final decision about the ownership and operational design of domestic satellites we think it is desirable to benefit from some exploratory operational experience."[5]

In 1969, a staff member working for Alaska Senator Mike Gravel read in a government report that NASA had time available on ATS-1. Senator Gravel thought this would be an excellent opportunity for Alaska,[6] and with other members of the Alaska Congressional Delegation he sponsored a meeting in Washington, DC, on July 31, 1969, to discuss a pilot satellite communications project for Alaska. In addition to Senators Gravel and Stevens and Congressman Pollock and members of their staff, attendees included a state delegation composed of the Director of Telecommunications, the Commissioner of Education, members of Alaska legislature, and members of Nome City Council; and representatives of the Office of Telecommunications Policy in the White House, the federal department of Health, Education and Welfare (HEW), NASA, Comsat, and RCA. The need for better communications to support health care was cited as a top priority: "It was stressed that it was difficult to be sick in Alaska except at certain times because, in general, telephone service is not available to reach Alaska's approximately 255 practicing physicians."[7] Alaskan medical authorities identified four principal health care needs that could be achieved by satellite: "provide 24-hour telephone service throughout Alaska so that emergencies could be answered, illnesses diagnosed, or transportation arranged; provide health education for the public; provide continuing education for health personnel; and provide for the utilization of a health data bank."[8]

The year 1970 was also significant for satellite policy initiatives at both the state and national levels. In Alaska, Governor Keith Miller signed into law Senate Bill 372, creating the Alaska Educational Broadcasting Commission (AEBC), which had a broad mandate to consider noncommercial needs for broadcasting: "While the members are not to be chosen to represent various special interests, the Bill states that the Governor shall give due consideration to representation from such fields as public health, public works, labor, commerce and the professions."[9] In the State Legislature, Representatives Gene Guess and Jalmar (Jay) Kerttula introduced a bill to allow the state to own and operate its own satellite communications system.[10]

In Washington, the Nixon White House established the Office of Telecommunications Policy. HEW also created an Office of Telecommunications Policy, which became a major funder of Alaska's satellite experiments and rural facilities. In May 1970, the Corporation for Public Broadcasting (CPB), representing several educational broadcasting organizations, including the newly formed AEBC, presented its case to the FCC to allocate the 2500–2690 MHz band (known as S-Band) for educational communications via satellite as well as for terrestrial instructional television. CPB noted that these were very efficient frequencies for space transmission and would permit development of ground terminals within the reach of educational budgets, perhaps from $150 to $750 in volume production.[11] These frequencies were later used to demonstrate video reception on NASA's ATS-6 satellite.

In 1972, Alaska Governor William A. Egan established the Governor's Office of Telecommunications (GOT or OT), which played a significant role in shaping Alaska's satellite and, more generally, telecommunications policies and strategies to serve all of Alaska's permanent communities during the 1970s.

The ATS-1 Biomedical Experiments

The Lister Hill Center of the National Library of Medicine, which had participated in President Johnson's task force, proposed a biomedical project on the ATS-1 satellite. The primary objective of the project was "to determine the degree to which satellite communications technology can be used for biomedical communications in remote areas where common carrier telecommunications services do not exist or are severely limited . . . and to conduct controlled experiments in effective methods of providing health care education and medical consultations to remote or geographically isolated locations."[12] They planned to use the VHF channel primarily for voice, but also wanted to experiment with nonvoice transmissions including ECG, facsimile, slow-scan TV, time-sharing computers from remote terminals, and teletype-to-teletype transmission. The Public Health Service, which was responsible for the health care of Alaska Natives, was invited to conduct the experiment in providing medical consultations via satellite between doctors at regional hospitals and village health aides. Technical operations were managed by the University of Alaska's Geophysical Institute; electrical engineering professor Robert Merritt designed the VHF earth stations using primarily off-the-shelf parts; the stations were built and installed with the help of his graduate students. Stanford University participated in some of the equipment trials, and Stanford's Institute for Communication Research and School of Medicine was contracted to carry out the evaluation of the rural biomedical experiments.

In August 1969, planning for the project began at an Alaska Conference on satellite communications. An Alaskan team led by Dr. Martha Wilson, then head of the Alaska Area Native Health Service (AANHS), a branch of the federal Indian Health Service (IHS), and an engineering team from the University of Alaska's Geophysical Institute, led by Professor Glenn Stanley, were the primary planners, and later implementers, of the experiment. In November 1969 the Alaska team submitted their formal proposal to NASA for time on the ATS-1 experimental satellite, which NASA approved.

In 1971, Alaska gained the opportunity to see if reliable communication using ATS-1 would help to improve health care for residents of rural Alaska. Since the 1950s, Alaska villages have been served by community health aides, local residents (primarily women) who receive basic medical training and provide first line care for the villagers. Health aides are supervised by medical staff in regional hospitals, which in turn are supported by specialists at the Alaska Native Medical Center (ANMC) in Anchorage. Although the health aide model was based on the assumption that the village aides would be supervised by doctors at the regional AANHS hospitals, the link between aides and doctors

An ATS-1 satellite antenna provides a single voice channel to a clinic in an Interior village. (National Library of Medicine)

was still high-frequency (HF) radio, which was highly unreliable because of ionospheric interference and poor maintenance. Twenty-six sites were chosen to participate in the satellite experiment, most of which were village clinics located in the Tanana Service Unit in the Alaska interior, one of seven service units operated by the AANHS. The Tanana Hospital had 26 beds and three doctors to administer health care to approximately 20 villages in an area of about 200,000 square miles. Communications to this group of villages had been less than 20 percent effective when attempted by HF radio.[13]

At a scheduled time each weekday, the physician in Tanana conducted a satellite "doctor call" with each of the village health aides in turn on the single ATS-1 audio channel. They presented the signs and symptoms of new patients and updates on the status of other patients. The physician could then propose a diagnosis or confirm the health aide's diagnosis and could confirm or modify the health aide's treatment plan. The impact of improved communication was dramatic: in the first year, the number of days in which a health aide could contact a doctor increased more than fivefold, and the number of patients treated with a doctor's advice more than tripled.[14] See Table 7-1. The Stanford evaluators also found that listening to the entire "doctor call" on the shared audio channel increased the health aides' medical knowledge, boosting their confidence in their work and making patients more apt to follow their instructions.[15]

In addition to conducting the formal evaluation of the ATS-1 biomedical experiment, Stanford University also produced a film about the project called *Satellite House Call* (now available on YouTube).[17] The purpose of the film was to show funders and decision makers in Washington, DC, what life was like in rural Alaska, and how reliable communications could improve the health and well-being of rural Alaska Natives. Some officials asked, "Why not use the telephone?," failing to grasp the isolation of bush communities and the difficulties of extending telephone service. Others wondered whether reliable voice communications would interfere with Native cultures. While cultural influence became a topic of debate when television later arrived in the villages, Alaska Natives were strong supporters of getting reliable voice communications for health care, emergencies, and staying in touch with family members. The film

TABLE 7-1   New Medical Episodes Handled by Teleconsultation

	BEFORE SATELLITE	AFTER SATELLITE	
	(1970–71)	1ST YEAR AFTER (1971–72)	2ND YEAR AFTER (1972–73)
9 Satellite Villages	47.1	184.6	290.0
4 HF Radio Villages	24.7	15.0	N.A.

showed consultations with village health aides on ATS-1 and an interview with an aide in a village without any reliable communication, and included endorsements from Native leaders Georgiana Lincoln and Willie Hensley, as well as Governor William Egan, and commitments from both Republican Senator Ted Stevens and Democratic Senator Mike Gravel to seek reliable operational satellite services for all of Alaska's communities.

Stanford evaluators Heather Hudson and Edwin Parker also published an article on the ATS-1 biomedical experiments in the *New England Journal of Medicine*.[18] Their goal was not only to reach medical professionals and researchers with data on the value of reliable communications in rural health care, but also to raise the ethical issue of involving Native Americans in a successful telemedicine experiment and then concluding, in effect, that "they could go back to dying in the old way" after the federally funded experiment was over.

Several other medical communications activities were piloted on ATS-1. For the first time, patients at Tanana Hospital could talk to their families in the villages. A doctor on the island of St. Paul in the Bering Sea could consult directly with specialists in Anchorage. And the University of Washington's medical school transmitted a genetics lecture to premed students at the University of Alaska. During periods when the audio channel was not assigned to Alaska, it was monitored by the NASA control stations at College (Fairbanks) or Mojave (in California). Health aides were able to reach these sites in an emergency; the control station staff in turn got in touch with physicians in Alaska by telephone, resulting in air medevacs of a resident with acute appendicitis from Allakaket and an injured snowmobiler from Huslia.[19]

The ATS-1 Rural Education Experiments

Educators in Alaska also wanted to experiment with satellite communications for education and community development. The newly established Alaska Educational Broadcasting Commission (AEBC) asked the Alaska Education Association for planning assistance to use ATS-1, which resulted in a study by Harold Wigren of the National Education Association and Henry Cassirer of UNESCO on the educational needs of Alaska villages and ways technology might help to meet those needs. Wigren and Cassirer concluded: "In many respects, a satellite was invented for Alaska because of Alaska's unique communication problems, lack of terrestrial communications facilities, mountainous terrain, harsh climate, and sparse population. These factors point to satellite communication as an ideal system of reaching all parts of the state on a real time (instantaneous) basis."[20]

Village schools typically had two rooms with a wood stove and diesel generator. Two teachers taught all grades through junior high, after which students had to leave the village to complete high school at residential schools. Most contact with the outside world was by mail and through visits from itinerant government officials. State educators proposed an ATS-1 pilot project including radio programs via satellite plus videotapes to be used in schools, and audio programs for teachers in remote communities to provide career development opportunities and for the general adult population in the villages as a means for community development.[21]

The educational content consisted of programs directed to rural teachers, audio materials to be used in village classrooms, student exchanges with distant schools, cultural programs for village adults, and radio broadcasts. The first demonstration of satellite communications for transmission of domestic radio programming took place on April 28, 1970. The University of Alaska's radio station, KUAC, originated a program featuring a lecture by visiting consumer advocate Ralph Nader, which was transmitted from UAF's recently constructed VHF terminal over ATS-1, received at NASA tracking stations and at Stanford University, and carried by Stanford's radio station KZSU and San Francisco's public radio station KQED-FM. Monitoring engineers reported that the technical quality was excellent.[22] There is no record of the various audiences' reaction to Nader's speech.

The educational project developed several different series of programs: three exclusively for teachers, five for classroom use, three for joint classroom and community use, and three designed for the community. Some 15 villages were in the target group, including several participating in the Tanana region biomedical experiments. The satellite radio was installed in a village classroom. In January 1973, the National Education Association (NEA) announced:

> Each Monday in nearly 20 frigid and far-flung Alaskan villages, teachers and other interested persons are taking part in unique training lessons bounced off a satellite 22,300 miles above the equator. Believed to be the world's first satellite seminar for teachers, the 15-week radio course . . . is titled 'Teaching Techniques for Rural Alaska' . . . An unscheduled feature of the initial broadcast was an NEA staff member's announcement to the remote villages of President Johnson's death that afternoon. The late President was a strong advocate of both space communications and education.[23]

Among the most popular community programs were those prepared with the help of itinerant broadcasters who transmitted from the villages programs including "Whaling in Barrow," "The Nulato Stick Dance," and "Keeping

Healthy Dogs." The Nulato stick dance broadcasts were a series of daily transmissions by participants in the stick dance over the modified taxi radio that served as the satellite gear. Producer Karen Michel remembers that one day, after hearing the program from Nulato, Rose Ambrose from Huslia left her bread rising to come to Nulato herself. A story exchange often involved playing tapes of traditional stories recorded by the Alaska Library Association's Alaska Native Oral Literature Project. Health aides and other listeners were then encouraged to contribute their own stories.[24] News of the Arctic Winter Games and Iditarod sled dog race from Anchorage to Nome was also popular: "Coverage of the Iditarod Race has become a tradition of the ATS-1 project because the villages of Ruby, Galena, Koyukuk and Nulato are on the race route and can participate as reporters."[25]

Classroom exchanges included school children in the Athabaskan village of Beaver and the Inupiaq community of Barrow talking with counterparts in Hawaii and in Wellington, New Zealand. However, such exchanges required extensive time and effort to arrange and coordinate. Organizational problems and funding delays resulted in the education experiments achieving less than had been originally anticipated. Nevertheless, the experience with ATS-1 contributed to further development of satellite content for village schools on later satellite projects and demonstrated that the conferencing capabilities were also valuable for education: "Due to the conferencing capabilities of the satellite radio it is possible for the school administrator to make a point to all of his teachers at the same instant, and for all of the teachers to interact with each other also. . . . [T]he satellite radio has provided the Tanana Area Schools with a dialogue mechanism usually unavailable for the urban schools of Alaska except for the cumbersome task of coming together in meetings."[26]

The shared audio channel also enabled students at distant boarding schools to be in touch with their parents in the village. And the reliable communications link could be a lifeline for village schools in winter: "In the past, generator failures, furnace blow-outs, or deficiencies in instructional materials would cause the whole school to shut down while a written message was mailed out. With the satellite radio, teaching conditions were greatly improved and teachers would call for immediate emergency parts during the critical winter months. The communications links between the schools and the area administration supported and encouraged the efficient operations of the schools."[27]

The ATS-6 Biomedical Demonstration

NASA's ATS-6 satellite, built by Fairchild Industries, was launched from the Kennedy Space Center on May 30, 1974. Senator Ted Stevens and Alaska broadcaster Augie Hiebert flew to observe the launch with former NASA rocket scientist Werhner von Braun, then vice president of Engineering and Development at Fairchild.[28] ATS-6 provided the first opportunity to use satellite communications for transmission of TV and multiple audio channels to low-cost earth stations. Known as ATS-F until its successful launch, ATS-6 had the capacity to transmit video and pioneered direct broadcast TV to small antennas. NASA also called ATS-6 "the world's first educational satellite"[29] because of its use for educational experiments in Alaska, Appalachia, the Rocky Mountain States, and later in India.

Alaska's experiments on ATS-6 built on the ATS-1 experience, primarily through the addition of video. The main purpose of the ATS-6 biomedical demonstration was to explore the potential of satellite video consultation to improve the quality of rural health care in Alaska. Like the ATS-1 audio experiment, the ATS-6 project was carried out in the Tanana Service Unit with specialist support from the Anchorage Native Medical Center in Anchorage. In addition to the Tanana Hospital, the participating village sites were Galena, staffed by a medex (comparable to a physician's assistant) and Fort Yukon, staffed by a nurse. Satellite earth stations with capability to transmit and receive video as well as audio were installed at the remote sites, with video receive-only plus two-way audio at Anchorage.

Although ATS-6 had advanced technology compared with ATS-1, there were also severe limitations. The satellite was available for medical consults only three hours per week because it was shared with projects in Appalachia and the Rocky Mountain region, as well as with the educational project in Alaska described below. Also, the U.S. experimental period was limited to nine months, because NASA had agreed to move the satellite to an orbital location over India for a year for India's Satellite Instructional Television Experiment (SITE). Consults for serious medical problems typically could not wait for the next scheduled transmission, and patients with less urgent issues were sometimes not available at the scheduled time. The number of patients treated was relatively small, and the clinics that participated were staffed by practitioners with much more training than the more numerous community health aides. The remote practitioners found the limited schedule and pressure to provide a patient for the teleconsultations frustrating at times. At one point, the exasperated nurse in Fort Yukon said, "One of these days, I'm going to put my dog on that table."[30]

However, the field trial did yield much valuable information about the difficulties and advantages of video teleconsultations and implementation in Alaska. The project demonstrated that small earth stations transmitting black and white television could provide signals of adequate quality to be useful for health care delivery. In fact, some dermatologists noted that images with good resolution in black and white were superior to poor color transmissions for diagnosis of some skin lesions. Approximately 325 video consultations were conducted on 104 scheduled transmission days. Health care providers at all levels of training were able to participate in the consultations, including some health aides who used the facilities in Fort Yukon and Galena.[31]

Although the video consultations were conducted for a wide range of medical problems, the unique capabilities of the video transmission played a critical role in only 5 to 10 percent of the cases presented. Otherwise, there was little measurable difference between the effect of video and audio consultation.[32] The Stanford evaluators also noted: "The health care providers involved in the demonstration generally felt that the video consultations improved the capabilities of the health care system, but questioned whether the improvement was worth the additional cost or inconvenience. They placed much stronger emphasis on implementation of reliable operational audio channels which they consider absolutely essential to delivery of health care in Alaska."[33]

As part of the experiment, a centralized computer-based Health Information System (HIS) was also introduced. The forms developed for the trial included all the information that the providers needed to include for the Indian Health Service plus diagnosis and treatment codes and evaluation questions for the project. The coded forms could be easily entered in a centralized computer system and the results made available at any IHS hospital where an Alaska Native might be treated or transferred. The health record system was judged by all participants to be a valuable addition to the health care delivery system, and they recommended that it be continued in the Tanana Service Unit and extended to other parts of the state.[34] However, an integrated computerized patient record system was not introduced for more than 30 years (see Chapter 17).

The ATS-6 Educational Satellite Communications Demonstration

In 1969, Robert Walp was managing the Internal Research and Development unit of the Space Systems Division at Hughes Aircraft and working on the design of a high-powered satellite transmitter using frequencies known as S-band (2500 to 2900 MHz), the same frequencies that were being used for some terrestrial instructional television networks. Coupled with a large antenna, the S-band package could transmit television to small, inexpensive satellite earth stations, which Walp and his colleagues thought could be ideal for developing countries such as Brazil and India. The Hughes team decided to propose an S-band experiment for Brazil in response to a request for proposals from NASA for its planned ATS-G satellite. When a NASA liaison informed Hughes that an international experiment could not be considered, Walp recalls looking at a U.S. map with an inset of Alaska and deciding "[Alaska] would be a good surrogate for Brazil because it also had a large, sparsely settled interior region with inadequate telecommunication services."[35] To convert the draft proposal from Brazil to Alaska, he contacted Frank Norwood, the executive director of the Joint Council on Educational Telecommunications (JCET), a Washington, DC, based organization representing the public television industry. Norwood put Walp in touch with Bob Arnold, then director of the Alaska Educational Broadcasting Commission (AEBC) in Anchorage. Arnold expressed interest and was willing to let his name and affiliation be included in the proposal. Walp modified and submitted the proposal to NASA, and a few months later Hughes was informed that NASA would fund the experiment.

Walp then visited Alaska to meet Bob Arnold and to learn more about rural Alaska and its communications facilities (or lack of them). In Fairbanks, he met Charles Northrip, then manager of KUAC, Richard Dowling, his chief engineer, and Robert Merritt, professor of electrical engineering at the University of Alaska, and a designer of the transmitters and antennas used in the ATS-1 experiments. At Hughes, Walp was also a mentor for Albert Horley, recipient of a Howard Hughes fellowship for graduate studies at Stanford, who had been working with Professor Bruce Lusignan on developing techniques for television transmission by satellite to small earth stations. Horley became an assistant secretary in the U.S. Department of Health Education and Welfare (HEW) in the Nixon administration.

Horley and Walp thought they might be able to get an S-band transmitter included on the ATS-F satellite, which had not yet been launched, and submitted a proposal. NASA responded that there was not enough time to consider a new application, but that they had instructed the contractors to include an S-band package on the satellite. Horley eventually managed to obtain $13 million from HEW for the project, which was expanded to include the Federation of Rocky Mountain States and the Appalachian Regional Commission. However, the S-band equipment for ATS-F (which became ATS-6 after launch) was supplied by Philco-Ford rather than Hughes, and NASA eventually canceled the ATS-G satellite. Walp decided to leave Hughes and was hired by JCET to assist with planning and implementation of the ATS-6 project.[36]

In addition to funding from HEW, AEBC also got some state support for the project. AEBC director Bob Arnold recalled: "The State of Alaska check of $43,000 . . . was made out to Robert B. Arnold. I remember calling Juneau and saying, 'Do I get to keep this? Or how do we deposit in the State Treasury?'"[37] Eventually, the newly formed Governor's Office of Telecommunications (OT) took over management of the project, initially under Charles (Chuck) Buck and then Marvin Weatherly, who had been chief engineer at AEBC, and previously worked for broadcaster Augie Hiebert.

While some plans had to be scaled back because of the limited schedule and relatively short duration of the U.S. experiments, Alaska's educational broadcasters were able to carry out a variety of demonstrations to produce and deliver content intended for rural children, Alaska Native adults, and rural teachers during 1974–1975. Satellite uplinks transmitted color video from Juneau and Fairbanks to 14 village receiving sites. Each village was equipped with a small earth station and single TV receiver that was placed in the village school. The transmissions also included four audio channels so that separate audio tracks could be used for translations into Koyukon, Athabaskan, and Yup'ik.

The content teams produced several innovative programs for their rural Alaskan audiences:

- *Amy and the Astros* was a program for five- to seven-year-olds set inside a space ship with two Astro children puppets and a robot puppet. An English-speaking Alaska Native woman discovered the spaceship and taught the Astro children to speak English. The 32 programs were shown twice a week.
- *Right On!* was a health education program for eight- to ten-year-olds set in a typical village health aide's home clinic. Millie the health aide gave advice to two animal puppets, a moose and a beaver. The 32 original productions were shown on Mondays; on Fridays, health films from commercial distributors were shown.
- *Alaska Native Magazine* was an hour-long public affairs program for adults, featuring news and current events with an Alaska Native host. A total of 31 programs were produced, including two 90-minute specials.
- *Tell and Show* was a series of 28 half-hour programs for rural teacher in-service training produced by the Alaska Department of Education.
- *Experiments of Opportunity* included a variety of programs such as "Politalk," with news about the state legislature, training programs for community librarians, and commercial films about fish and game.[38]

The project encouraged village involvement through consumer committees and utilization aides, who were residents in each village hired part time to take responsibility for technical operation of equipment, to fill out logs on reception quality and audience size, and to encourage audience interaction.[39] However, the tight production schedule and varying participation of consumer committees made community participation and feedback difficult. Project cinematographer Mark Badger said that while the consumer committee for *Alaska Native Magazine* served as a formal representative structure, there was no process for involving villagers. "In practice . . . it was up to the production crew to ascertain what villagers thought of the project and whether it met their needs,"

ATS-6 *Right On!* program. Millie the village health aide with Rex Moose and Charlie Beaver. (Heather Hudson)

and the tight schedule was a "meat grinder" that did not leave enough time for the production crew to learn from and work with the villagers.[40]

A policy review pointed out that the education projects had been plagued "with outrageous troubles beginning with too little time and money to do the planning that was necessary."[41] The early stages were hampered by funding uncertainty and management confusion at the federal level. HEW and NASA required that the sites be selected long before Alaska management could take adequate steps to make selection more responsive to local wishes. And while the satellite had multiple channels to receive audio, participants were forced to use the ATS-1 satellite for interactive responses because the Department of Defense was using the frequency intended for interactive audio on ATS-6. Participants needed to change audio channels on the television receiver to avoid feedback, and broadcasters found it hard to understand responses from children over ATS-1.

Project reviewers Theda Pittman and James Orvik made several recommendations concerning rural Alaska media services, local control of media, and television programming. Among their suggestions were that videotaping equipment be provided in each community with TV reception so that programs could be viewed at the residents' convenience. They also recommended that consumer committees should be given funds to purchase programming on behalf of their villages.[42] Later, rural television projects LearnAlaska and RATNET actually did adopt versions of these practices. (See Chapter 10.)

Pittman and Orvik pointed out that some village high schools were being built, "however, villages have neither the expertise available nor the educational materials to deliver full quality high school education."[43] They advocated that "Telecommunications in Alaska should take as its mandate: solution to the 'high school problem'"[44] by augmenting the village high school curriculum through distribution of existing programming, teacher sharing via audio presentation and interaction, and new programming on Alaska Native history. The need for high school content and teacher support increased dramatically after the settlement of the law suit known in Alaska as the "Molly Hootch Case," for the Yup'ik girl whose name headed the original list of plaintiffs who sued the state of Alaska in 1972 for failing to provide village high schools, thereby forcing village children to attend distant boarding schools to finish high school. The settlement in 1976 of the case officially known as *Tobeluk v. Lind* consisted of a consent decree providing for the establishment of a high school program in all of the 126 villages covered by the litigation, unless the village decided against having its own high school.[45]

Despite the shortcomings, the reaction of villagers to the ATS-6 community education projects was generally very positive but raised expectations for television and Alaska content without any follow-on plans. Pittman and Orvik emphasized that telecommunications demonstrations should be undertaken in rural Alaska "only when there are resources and commitment for putting aspects of the demonstrations which users deem successful directly into operations."[46]

Lessons from the ATS Experiments

At the invitation of broadcaster Augie Hiebert, Wernher von Braun visited rural Alaska during the ATS-6 experiments, where he watched a telemedicine consultation from Galena to Tanana and then spoke to Tanana school children. Marvin Weatherly remembered:

> Wernher and I were sitting on the bank of the Tanana River, and he was skipping rocks on the water . . . and the principal of the school came

down and said, 'Dr. Von Braun, would you mind talking to one of our classes?' . . . and Wernher said, 'Ah, no, I'd be happy to.' I said, 'Boy. Here you have the most renowned scientist in rocket communication talking to a sixth grade class in Tanana, Alaska. . . .' He drew this circle representing the earth and then he drew the satellites that were in synchronous orbit around the earth and then the earth stations, the Tanana earth station, and explained how it worked and everything. . . . Well, you know these kids are awestruck. There was one little boy in front (who) raised his hand tentatively . . . : 'Dr. Von Braun, I understand everything you're saying to me. What I don't understand is what can we do with this?' Wernher, instead of going into some long diatribe about the technical aspects of the satellite, looked at this young boy and said . . . 'What do you want to do with it?'

Weatherly reflected on how von Braun's response resonated with him: ". . . years later, the engineers that worked for me and in projects I have been involved in, that was always the first question: What do we want to do? Not what *can* we do, what do we *want* to do with the project? Not how much does it cost? What do we want to *do*?"[47] As Augie Hiebert's daughter commented: "The doors of the imagination were flung open. . . ."[48]

In a sense, the ATS experiments had already opened the doors of imagination of Alaska's policy makers and providers of rural health care and education. ATS-1 was to play a key role in helping Alaskans learn about the potential of satellites to bring reliable communications to isolated communities and in persuading Alaska's leaders to seek long-term operational (as opposed to experimental) satellite communications for Alaska villages. As discussed in the following chapter, state government officials and consultants built on the ATS-1 experience to develop strategies to bring reliable voice communications to all permanent Alaska communities, primarily by satellite.

Both health care providers and educators found the shared audio "conference" circuit on ATS-1 valuable for learning from other participants and for sharing information with all the sites. The Public Health Service specified audio conferencing as a requirement of the statewide operational satellite network they later funded. Audio teleconferencing for public participation in state governance was later introduced throughout the state. (See Chapter 10.)

ATS-6's impact was less significant for rural health care because of the severe time constraints on the satellite and the expense and relatively limited incremental value of interactive video, where health aides had minimal training and many village clinics lacked running water and electricity. However, the educational television programs on ATS-6 did serve as lasting models for future content on LearnAlaska and some of the later public broadcasting programming for statewide audiences. And state oversight of the ATS-6 experiments resulted in continued opportunities for planning operational satellite services for Alaska. As the Governor's Office of Telecommunications had stated, a goal of the project was to "install and operate an experimental satellite communications system so that state planning and requirements for future operational communications could be grounded in actual experience."[49]

Chapter 8

From Satellite Experiments to Commercial Service

We can have the finest, most sophisticated communications available anywhere. . . . It's our right, but we have to seek it.

Senator Ted Stevens in the film *Satellite House Call*[1]

Building on the NASA Experiments

Following the successful demonstration of reliable voice communications on NASA's ATS-1 satellite, rural health care providers, Alaska Native organizations, community representatives, and many state politicians concerned about constituents in rural Alaska wanted reliable voice service to be continued and expanded, and Senators Gravel and Stevens vowed to provide bipartisan support for a long-term solution. Alaska broadcasters were also eager to implement rural satellite TV distribution that was demonstrated by Comsat, and educators were enthusiastic about the potential of audio and video for distance education from their experience with ATS-1 and ATS-6. The experimenters, evaluators, practitioners, and politicians were drawn into a lengthy political process to find a means of providing reliable communications to all of Alaska's communities.

The first satellite used for domestic (rather than international) communications anywhere in the world was the Canadian satellite called Anik (the Inuit word for "little brother"), launched in November 1972. Anik brought reliable telephone service and television to the Canadian North for the first time. The FCC's "Open Skies" policy enacted in 1972 led to the launch of multiple U.S domestic satellites systems in the 1970s, with the first being Western Union's Westar in April 1974. Although as discussed in Chapter 6, Comsat was eager to launch a satellite that RCA could use to serve Alaska, RCA decided to launch its own domestic satellite in December 1975, with Comsat following in 1976. From 1973 to 1975, RCA and the state of Alaska tangled repeatedly over the design of the satellite and the earth stations for Alaska villages.

In 1973 and 1974, the Stanford team that had been involved in the ATS-1 telemedicine evaluation continued to support the transition from experiment to operational service. With funding from the Lister Hill Center for Biomedical Communications and NASA, they produced plans for a network of small, inexpensive earth stations that could provide affordable communications services, including telemedicine services to villages in the Alaska bush. The Stanford engineering team, directed by Professor Bruce Lusignan, proposed a 10-foot-diameter earth station with a one-watt power supply for telephone service in rural locations that they estimated would cost about $15,000 each in quantities of 100 or more. The earth stations were designed to be flown into villages in small bush planes and maintained by local residents. However, use of such low-power earth stations would require a significant increase in the power amplification capability of the satellite.

At the time, all commercial satellites used large, expensive, and powerful satellite earth stations for high-volume traffic between two distant points, such as the Comsat Talkeetna earth station, which was 100 feet in diameter. A "small" earth station was 30 feet in diameter and cost about $500,000. The concept of using much smaller, low-cost and low-power earth stations for "thin route" rural communications with only a few telephone circuits per location was completely novel. In a network with only a few earth stations, total system cost was minimized by using large, expensive earth stations because they reduced the cost and power requirements of the satellites. However, when a large number of earth stations was required, such as to serve Alaska villages, it was more economical to design a satellite with higher power amplification in order to reduce the size and cost of the earth stations.

The Federal Communications Commission (FCC) granted waivers of its regulations to permit RCA to begin construction of the satellite at its own financial risk, pending FCC approval of its technical and financial plans. However, the initial design proposed by RCA was not considered suitable for Alaska because its coverage and power levels would make the size and cost of earth stations prohibitively expensive to serve rural communities. The technical plan that RCA filed with the FCC was, in Lusignan's opinion, close to ideal for providing commercial services in the lower 48, but not suitable for Alaska. In addition, the financial plan for the satellite proposed that slightly more than 50 percent of the satellite's cost be allocated to Alaska communications, apparently subsidizing the carrier's services in the competitive lower 48 market with revenues from Alaskans.

The Point Reyes Demonstration

At the request of the Stanford team, Senator Ted Stevens requested RCA to allow a technical demonstration of a 10-foot-diameter earth station at the Point Reyes, California, site of an RCA international communications satellite facility with a 100-foot earth station. Because no U.S. domestic satellite was yet in orbit, the demonstration used Anik, the recently launched Canadian domestic satellite. With funding from the National Institutes of Health (NIH), Stanford rented the electronics from California Microwave, then a small Silicon Valley company, and the 10-foot antenna from Westinghouse Electric. In February 1974, Senator Stevens and a group of Alaska state legislators, including Richard Eliason, Marlo Fritz, and Larry Peterson, spoke through the satellite link, with their voices transmitted through the 100-foot earth station to the satellite and received by the 10-foot earth station. The contrast between the small and large earth stations was dramatic.[2]

Following the February 1974 small earth station demonstration, RCA senior vice president Howard Hawkins requested a private meeting with Governor

Participants at the demonstration of a small satellite antenna at RCA's satellite site in Point Reyes, California, in February 1974. From left: State Representative Richard Eliason, Sitka; State Representative Milo H. Fritz, Anchorage; US Senator Ted Stevens; Stanford Professor Ed Parker; State Representative Larry Peterson, Fort Yukon. (Heather Hudson)

Egan to explain why the changes Alaska was requesting were not technically or economically feasible. The outcome of the private lunch in the governor's mansion in Juneau, to which Professor Parker was also invited, was that the governor did not accept RCA's assertions, and the parties agreed to establish a joint RCA-Alaska-Stanford technical task force to attempt to reach an engineering consensus.[3]

RCA also provided a briefing to Senator Ted Stevens in Washington in early March of 1974, which Professor Parker also attended. Following that meeting, Senator Stevens wrote to Parker: "Thank you very much for taking time to travel back here to the Nation's Capital and attending the meeting Monday in my office regarding communications in Alaska. . . . I've written to the Governor, asking that he encourage outside members on the task force; also, I've requested to both he [sic] and Howard Hawkins that the 30-day time limit possibly be extended, to assure adequate time for other sources to contribute."[4]

On the same day, Senator Stevens also wrote to Governor Egan:

> Moving to the areas of low-cost earth stations and your task force, I'm disturbed at a statement Mr. Hawkins made at the beginning of the meeting—the theme of which he carried through the ensuing discussion of RCA's plans and goals. He said there is a difference between RCA's basic satellite program and the possible utilization of small earth stations. This seems to me to again view the bush as the step-child of the modern system RCA proposes, and I hope we, you and I, can take the necessary steps to prevent anything less than the most modern system over the entire state. . . .
>
> The summary of Alaska's needs and goals, could, if you so desired, be submitted to the Federal Communications Commission along with RCA's application for a domestic satellite construction permit for consideration by the commissioners. This was recommended last fall, as you'll remember, by the then-director of the FCC's Office of Plans and Policy, Walt Hinchman. (Walt is now head of the Common Carrier Bureau.) Since the task force report is now in its formative stages, this course of action might be an excellent one to consider seriously. Walt Hinchman told my office that the RCA construction application for the domestic satellite is in the final stages of being approved, and if there is no additional input, either from the State or the APUC, it may well be too late to require any modification in the system. . . .[5]

There was also considerable federal government interest in the work of the task force. Senator Stevens had asked Dr. Albert Horley[6] (Assistant Secretary of Health, Education, and Welfare) and General Lee Paschall (then director of Command, Control and Communications for the air force) "to lend some people to the task force." Clay T. "Tom" Whitehead, director of the Office of Telecommunications Policy in the Executive Office of the President, also suggested that two of his staff join the task force.

Stanford and RCA members of the task force then met to discuss a variety of options for satellite and earth station design for both telephony and TV reception, but the RCA team appeared reluctant to consider seriously any modifications of the satellite because its construction was at their own financial risk, and the FCC was unlikely to approve any revisions to their rates to cover additional costs. They also wanted to avoid any changes that might delay the satellite launch and start of service. On April 5, 1974, RCA sent a report to Governor Egan summarizing the agreements and disagreements between the RCA and Stanford engineers.[7] RCA proposed using the unmodified satellite to provide communication to Alaska as quickly as possible, using 15-foot earth stations with estimated costs of $70,000 per earth station. They claimed this would avoid an $8 million cost and one-year delay in launch of the satellite. They also proposed to proceed with telephone service to 60 villages by the end of 1974 and an additional 60 villages in 1975, "depending on immediate State and FCC approval and ability to procure hardware" and a firm arrangement for revenue support. RCA's plan assumed that the earth stations would initially operate with the Canadian Anik satellite and later be repointed at the RCA satellite after its launch.

Although this option appeared attractive in its proposed commitment to serve 120 villages starting before the RCA satellite was launched, the caveats and changing cost projections concerned the Stanford team. Professor Parker wrote to Governor Egan: "In general, the recent task force activities convinced the non-RCA participants that RCA cost estimates were not to be taken at face value: they changed the numbers every day, including the day of the presentation to you after the task force had finished . . . My recommendation would be to try to block RCA's construction permit at the FCC until they propose to modify their satellite to provide higher power. . . ."[8] Meanwhile, Governor Egan had written to Howard Hawkins in his role as chairman of both RCA Globcom and RCA Alascom: "We know that the small 'Stanford type' station can immediately provide single channel voice service using RCA's leased capacity on the Canadian Anik satellite and also has the capability of expanding to multiple voice channel and television receive capability with the availability of the modified satellite. . . . In view of the urgent requirement for 24-hour-a-day telephone service to so many Alaskan communities, I urge you to begin installation of these small earth stations at the earliest possible date. Similarly urge you to immediately initiate modification of RCA's planned satellite to include a high power 4 GHz transponder. . . ."[9]

In early May 1974, RCA presented a new proposal for Alaska satellite services to Governor Egan, who responded very favorably in a letter to Howard

Hawkins: "Your assurance that complete and modern telephone service and television capabilities can be made available to 50 Alaskan villages by the end of 1974 is indeed well received. It is to be clearly understood by all concerned that the overall plan as presented by RCA Alascom in constructing satellite ground stations will provide quality telephone service and television capability to a total of 170 bush or village communities throughout Alaska."[10] The Stanford team's analysis of RCA's May 1974 plan was less favorable, pointing out equipment costs that were left out of the proposal, high-cost estimates for modifications to the satellite, and poor signal strength for western Alaska and the Aleutians from a satellite antenna optimized for coverage of the contiguous 48 states.[11] In a similar letter to Senator Stevens, they also suggested that "if RCA continues to be uninterested in providing services to smaller communities, it would seem to be appropriate to invite offers from other satellite companies."[12]

By late June 1974, Governor Egan realized that RCA was not going to deliver on installing telephone service in some rural communities by the end of 1974; he communicated his disappointment and frustration in a letter to RCA's Howard Hawkins:

> I have just recently learned to my dismay that RCA-Alascom has no plans to implement remote area satellite telephone service in 1974. In the light of the comments made by RCA during your last meeting in my office, I must confess I am greatly disturbed. . . . In addition to the indefinite delay of the program, I am also greatly disturbed by the escalation of costs for the small village installations. According to the latest information available to me, the cost of a small earth station to provide a single telephone circuit is now up to $73,000, a sharp contrast to the $20,000 stations we were discussing in March and April of this year. . . . I have no alternative but to inform the Federal Communications Commission that RCA's current plans for a domestic satellite system will not adequately meet the needs of Alaska's rural areas.[13]

Hawkins replied to the governor that "RCA does have firm satellite telephone service plans and is moving ahead as rapidly as possible to implement them." He also pointed out that implementation of RCA Alascom's plan would require regulatory approval by the FCC and the Alaska Public Utilities Commission (APUC), and added: "We hope that these approvals can be granted promptly."[14]

Governor Egan also sent a letter to Walter Hinchman at the FCC, outlining Alaska's communication requirements, stating "The first and most urgent of these is the requirement of a minimum of a single duplex voice channel of

24 hours per day telephone service to all communities in Alaska with a population greater than 25. Obviously, to serve such small communities the cost and complexity of the equipment must be held to the absolute minimum."[15]

The Sale of White Alice

While debates continued about design of a satellite for communication within Alaska, much of civilian communication continued to depend on the White Alice tropospheric network (known as WACS) operated by the air force (see Chapter 3). RCA had finalized purchase of the Alaska Communications System (ACS), but WACS was still owned by the Department of Defense. The air force had determined that the WACS "A route" from Anchorage through Ketchikan to the contiguous 48 states was surplus to their needs and would be discontinued effective June 30, 1974. Although the air force was willing to sell the WACS to RCA, the state concluded that the proposed price would make long-distance calls extremely expensive for Alaskans because, under the monopoly regulatory structure of the time, RCA would be able to put all of the White Alice capital and operating costs in their rate base, which in turn would be used to set the prices they charged Alaskan customers.

Senator Ted Stevens addressed the issue in letters to the chairmen of the Senate and House Armed Services Committees in March 1974:

> Much of WACS, while the most modern of the time when installed over 15 years ago, is now obsolete; its operation and maintenance are extremely costly with repair and part-replacement near impossible. . . .
>
> Controversy arose early last year, when RCA wanted to add more circuits to some basic trunk lines along the White Alice system, to accommodate increased civilian demand. The military essentially approved the overbuild, planning with RCA to turn over to the private carrier some of the WACS in the future, consistent with DOD objectives to leave the communications-provider role. Any acquisition of WACS by the common carrier must, according to the ACS disposal act, have the approval of the Alaska Governor. Governor Egan, however, would not give his okay, fearing the inclusion of the antiquated system in Alaska's rate base. He felt the common carrier should not take on equipment already overdue for replacement by satellite-age technology.
>
> With the situation stymied by the Governor's inaction, RCA had to seek alternative methods of providing services. This has forced RCA to accelerate the schedule to provide modern communications by satellite to the 49th State.

Last December, a proposal by Air Force Deputy Assistant for Transportation and Communications, John Perry, forced a move from dead center of the communications situation in Alaska. Mr. Perry announced Air Force plans to cease operation of the WACS A Route, which provides communications through Southeaster Alaska to the contiguous U.S. . . . .

Mr. Perry did offer the alternative to the State of having RCA purchase the A Route, thus insuring communications along this route to continue. Both Governor Egan and I strongly protested this course of action, again reluctant to allow the antiquated equipment to go into the rate base. Subsequently, Mr. Perry . . . has stated that DOD would be willing to lease portions of the WACS A Route to the carrier, in order that communications could be continued until the satellite system is completed. . . .

Mr. Perry, drawing his conclusion from the language of ACS Disposal Act, stipulates that RCA must lease the A Route of WACS for the fair market value of the system. For "fair market value," the military uses a yardstick of its investment, operation and maintenance costs. This is hardly a reasonable estimate of the "fair market value" of a system—which were it in the hands of private communications carrier—would have been written off long ago. The life-span of modern technology is seven years, so the primitive tropospheric scatter system of much of Route A has been not only doubly amortized, it is, by the military's own admission, totally excess to present DOD needs.

Why then must the few Alaskan taxpayers be forced to pay for a military mission long ago accomplished? The Governor has proposed, and I concur, that a lease fee of $1 be paid by RCA to the military for the use of the A Route after June 30, 1974. RCA would operate and maintain the system until such time as the company can provide a completely modern alternate routing system; RCA says it can have such a system in by the end of this calendar year. Mr. Perry has informed me that if both the House and Senate approve the $1 lease arrangement, the Department of Defense will follow the desire of Congress. . . .[16]

After protracted negotiations, RCA Alascom agreed to lease White Alice from the air force with the option to purchase segments required to support the satellite system being built at the time. In turn, the air force leased back facilities needed to support its Alaska bases and sites. In 1974, Alascom took over the operation and maintenance of the segment between Palmer and Ketchikan with circuits leased back to the military. Two years later, Alascom assumed control of the rest of the White Alice system from the air force. By 1981, all troposcatter facilities had been replaced with satellite terminals, and the troposcatter sites were turned back to the Alaska Air Command for disposal. Acting Secretary of the Air Force Tidal W. McCoy at last signed the White Alice sale agreement on August 25, 1983, finally ending military ownership of long-line communications in Alaska.[17] The last WACS tropospheric link, from Boswell Bay to Neklasson Lake, was used until January 1985 to connect Middleton Island to the network; the links were replaced by a satellite earth station.[18]

In addition to the obsolescence of the White Alice system, other factors were driving change in Alaska communications in 1974. The Alaska economy was booming, with the Trans Alaska Pipeline System (TAPS) under construction, bringing thousands seeking jobs to build the line or to work in the numerous other businesses that sprang up to support the oil industry. Demand for communications grew dramatically to serve these new pioneers; RCA Alascom also had the contract to provide communications at Prudhoe Bay and at the pump stations along the pipeline. The carrier was hard-pressed to meet the pent-up prior demand for telecommunications services; keeping up with the rapidly expanding new demands was almost impossible.

Also in 1974, Governor Egan was running for re-election. It is not surprising that he was eager to accept any RCA promise that would bring service to rural Alaska in the summer of 1974 construction season, prior to the November election, and that he was not pleased when he learned that RCA would be unable to keep that promise. Tangible results in the form of telephone service to rural Alaska would demonstrate his commitment to Alaska Natives, who were a significant voting bloc.

On June 28, 1974, both the *Anchorage Daily News* and the *Anchorage Daily Times* published stories of an attack on Governor Egan and RCA by former Governor Walter J. Hickel, who was again running in the Republican gubernatorial primary. The *Daily News* reported that Hickel had sent a letter to Walt Hinchman at the FCC asking the FCC to deny RCA's satellite proposal for Alaska until it could be upgraded to serve the bush. It quoted Hickel as saying, "Our governor should be demanding this satellite serve all villages of a certain size. Instead he is letting it drift, which could cost our Native people up-to-date communications for more than a decade."[19] The *Times* quoted Hickel's letter to the FCC as saying that the RCA plan "will exclude the vast majority of small communities and villages in Alaska—and these are the people who need it the most."[20] Ironically, Hickel's goals were also Egan's. In response, Egan denied Hickel's charges and released his letters to Howard Hawkins and to Walt Hinchman. The satellite issue apparently did not help Hickel; he lost the primary to Jay Hammond, a bush pilot and trapper, and former member of the state legislature and mayor of the Bristol Bay Borough.

Others Become Involved: The Lister Hill
Center, the APUC and HEW

In May 1974, Dr. Albert Feiner, then director of the Lister Hill Center for Biomedical Communications, sent a memo to the Alaska Native Health Service recommending implementation of a health communication network in the Bethel District of Alaska. He suggested that the specifications for earth stations should be compatible with RCA requirements in order to facilitate transition from the Canadian Anik satellite to RCA's system. Feiner's memo cited the Lister Hill Center support for the ATS-1 experimental network that proved so successful in improving health care in rural Alaska and its support for the February 1974 Point Reyes demonstration.[21] Senator Stevens soon sent letters to Dr. Emory Johnson, director of the Indian Health Service, and to Art Lusk of the Alaska Area Native Health Service endorsing the recommendations of the Lister Hill Center.

Although approval of a construction permit for RCA's satellite was a matter for the FCC, there was an issue that could also require state regulatory approval. Professors Parker and Lusignan wrote to Gordon Zerbetz, then chair of the APUC, explaining that if the satellite were not modified to increase its amplification, the costs of earth stations would need to be increased significantly, and the voice transmissions would require a "double-hop" through the satellite from a small station to a larger one and back to the second small earth station, an inefficient use of satellite capacity (and an annoying feature for callers who would experience the delay of the signal transmitted twice through the satellite, a distance of nearly 90,000 miles). The issue for the APUC was that RCA Alascom would likely propose to include the extra costs of these inefficiencies in the rate base from which their prices for Alaska long-distance service would be derived. Parker and Lusignan therefore recommended to Zerbetz "that the APUC write to the Federal Communications Commission requesting that the construction permit for RCA Global Communications satellite system not be approved unless and until the APUC has had a chance to review their plans. If RCA does launch its satellite system without APUC prior approval, we recommend that your commission refuse to allow any portion of the satellite costs to be included in the Alaska rate base. Instead, RCA Alascom could be required to lease satellite services from whichever domestic satellite carrier could serve Alaska at the lowest competitive rates."[22]

The Federal Department of Health Education and Welfare (HEW), which had funded both the ATS-1 audio telemedicine experiment (through NIH) and the ATS-6 video transmission experiment in Alaska, continued its strong interest and support. A June 1974 draft staff report concluded that "DHEW should provide continuing technical, analysis, and liaison support to the Alaska telecommunications project . . . to assure that any Alaskan telecommunication system meets the health, education, and other social service needs of the people of Alaska in fulfillment of DHEW's relevant responsibilities . . . [and] so that the technological and service experiments and experience can yield benefits to other potential users, including statewide, regional or national social service systems." It concluded that "DHEW's interests are congruent with Alaska's overall interests seeing that an adequate telecommunications system exists to service their needs . . ." and that "an integrated system serving all users is the most cost-effective approach."[23]

Challenging RCA's "Red Book"

In July 1974, FCC Common Carrier Bureau Chief Walter Hinchman wrote to Howard Hawkins of RCA, requesting detailed information about RCA's communications plans for Alaska for the period 1974–1980:

> As you know, the special role of RCA as the only currently authorized long-lines carrier for Alaska has been a major factor in the Commission's consideration of RCA's plans to enter the domestic satellite field. Accordingly, RCA's plans for long lines service to and within Alaska via satellite are of major significance in the Commission's consideration of RCA's pending applications for satellites and earth stations. It is of particular importance therefore that we have a clear, complete and comprehensive understanding of those plans, including the basis for the plans, the systems and services to be provided, and the economics involved. Given an understanding of the total plan we will be able to relate the individual applications to it and make the necessary finding for authorization. This will also enable us to deal appropriately with the several comments and complaints we have received from the State of Alaska concerning specific aspects and desired documentation of the future plans for Alaskan communications.[24]

A major reason for the FCC's request was that if capital costs were higher than necessary, those costs plus a rate of return (or profit) could be passed on to customers. Further, the proposed division of costs between Alaska services and lower 48 services could result in cross-subsidies to the potentially competitive interstate services from the monopoly intrastate Alaska services. In the 1970s, U.S. telecommunications policy was in the early stages of transition from regulated monopoly services to competitive services. The federal "Open Skies" policy authorizing competition in satellite services was an early step in a continuing transition to competitive services, which would include interstate long distance by the end of the decade. Consequently, the state of Alaska attempted to use both competition and regulation to make services affordable for Alaska businesses and residents.

Telephone services remained a regulated monopoly in which a single monopoly carrier, in the case of Alaska, RCA Alascom, provided long-distance service with interstate rates approved by the FCC and intrastate rates approved by the APUC. In both jurisdictions, the rates (prices) were proposed by the carrier, which was entitled to get sufficient revenue from its revenues to recover all of its operating costs and capital investment costs plus a rate of return based on a percentage return on its capital investments. Carriers thus had little incentive to control costs because their monopoly prices would recover their costs; they also had significant incentives to increase their investments because the larger their investments, the more profit they would be allowed to make. Much of the regulatory battle fought between the state of Alaska and RCA during this period was about how much of the RCA satellite investment would be put in the Alaska "rate base," that is, what capital costs RCA would be allowed to charge to Alaska telephone customers along with a percentage profit allowance based on those costs. This "rate of return" regulatory system functioned like "cost plus" contracting, where the contractor determined the base costs and the regulators determined what percentage profit was allowed on that base.

RCA submitted its communications plan to the FCC on September 20, 1974. Governor Egan asked several experts to comment on the RCA plan, known as the "Red Book," including broadcaster Augie Hiebert, Walter Parker (who later became Commissioner of Highways), Fred Brown (member of the State Legislature from Fairbanks), Professors Glenn Stanley and Robert Merritt from the University of Alaska, who were consulting for Brown's legislative committee, Stanford faculty Edwin Parker and Bruce Lusignan, and Robert Walp, who was in Alaska to work on implementation of the ATS-6 satellite experiment.[25]

Robert Walp remembers:

> RCA Alascom's Red Book described a plan that placed large satellite earth stations at Alaska's regional centers. They were to supply telephone trunk service to and from an earth station at Talkeetna, Alaska, which connected to a microwave radio system linking it to Anchorage and Fairbanks. . . . Telephone service to the small communities that were clustered around the regional centers would be carried by VHF radio using mobile telephone equipment of the early 1970s. [RCA had installed VHF telephones in some villages on the Seward Peninsula starting in 1970.]
>
> The consensus of the governor's advisors was that the RCA plan was limited because of its use of VHF radio, which could not carry quality

telephone service compatible with the North American system. Further, there would be no way to bring television to the small communities by satellite. We had no authority and could only persuade RCA, using the stature of the governor's office or, possibly, via the FCC. Although the clock was ticking toward 1975, when an RCA satellite was to be launched by Globcom, the RCA plan had not been endorsed by the State of Alaska. We were concerned that Alascom would go ahead with its plan, relying upon tacit approval of the FCC.

A parallel concern was the business relationship between Alascom and its parent, Globcom. Alascom was a monopoly carrier in a regulated environment, customarily entitled to a fixed return on its total investment, with rates being set by the regulatory authorities to guarantee a reasonable return. Its parent was set up to supply telecommunications services in the competitive market in the contiguous lower 48 states. Unless regulatory safeguards were installed, Alascom as a monopoly could subsidize Globcom, which would be competing in the domestic satellite market. RCA's plan was to have Globcom and Alascom jointly own their satellite system, with the portions based upon usage."[26]

The state filed a detailed response to the FCC on October 21, 1974, appending a technical review of RCA's plans prepared by Professor Bruce Lusignan of Stanford. In his covering letter, Governor Egan concluded: "The RCA plan falls short of meeting the requirements of the State of Alaska for low-cost telephone and television service to remote communities." The state's response included a 12-page statement of the state's requirements for low-cost telephone and television service to all communities in Alaska with a population of 25 or more, and stated: "The RCA-Alascom statement of requirements does not satisfactorily specify the requirements for communication service for Alaska, especially for service to remote areas."[27] The technical comments also pointed out that less-expensive, proven alternatives were available, such as satellite telephony provided by GTE in Algeria and Brazil (using capacity on Intelsat satellites).

The Alaska response raised the additional issue of cost separation between interstate and intrastate jurisdictions: "The RCA plan does not separate costs and revenues associated with intrastate traffic from those associated with interstate traffic."[28] It also commented on separation issues between RCA Globcom and RCA Alascom, noting that, according to the RCA plan, 51 percent of the costs of the proposed RCA satellite and earth stations was to be charged to Alascom and therefore Alaska ratepayers, with the balance allocated to RCA Globcom. It found that amount excessive, and concluded: "RCA-Alascom should therefore be restricted from financial transactions with RCA affiliated companies unless these transactions are entered into as a result of a competitive bidding process. No shared ownership of facilities should be permitted. . . . "[29]

The state also pointed out that RCA's right to engage in common carrier business in Alaska, as a condition of its purchase of ACS, was to provide service to rural Alaska, specifically to 142 communities. It noted that the APUC had recently found that RCA was apparently in default of that agreement and liable to have its certificate revoked for noncompliance.

While the struggles with RCA continued, after a recount following the November 1974 election, Republican Jay Hammond defeated incumbent Democrat William Egan and became the next governor of Alaska.

Chapter 9

Telephone Service for Every Village

We had no authority. We just were right.

Bob Walp, about the Governor's Office of Telecommunications[1]

A New Administration and a New Strategy

After Governor Hammond assumed office, he quickly took charge of tele-communications issues, including RCA's satellite plans. Hammond appointed Marvin Weatherly, formerly director of the Alaska Public Broadcasting System (APBC), as director of the Governor's Office of Telecommunications (OT), began a discussion with the Alaska legislature to achieve consensus on the policy issues, and established a Telecommunications Working Group headed by Highways Commissioner Walter Parker. Stanford professors Edwin Parker and Bruce Lusignan were again engaged as consultants, as was Robert Walp, who had worked at Hughes Aircraft during the NASA procurement of the ATS-6 satellite. Walp would later become director of OT.

RCA was quick to engage with the new administration, presenting pro-posed modifications of their September 1974 plan to the state in January 1975. However, Governor Hammond concluded: "the State cannot endorse the RCA plan, as it is not based on a system designed for Alaska's unique needs." Concerning the comments submitted by the Egan administration in October 1974, Hammond stated: "This Administration agrees with the objections expressed in the October comments, confirms their validity, and fully endorses the concepts and recommendations expressed in them."[2]

In July 1974, Edwin Parker had counseled Governor Egan: "I have become convinced that the only strategy powerful enough to force RCA compliance would be to have an alternative that could work without RCA cooperation and to convince RCA that you will act if they don't. If RCA senses that you don't have the will to carry out an alternative plan, they may be less likely to provide it themselves. Conversely, and paradoxically, the more committed you are to moving independently of RCA, the more likely it will be that RCA will comply and that you won't have to carry out the threat."[3] Parker's advice was prophetic. Robert Walp and Marvin Weatherly came to the same conclusion six months later at the beginning of the Hammond administration.

In January 1975, Marvin Weatherly, as the new director of OT, decided to hold a public review of the RCA plan in Anchorage. RCA was represented by Steven Heller, president of RCA Alascom, and several upper-level corporate executives from New Jersey, with the state represented by Weatherly, Robert Walp, and a few other advisors. Walp remembers:

> We may have lacked experience, but our plan proved to be sound. We argued over the practicality of putting small earth stations in the villages and even the feasibility of ever using small earth stations. Costs, timing and schedules were discussed. RCA's approach was based upon equipment that was in use at that time; this was conservative and relatively risk free. But it was based upon practices used for trunking heavy traffic between international centers, not for thin-line circuits to tiny towns and used the satellite inefficiently. We had spent a lot of time analyzing the small earth station in each village model and knew that it was economically feasible. . . .[4]

Walp recalls that over lunch at the invitation of Heller at the Petroleum Club, an exasperated Weatherly erupted: "'Dammit, Steve, if RCA doesn't put in small earth stations, the State will!' Steve replied, 'Be my guest!' After lunch, Marv and I walked back to his office . . . a few blocks away. By the time we got out of the cold, we had concluded, 'Why not?' starting a chain of events that eventually put a small earth station in every Alaska village."[5]

Weatherly quickly contacted selected members of the State Legislature and received strong indications of interest and support for state-owned earth stations if that was the way to get communications to rural Alaska. On February 21, 1975, the state issued a press release announcing that it intended "to purchase a number of small earth stations for installation in 1975 and 1976. . . . These stations are needed to provide emergency and medical communication in the bush. They shall also be fitted to supply message toll telephone service. These stations will work with existing or planned domestic satellites employing standard C-band transponders. The program calls for a quantity of 100 to 150 stations."[6]

State Funding of Satellite Earth Stations

Robert Walp describes the state's activity behind the scenes:

> There were well over a hundred unserved communities at the time, we estimated. To get things started we decided to request funds for 100 small earth stations and worry about their location later. Because the legislature is mandated to end its session in the spring, there was no time to analyze needs in a methodical manner. I estimated that an average station would cost $50,000. . . . We prepared a budget request for $5 million to procure equipment for 100 small earth stations, twenty to be operational by year-end 1975. . . .
>
> As this was at the beginning of the oil pipeline boom, money was easy. After a month or two without final action, I was in Marv's office when Frank Ferguson, the senator from Kotzebue . . . phoned. Frank . . . wanted to know if these earth stations could receive television as well as supply telephone service. I made some "back of the envelope" calculations . . . and called the senator back and told him we would be able to do it soon, if not immediately. He said, "You've got your five million dollars!"[7]

Marv Weatherly also remembered the efforts to get funding from the legislature: "I needed $5 million. That was the magic figure. I went to [Democratic senators] Jay Kerttula and Chancy Croft. RCA was all set up to kill this thing in the Senate—my request for the $5 million. But the House—that was no problem. The Senate was the battleground. Now here we have a Republican governor and Republican cabinet. I was a Republican. My opposition in the Senate was from the Republicans. . . ."[8] But Croft and Kerttula managed to maneuver the bill through the Senate. Weatherly remembered: "Jay called me on the phone. He says, 'Are you sitting down?' I said, 'Hell, I'm in bed.' . . . and he says, 'You got your $5 million. Now go out and build it.' And . . . you could hear my yells from one end of Anchorage to the other. I took the $5 million and started procurement."[9]

The appropriation bill was passed in May 1975 with House Concurrent Resolution No. 60, which called for the establishment of a state-owned system of small satellite earth stations. The legislature found that although it did not wish the state of Alaska to become extensively involved in the operation of the state's long-lines network, "it does find State ownership of satellite earth stations necessary and in the public interest."[10]

Meanwhile, the Alaska Area Native Health Service continued to seek a solution to connect Native health aides in remote villages to their regional hospitals. In March 1975, RCA Alascom responded with a proposal for a system based on wireless terrestrial links to existing communications facilities (and in one case to a shared satellite earth station).[11] It was clear to all that RCA was not on board with the state plans for small earth stations to provide communications to and from Alaska bush communities.

On April 9, 1975, Professor Glenn Stanley of the University of Alaska's Geophysical Institute sent a memo to Governor Hammond recommending that "the State of Alaska should procure and cause to be installed a reasonably large number of earth stations capable of providing immediate 24 hour/day emergency service to rural Alaska. . . . these stations should be leased to local utilities where present and in the absence of a local utility, to the common carrier until such times as a local utility is available."[12] On the same day, Walter Parker, Alaska Commissioner of Highways and head of the cabinet-level Communications Working Group, sent a memo to Governor Hammond with recommendations concerning the purchase of small earth stations listing the following objectives developed by the working group:

1. Small earth station installation must begin in the summer of 1975.
2. The state must preserve its options for obtaining future communications services, especially for education and health delivery systems.
3. All business arrangements between RCA Alascom and other RCA companies must be specified in contracts of public record, subject to prior approval and monitoring by the APUC and the governor's office.

Commissioner Parker's memo listed several recommendations, including that the state seek enactment of legislation for the purchase of small earth stations and proceed with the procurement.[13]

Concerning economic issues, as part of his consultancy to Governor Hammond, Edwin Parker prepared a 10-year cash flow analysis, demonstrating the financial feasibility of a state network of 100 earth stations for rural service. At a meeting with Robert Walp and the Stanford consultants, RCA Alascom's vice president and treasurer Robert Posma admitted that no revenue analysis for potential small bush earth stations had been made until a few days prior to the meeting. Walp then sent a memo to OT Director Marvin Weatherly, concluding that the revenue estimates for small bush earth stations provided by RCA show that " the small earth station makes good business sense and is certainly not the 'loser' RCA has called it in the past."[14]

Opposing FCC Authorization of the RCA Satellite

With a strong consensus throughout his administration, from the legislature and from his various advisors that small earth station service was technically feasible and economically viable, Governor Hammond sent a letter on April 11, 1975, to FCC chairman Richard Wiley stating:

> After intensive negotiation between RCA and representatives of my office, it became clear today, that it will be difficult and time consuming to reach a negotiated agreement with RCA on communications policy issues affecting the State of Alaska.
>
> Therefore, we must reaffirm the objections to the RCA plan stated in my February 14 letter to Walter Hinchman and the objections contained in statements of the previous administration to the FCC. The RCA plan of September 20, 1974 is unacceptable to the State of Alaska. . . .
>
> We strongly urge the FCC to not award a satellite construction permit to RCA Alascom (or jointly to RCA Alascom and Globcom). Such action by the FCC would create significant harm to Alaska by burdening the State ratepayers with costs which we consider inappropriate and excessive. We do not wish a monopoly service in Alaska to subsidize a competitive service in the remainder of the United States.[15]

The first major success in the state's campaign to get RCA to improve service to rural Alaska came on April 21, 1975, when RCA filed an amendment to their domestic communications satellite system application with the FCC, stating: "This is to inform you that the RCA applicants have made a number of recent changes in the spacecraft to be used in their proposed system. The net effect is an improvement in the performance of certain of the transponders making possible provision of a better service to the public." RCA proposed to increase the gain by four decibels (dB) in two transponders, thus significantly reducing the power requirements and therefore the costs of small earth stations, although not by as much as the Stanford team had recommended. They also proposed to move the beam of one of the transmit-only antennas for TV .5 degrees westward toward Alaska, to increase the transmitted power over much of Alaska, including the Aleutian chain. This adjustment would make possible reasonable quality TV reception with 15-foot diameter antennas.[16] During much of April and early May of 1975, the governor's staff and consultants reviewed and suggested changes to various drafts of a proposed agreement between RCA Alascom and the state of Alaska concerning satellite earth stations. However, RCA refused to enter an enforceable agreement to install and operate earth stations that would be cost effective for telephony and television service in Alaska villages.[17]

In early May 1975, the telecommunications trade press reported that RCA Globcom and Alascom had applied for a construction permit for a $6 million earth station in New York.[18] Edwin Parker brought this new development to the attention of Marvin Weatherly, director of OT, recommending that the state ask its counsel in Washington, DC "to file in opposition to this application, opposing the joint ownership feature and recommending that if it is approved, it should be approved solely as a Globcom application. If the two RCA companies are permitted joint ownership of any facilities, whether in space or on the ground, they will be able to make arbitrary transfers of cost allocations between the two companies, thereby defeating the regulatory intent of having separate companies."[19]

The state engaged economist Professor William Melody (formerly chief economist at the FCC) to work with Edwin Parker and Marvin Weatherly on the state's objection to RCA's earth station application. On May 30, Jack Pettit, counsel for the state of Alaska and former general counsel for the FCC under Chairman Dean Burch, sent a letter to FCC Chairman Wiley, all of the other FCC commissioners, and Common Carrier Bureau Chief Walter Hinchman, summarizing the state's position:

> A member of your staff was kind enough to furnish me a copy of the letter to you from Howard R. Hawkins, Chairman, RCA Alascom and RCA Globcom, dated May 8, 1975. This letter purports to be a 'progress report' regarding communications services for Alaska. I might add that RCA did not contact the State of Alaska in the preparation of this progress report, seek its approval, or indeed send a copy to anyone to my knowledge in the Alaska state government. . . .
>
> Mr. Hawkins' letter encloses a copy of the proposed agreement between RCA Alascom and the State of Alaska. Mr. Hawkins notes that 'the signing of this agreement has been deferred while the State legislature considers a proposal for creating a State-owned and operated communications system for Alaska.' The fact is that this proposed agreement was not and will not be signed, as evidenced by the enclosed copy of House Concurrent Resolution No. 60—adopted by the Alaska Senate on May 9, 1975—which clearly rejects the proposed agreement referred to by Mr. Hawkins.
>
> The State has had a continuous series of problems over the past four years in attempting to get RCA to commit itself to a communications system that will serve the needs of Alaska citizens. For example, throughout this period the State has on numerous occasions attempted to get RCA to establish a communications system plan for Alaska under a corporate structure that will permit effective regulation by the State. RCA has flatly refused to do this. . . .
>
> One current focal point of debate is ownership of small earth stations. Alaska has decided to seek authorization for state ownership of small earth stations because it has concluded that RCA is not responsive to the needs of the bush communities. Throughout some four years of negotiation, RCA has stated and continues to state publicly and to the FCC that there

are no real differences between the State and RCA. Unfortunately, when these assertions have been examined in detail, the differences have proved to be substantial. . . . Contrary to the assertion of Mr. Hawkins, the FCC should be aware that RCA has not demonstrated a 'firm and unstinting commitment to provide necessary communication to all of Alaska.' This is precisely why Alaska has felt compelled to seek authorization for earth station ownership.

Mr. Hawkins also states that 'state ownership of small earth stations is unnecessary. . . .' Mr. Hawkins is certainly entitled to his view. I would point out, however, as is dramatically evidenced by House Concurrent Resolution No. 60, that the elected representatives of the people of the State of Alaska have determined that 'state ownership of satellite earth stations [is] necessary and in the public interest', have strongly supported Governor Hammond's determination to proceed with the State's procurement of 100 small earth stations for rural Alaska, and have backed that determination with an appropriation of $5,000,000.

The State of Alaska will be seeking within the next several weeks the necessary authorizations for the procurement and construction of 100 small earth stations in rural Alaska. In addition, the State intends to oppose those applications of RCA Alascom and RCA Globcom, the granting of which would or could prove detrimental to the interests of the people of Alaska. . . .

A major goal of the State of Alaska is to require RCA to deal at arm's length with RCA Globcom. The State does not want RCA Alascom's service in Alaska to subsidize RCA Globcom's competitive service in the remainder of the United States. In particular, the State wishes to prevent RCA Alascom from owning any facilities jointly with RCA Globcom, including satellite and satellite ground stations.

As long as RCA Globcom insists upon pursuing its satellite plans under joint arrangement with RCA Alascom, it will be impossible for Alaska—or for that matter the FCC—to regulate effectively the division of investment, expense, revenues and profits between Alaska and RCA's other services. For this fundamental reason, as well as others, Alaska opposes all RCA applications for earth stations that would be jointly owned by RCA Globcom and RCA Alascom.[20]

An Interim Agreement with RCA

The Joint Resolution of the Alaska legislature and the appropriation of $5 million for small earth stations referred to in Pettit's letter to the FCC set the stage for the next battle with RCA. The state of Alaska proceeded with procurement of 100 small earth stations capable of telephone communications from small earth station to small earth station using the recently amended specifications of the RCA satellite. The $5 million appropriation for 100 earth stations meant that the price had to be under $50,000 per earth station. The cost was known to be higher than the early Stanford estimates because RCA did not modify the satellite as much as had been requested and because the state specified 15-foot antennas. The latter change was made to facilitate television reception from the modified RCA satellite. The winning bid came in at $37,500 per earth station.

The technical design of the earth stations began, with the state adding as an advisor William B. Pohlman, a retired Army colonel with a doctorate in electrical engineering who had worked in the aerospace industry. Bob Walp comments:

> Never before had earth station antennas smaller than 10 meters diameter been proposed. Satellite spacing was based on the narrow beamwidth of the larger antennas and it was assumed that earth stations with small antennas couldn't discriminate between adjacent satellites because their beams would overlap. Proving that this isn't a problem in the Alaska case required complex mathematics due to the many earth stations, each potentially contributing a bit of interference from the ground to a nearby satellite. Bill Pohlman's calculations in this and other areas not only enabled our application to be successfully filed and processed, it paved the way for the plethora of small earth stations now seen everywhere.[21]

In June 1975, the state filed applications with the FCC to install and operate the first six of the small earth stations. In its application the state indicated that these represented the first of approximately 100 applications it intended to file, and stated: "The State of Alaska does not intend to operate the proposed small earth stations. . . . One possibility . . . would be to lease the stations to RCA Alascom for operation in connection with its long lines services. Other possibilities, however, include local operation of the earth station by either the local telephone company or by a native corporation."

RCA opposed the state applications and filed competing applications for the same size and type of earth stations at the same sites. The state's representatives soon realized that contesting ownership at the FCC would be a very slow and expensive process and wanted to expedite getting service to rural Alaska. Walp explains:

> The state's efforts were driven mainly by the need for acceptable communications in rural Alaska; at the beginning it was a given that this was RCA Alascom's responsibility. We would still meet our original objectives, even

if Alascom built and operated the system. . . . We wanted to expedite the process which was taking longer than we had estimated. [Attorneys Jack Pettit and Dick Edge][22] arranged for RCA and the State of Alaska to use the good offices of FCC Commissioner Abbot Washburn to work out a compromise and avoid lengthy hearings to determine who was best qualified to serve the public. On July 11, an [interim] agreement was reached in which the State would own the earth stations with RCA Alascom installing them and operating them as part of its system on an interim basis, pending ultimate resolution of all issues. Thus, we were spared the job of setting up a duplicate organization in the bush, yet would eventually earn revenue from our investment in the hardware or be reimbursed for its cost. That would be addressed later, we assumed, thankful that an impasse had been averted. (RCA badly wanted to have outright ownership of the village earth stations because they would add to their rate base, increasing their revenues. Also, there were subsidies based on the assumed high cost of providing rural service. In Alaska these subsidies were immense. OT, and, of course, the small local exchange companies that did exist, wanted local ownership. We thought we could sell the earth stations to the locals, after operation began and things settled down. . . .) Steve Heller's exclamation had initiated a chain of events that got Alascom to put small earth stations in the bush!

A High Stakes Poker Game

Meanwhile, the state continued to oppose RCA at the FCC. Bob Walp explains:

. . . the standoff between RCA and the State came to a head in late October 1975 when the State petitioned to deny the launch of the RCA F-1 satellite, planned for December 12. The State contended that the RCA applicants had not demonstrated compliance with requirements of the ACS purchase agreement, that the joint satellite proposal would adversely impact cost and quality of communications service in Alaska, and that the proposed ownership arrangements appeared more expensive that a lease arrangement and that RCA Globcom was apparently attempting to use the joint ownership arrangements as a source of cross subsidy for its competitive services. The State did point out that it did not oppose launch of the satellite should Globcom elect to proceed in its individual capacity and would not object to a lease arrangement pending resolution of issues raised in its pleading. AT&T then filed its petition to deny a few days later. Both petitions were opposed by RCA almost immediately.[23]

In an order on November 6, the FCC set out conditions for RCA to launch their F-1 satellite that would prevent unnecessary investment by Alascom and preclude cross-subsidization from Alascom to Globcom. However, there was no guarantee that the satellite would carry Alaska traffic. On November 14, 1975, RCA asked the Commission for immediate reconsideration of its decision to defer a resolution of the issue of whether the AT&T or the RCA satellite should be used for Alaska service, stating: "Before the RCA applicants can irrevocably commit this investment by launching their first satellite on December 12 the basic question must be favorably resolved as to whether RCA Alascom will be authorized to continue to keep its interstate and intrastate MTS and other traffic on the RCA applicants' domestic satellite system."[24] In other words, RCA was threatening not to launch its satellite. As Walp pointed out, this was becoming a high stakes poker game.

As the launch date approached, RCA's Howard Hawkins tried to get Governor Hammond to meet with him in Washington, DC, and sent a letter to FCC Chairman Wiley on November 24, restating the RCA position. In the next ten days, Walp remembers that at least seven letters to the FCC addressed the issues from various parties including the state of Alaska, Western Union, and AT&T, as well as from RCA. Finally, on December 10, FCC Chairman Wiley sent a letter to RCA, stating that the FCC found that a series of letters from RCA on December 8 constituted adequate acceptance of conditions set forth in the FCC November 6 Satellite Order and that launch of RCA's Satcom F-1was authorized—for December 12, just two days later.

Telephone Service Begins

Telephone service to 100 bush communities with the small earth stations commenced in 1975 and 1976, with RCA Alascom installing and operating the state's 15-foot-diameter earth stations. Service began using leased capacity on the Canadian Anik satellite, later transferred to the RCA satellite. The experimental telemedicine network that began the process continued on NASA's ATS-1 satellite until medical communications were transferred over to the RCA satellite and expanded to include communities with the new small earth stations. An updated version of this telemedicine service remains in operation to this day (see Chapter 17). Rural television delivery would have to wait for additional funding from the state legislature, as discussed in the next chapter.

Bob Walp later reflected: "I was surely not well liked by RCA. . . . Here we were their best friends, we really were. We brought them in, dragged them kicking and screaming into the modern satellite age, and they dragged their feet. And we had great aims for building up the local exchange business."[25]

RCA Brings the Telephone to Alaska's Bush

Alaska's remote villages
are getting their first link
with the outside world
through Alascom's Bush
Telephone Program

Since July, the 89 inhabitants of Little Diomede, Alaska, have been getting accustomed to an unfamiliar instrument—the telephone.

A rocky Bering Sea island barely 25 miles from Siberia, Little Diomede is the farthest point yet reached by the Bush Telephone Program, an RCA Alascom project designed to bring long distance, direct dial telephone service to 142 remote villages of 25 or more people. While Bush Telephone is the most recent—and unusual—facet of RCA's role in Alaska, the Company for years has been instrumental in providing long-lines communications to the forty-ninth state.

"Before we put in the phone," says Bill Piotter, one of three field technicians who made the installation, "Little Diomede was accessible only by skin boat (25-foot walrus hide vessels), float plane, or helicopter." Other Bush telephone locations—there are now about 50 in operation—are either on the vast, barren plains above the Arctic Circle or in the massive state's virtually impenetrable woods and mountains.

Stephen D. Heller, who frequently visited Alaska as an RCA Service Company executive before joining Alascom as President last year, sees the Bush Telephone Program as a great asset to Alaskans living in remote villages. "Now the people in the villages can really communicate for the first time," he says, "with a world that could only be reached before by such primitive means as dogsled messengers or the most rudimentary radio service."

Calls from Bush Telephone locations are transmitted through a series of microwave or UHF and VHF radio in-

A student at Moravian Mission, near Bethel, makes his first call, which is being processed by Alascom long distance operators at automated Toll Service Desks (right) in Fairbanks and Anchorage.

Village telephone service: making a first telephone call. (undated RCA newsletter)

The success of the Governor's Office of Telecommunications and team of consultants in reaching an agreement to provide telephone service to rural Alaska was not the end of their involvement. In the fall of 1975, Marvin Weatherly became a commissioner of the APUC, and Robert Walp became director of the Office of Telecommunications, which Governor Hammond moved from Anchorage to Juneau. George Shaginaw became deputy director of OT, and Richard Dowling, who had been chief engineer for KUAC in Fairbanks, became chief engineer at OT. (Shaginaw eventually became president of Alascom after it was sold by RCA, and Dowling became a senior vice president of GCI.) Professors Parker and Melody continued to consult for the state, joined in 1976 by Heather Hudson, Douglas Goldschmidt, and Aileen Amarandos. University of Alaska professors Stanley and Merritt continued to consult for the Select Committee on Telecommunications chaired by State Representative Fred Brown.

Chapter 10

Broadcasting and Teleconferencing for Rural Alaska

We formed the Rural Alaska Television Network without any thought as to what its acronym was. . . .

Bob Walp, former director of the Governor's
Office of Telecommunications[1]

The Growth of Public Broadcasting

Government-supported noncommercial radio, and later television, were the means to expand broadcasting in much of rural Alaska. Radio station KUAC was established at the University of Alaska Fairbanks in 1962, but public broadcasting did not take off until the following decade. Two national initiatives in 1967 on educational and public broadcasting influenced Alaska: the Carnegie Commission Report on Educational Television, and the Public Broadcasting Act of 1967. Following the passage of the Public Broadcasting Act, Congress made funds for facilities available, and Governor Keith Miller appointed a Public Broadcasting Authority, which later became the Alaska Educational Broadcasting Commission (AEBC) with responsibility for coordinating and encouraging educational broadcasting. Under KUAC station manager Charles Northrip, who was appointed the first AEBC director, by late 1969, AEBC had "drafted legislation to establish itself by statute, begun to consider noncommercial radio as a basic information service to rural Alaska, convinced the University of Alaska to establish a television station, met with educators to determine goals that might be achieve through electronic media, and submitted a proposal to NASA for experimental satellite use."[2]

As discussed in Chapter 7, the AEBC, which moved to Anchorage in 1971 under Executive Director Robert Arnold, played a key role in the ATS-1 and ATS-6 educational satellite experiments. It also began to support the establishment of public radio stations with funding from the state and from the federal Department of Health, Education and Welfare (HEW). KYUK in Bethel went on the air in May 1971. Also in 1971, the legislature appropriated $50,000 and an additional $89,000 during the 1972 session to begin construction and cover operating costs of KOTZ in Kotzebue. KOTZ suffered several setbacks but survived to become the voice of Kotzebue and the surrounding region. Two weeks after KOTZ went on the air, an electrical arc on the makeshift antenna burned down the transmitter shack, but the station was back on the air two months later. A barge carrying construction material for a new studio sank in 1975, but a second shipment arrived in time for fall construction, and KOTZ moved to new studios in April 1976.[3]

In January 1973, state Senator Willie Hensley introduced a bill in the legislature to fund radio station construction at Barrow, Dillingham, Kodiak, and Petersburg. Applicants in Juneau received a capital grant from AEBC and additional funding from HEW to establish KTOO. An alternative to commercial stations in the state capital, KTOO went on the air in January 1974, and Ketchikan's KRBD Rainbird Community Broadcasting followed in 1976. Wrangell and Unalaska public radio stations began operations in 1977, with Unalaska initially rebroadcasting the AFRN feed after the AFRN repeater was taken off the air at the White Alice site in 1975.[4]

Some noncommercial stations were established with other sources of funding. Nome's KNOM, the oldest Catholic radio station in the United States, was licensed to the Catholic Bishop of Northern Alaska and began broadcasting in July 1971.[5] KMTE at Mount Edgecumbe near Sitka, then a Bureau of Indian Affairs (BIA) residential high school for Alaska Natives, was built with federal funds from the school budget. This was the only site where the Department of the Interior authorized a low-power "educational" FM station.[6] KMTE also turned out to be one of a kind as a radio station funded by a federal agency (other than the State Department or Department of Defense). The Federal Communications Commission (FCC) concurred with the Department of Interior's proposal for the station, setting conditions that it would not cause interference and would not impede private broadcasting development, but urged federal agencies in future to use "private, state or local government entities which could be licensed by the FCC to meet their educational broadcasting requirements regardless of whether federal funds were involved."[7]

Former AEBC director Bob Arnold reflected on the value of the radio stations: "very broadly the idea of making information available to people who didn't have it before was enormously satisfying. Speaking of villages . . . we tried—and found some success—in obtaining information from the villages."[8] As was the case with Alaska's first commercial radio stations, the nonprofit stations in rural Alaska served a vital function of getting messages to and from the villages, especially before telephone service was widely available. Arnold recounted the critical role of KYUK in Bethel on election day: "the mayor or the election official of a small village had been stuck by a blizzard in Bethel, and the ballots were in the city safe. And so he went on the air in Bethel to announce the combination to the safe so that they could get the ballots out and have the election."[9]

The message service at KOTZ in Kotzebue was called Tundra Telegram. Alex Hills remembered messages sent from KOTZ before the villages in the region had telephone service. "It was probably the most listened-to program of the day because it was like listening in on the party line . . . and the senders of the message would try to figure out secret little ways to say things that no one else would understand. . . . Bush pilots would bring in notes [from the villages] or maybe the message would be relayed by someone's short-wave radio to get it as far as the radio station. . . . I remember someone left her false teeth on the dresser at home and was asking that they be sent in."[10]

Alaska's first public television station, KUAC-TV, went on the air at the University of Alaska Fairbanks in 1971. It was followed by KYUK-TV in Bethel

Living color in the bush!
Television debuts in Bethel – and it's a sellout

A new line has been added to the inventory of Swanson's Department Store in Bethel — television sets — and they're selling faster than the store can keep them in stock.

There are no televisions to be bought at Bethel's other emporiums, either, because last week KYUK-TV went on the air.

As many as 80 people jammed into the recently-completed station on opening night, general manager Andy Edge said. And, he maintained with a grin, his 23 by 23 foot house was jammed with 63 people Saturday night — all craning their necks to get a look at the show being broadcast in living color.

"Believe me, it's beautiful," Edge told the Alaska Educational Broadcasting Commission, which oversees the operation of Bethel Broadcasting, Inc., at its meeting Monday. (Story, Page 2)

Edge reported that there are approximately 100 sets in Bethel and the surrounding area at the present time, and that the stores — Swanson's, NC and the Steak House, which got in the television business for the occasion — are sold out.

"Sears opened up a store for a couple of days and sold 20 sets, then packed up and left with orders for 50 more," Edge said.

The station will be on the air five days a week for an average of seven hours a day. Edge said. Wednesday night the first live news will be broadcast, he said, adding that about 20 hours of locally-produced shows are in the can waiting for additional equipment before they can be shown.

For many of the Yupik Eskimos in Bethel, last week's broadcast was the first they had ever seen. The station is televising at low power, reaching only the immediate vicinity of Bethel, but later this year the broadcast will be increased, reaching most of the isolated villages in the Kuskokwim and Yukon River region.

Larry Fulton

Youngsters gather round a TV set — one of the few in Bethel.

Bethel residents watch TV via satellite for the first time. (*Anchorage Daily News*, September 19, 1972)

in 1972. Anchorage's public television station, KAKM, did not begin broadcasting until 1975. KTOO-TV brought public television to Juneau in 1978.

Project Wales

Chapter 6 described Comsat's technical demonstration of small satellite earth stations for TV reception in May 1972; the equipment was moved from Juneau to Kodiak, Bethel, Nome, Barrow, and Fort Yukon. During the following year, a six-month experiment conceived by Commissioner James Hendershot of the Alaska Public Utilities Commission (APUC) provided taped television programs to the Inupiat village of Wales. With a population of about 150, Wales is the westernmost community on the American continent, some 110 air miles from Nome, but only a few miles across the Bering Strait from Siberia.

The homes in Wales were connected with cable that provided two channels, offering a choice of educational content or entertainment programming; the project also provided TV sets to residents. Researchers found that the school children gained in auditory processing abilities of spoken English (compared to the control village of Kivalina).[11] In general, the residents preferred entertainment to educational programming, and watching TV to some other social activities. "A surprise finding was that television tended to reduce alcohol consumption because of the apparent impaired ability to remember the previous evening's programs, a popular conversational topic the following day."[12]

Unfortunately, perhaps less surprising was the lack of any plans to continue providing television programming to Wales after that six-month period in 1973. As with the Comsat demonstration and the ATS-6 educational communications project, when the experiments ended, so did television for the participating communities.

Mini-TV for Rural Alaska

As discussed in Chapter 4, in 1959, a mountaintop translator and low-power local TV transmitter brought television to Suntrana, the Usibelli mine, and Healy. Translators later extended the reach of television through microwave repeaters from Fairbanks and from Anchorage down the Kenai Peninsula. In 1972, the Alaska Public Broadcasting Commission (APBC, the successor to the AEBC) received permission from the FCC to test videotape as a program source rather than an off-the-air signal from a translator as a means of providing television to rural communities that could not be reached by terrestrial repeaters. In 1973, the APBC began the test by mailing videotapes to three villages. Each village played the tapes on a videotape player connected to a low-power TV (LPTV) transmitter with a signal that covered the community. After the successful trial, LPTV, or what became known as mini-TV in Alaska, became a means of providing television for isolated villages. Marvin Weatherly, then director of the APBC, described the process of getting the FCC licenses:

> . . . something I wanted to pursue on the broadcasting commission [was] the low-power satellite television. It was my sense that if we could get these in operation in various sites around the state that it would help our overall plan of telecommunications. The problem [was] that the FCC was [granting them only] experimental licenses and there was a finite life to these things, so I wanted to regularize these licenses. I think it was Ted Stevens, or his office, that I used to gain entrée to [Commissioner] H. Rex Lee at the FCC. And I explained to Rex what we wanted to do, and I said, 'I want a regular license.' And he turned to his people and he said, 'You see any reason why we can't do this?' 'No.' 'Well, why don't we do it?' So we got a procedure for regular licensing of these [LPTV transmitters].[13]

In order to allow the LPTV sites to broadcast tapes from the commercial networks as well as from PBS, APBC arranged for each station to be licensed as an affiliate of all four networks. Michael Porcaro, who became assistant director of the APBC in 1975, and then executive director the following year, helped to obtain the special permission necessary from the FCC to create the network of mini-TV stations, and along with commercial broadcaster Augie Hiebert, gained permission from the networks to allow their copyrighted content to be retransmitted across the Alaska satellite system.[14] Alaska had pioneered the beginning of what became LPTV licenses for local broadcast retransmission throughout the United States.[15]

Between 1974 and 1976, the builders of the Trans Alaska pipeline adopted the LPTV model to provide television to workers at their construction camps. Northern TV and Midnight Sun Broadcasting applied for 18 identical LPTV licenses from the FCC so that mini-TV stations could transmit CBS and NBC programs.[16] The FCC authorized operation of the 10-watt mini-TVs in all of the construction camps along the 800-mile route from Prudhoe Bay to Valdez. Programs from Northern TV and Midnight Sun Broadcasting were taped in Fairbanks and flown in daily by the Alyeska Pipeline Services Company.[17] In 1975, Northern TV also applied for a license for an FM translator in Prudhoe Bay to rebroadcast KYBR AM from Anchorage via satellite, pointing out to the FCC that "things are different in Alaska." It became the first 100-watt FM translator in the country rebroadcasting an AM station received by satellite.[18]

The Satellite Television Demonstration Project (TVDP)

While the strategies to get reliable voice communications in the mid-1970s were intended to find near-term solutions for all communities, there was also great interest from rural communities and their elected representatives to get television to the bush, but no comparable funding commitment for TV. At the time, the Alaska network affiliates got most of their programs on tapes shipped from the mainland first to Hawaii (seen on one-week delay), and then Anchorage, Fairbanks, or Juneau. This system of "bicycling videotapes" was inexpensive and allowed local affiliates to insert advertising, but it was cumbersome for the stations and frustrating for viewers. However, the cost of satellite distribution was about $1000 per hour, so bicycling tapes was preferred.[19] The taped nightly network news was shipped on commercial flights from Seattle, usually arriving in time for late evening broadcasts.

In late 1975, the Governor's Office of Telecommunications (OT) called Alaska's broadcasters together to determine how to distribute programs from Anchorage network-affiliated TV stations to the village mini-TV sites via satellite. Robert Walp, then director of OT, described the problems they faced:

> To start with, there was no funding to pay for a satellite transponder to carry a television signal. Naturally, RCA had no incentive to donate one; after all, the satellite had to earn revenue. Then, there were four television networks: ABC, CBS, NBC and PBS, all wanting to get their programming on the system. Next, there was the problem of getting television to Alaska in the first place, since the only way at the time was to "bicycle" videotapes from outside. . . . If the networks didn't have their programming brought in live they said that they wouldn't be able to release it to the bush. To get over that hurdle, we decided to use the not yet funded transponder to transmit TV to the bush during afternoon and prime time; then, it could be used to bring television to the State from the lower 48 over the remaining hours. Programming would have to be recorded before transmission to the bush as it wouldn't be possible to simultaneously import it from the lower 48 and transmit it to rural Alaska.
>
> While logistics issues were being attacked, we tried to find a way of paying for a TV transponder. With a chance of funding from the state, the broadcasters discovered they were too poor to make any contributions, even proposing that they be paid for supplying their programming to the bush. Negotiations went on for a long time with OT trying to get a low transponder price from RCA . . . while reasoning with the broadcasters. The networks were no help, either, maintaining that TV in Alaska wouldn't appreciably increase their advertising revenue.[20]

The state legislature finally came to the rescue, and appropriated $1.5 million to OT in its 1976 session to lease a satellite transponder and modify 23 of the earth stations installed for telephony to provide TV for a one-year period, extended with funding for an additional six months by the legislature in 1977.[21] However, Walp continues:

> Governor Hammond wasn't happy to have the State to pay for the commercial networks' programming, nor were we, only the Governor was far less happy and didn't want to sign the funding bill. He convened a meeting with the broadcasters, some legislators and a few advisors. The Governor asked for statements from all, hoping to find a way to get television without state funds. Pressure from all over the State for live TV was growing and the bush, especially Senator Frank Ferguson [from Kotzebue], was adamant. The commercial broadcasters had implied that they could eventually support the system, just not now. The meeting wasn't going anywhere until Representative Fred Brown from Fairbanks made an eloquent plea, citing the importance of timely news, education and culture in the bush. And I assured the Governor that before long State funding would not be needed because the broadcasters would eventually find a way to contribute. Governor Hammond reluctantly agreed to sign the bill.[22]

OT had already set up a TV Advisory Committee representing a cross-section of industry and government to establish technical standards for use in transmitting television via satellite. Its calculations had shown that small earth stations were capable of receiving usable signals from the satellite despite the fact that their sensitivity was only one-tenth of the larger 10-meter antenna stations.[23] Low-power TV transmitters were used to transmit the signal in the villages.

OT issued a request for proposals for quotes for a television distribution system and got responses from RCA Alascom using its RCA satellite and Robert Wold using a Western Union satellite. It found that Wold was cheaper for the interstate links but did not include the required intrastate communications. OT also concluded that state-owned earth stations for TV reception were much cheaper than RCA prices for TV to the bush. In its report to the governor, OT quoted House Concurrent Resolution No. 60 (1975) in which the legislature declared that "state ownership of satellites earth stations (is) in the public interest" and urged that "steps necessary to facilitate acquisition and installation of state earth stations *including but not limited to television reception capability* be taken"[24] (italics in report). OT estimated the total cost for interstate, intrastate public interest, and intrastate entertainment television including amortized capital investment and annual operating and maintenance costs at $4.32 million or about $10 per capita statewide.[25]

After extensive negotiations, the state contracted with RCA Alascom in December 1976 to provide the services including a one-year lease of a full-time transponder from Alascom for $500,000, installation of earth stations, and low-power TV transmitters in 23 villages.[26] OT's chief engineer, Richard Dowling, and an engineer from RCA devised a means of eliminating interference to allow the satellite transponder to carry two television signals, making it possible to bring two programs into the state at the same time. Walp comments: "Doubling the amount of programming for the broadcasters made them happier."[27]

In January 1977, Augie Hiebert, representing the commercial broadcasters, and George Shaginaw from OT met with the commercial networks in New York to finalize agreements for delivery of the entertainment programming. They stated that if any abuses occurred, the agreement would be canceled and tests terminated immediately, and that they would return in one year with a report "at which time the networks would decide if the concept could be implemented on a continuing basis."[28] They then proceeded to Washington, DC, to ensure that the licenses for the local transmitters to receive the satellite programming had been issued. Hiebert found that some of the state's applications were still in an FCC in-basket. He recalled: "This was on Friday, January 12, and the service was supposed to start on Monday, January 15! . . . The chief of the FCC Translator division sent a telegram that Friday to the State Division of Communications authorizing them to start the demonstration the following Monday." Hiebert called the helpful FCC official "a bureaucrat with a big heart."[29] The transmissions began on schedule on January 15, 1977; the first rural satellite-fed TV station to go on the air was Tenakee Springs on Chichagof Island in southeast Alaska (population about 100).[30]

The Rural Alaska Television Network: RATNET

Under the auspices of the Alaska Federation of Natives (AFN), a Telecommunications Committee comprising representatives from each of the 12 Native corporations was established to select sites and programming. Those selected ranged from Anaktuvuk Pass in the Brooks Range to St. Paul in the Pribilof Islands, King Cove in the Aleutians, and Tenakee Springs in southeast Alaska. It was not possible to get a satellite station installed on Diomede, so the residents of the closest community to Soviet Siberia had to rely on tapes sent on infrequent flights. (The island of Diomede at that time relied on phone service via VHF radio from the Tin City radar station near the village of Wales on the mainland.) When the one-phone and one TV channel service reached

Mountain Village, the local district school superintendent announced the phone number, advising callers to phone early in the day as "it takes me three minutes to run down there to take the call."[31]

The committee that managed the channel became known as RATNET— the Rural Alaska Television Network; it selected content for one video channel including a variety of entertainment programs from the three commercial networks, plus educational content. Initial broadcasts included several hours of instructional TV. In urban areas, the commercial broadcasters were able to receive two channels per transponder, but required 10-meter or larger antennas that would have been prohibitively expensive for the bush. In general, the hours from 5 pm to midnight were reserved for delivery of programs to rural locations; live programs were sometimes transmitted during the evening to both urban and rural viewers. The system was not flawless but worked far better than many thought possible. The weakest link turned out to be the off-air pickup of programs in the lower 48 states that were then uplinked to Alaska; problems included electrical interference, operator errors, and program changes.[32]

In the first year, more than 5800 hours of programming were carried by satellite, including 2500 hours dedicated exclusively to network programs for rural communities. There were an additional 900 hours for educational, public service, and special intrastate programs, plus more than 2400 hours for interstate delivery of network programming to urban broadcasters. The largest non-network user was the Department of Education, which transmitted 10 hours per week, later expanding to 20 hours, of educational programs, including a national series for GED preparation in which more than 80 rural adults enrolled.[33] The APBC requested a block of time and produced many programs, building on its experience with the ATS-6 projects. The Alaska Area Native Health Service also expressed interest but could not obtain funding to produce health-related programs.

The satellite-delivered programs reached an estimated 5000 rural residents in the 23 communities, plus urban viewers.[34] Like their predecessors in Wales, the rural viewers preferred entertainment; their favorite programs included *The Six Million Dollar Man*, *Charlie's Angels*, *Hawaii 5-0*, and *The Bionic Woman*.[35] According to the project manager, one of the more frequently heard observations from urban residents was: "Those people in the bush get better television than we do!"[36]

Broadcaster Augie Hiebert was enthusiastic about the TVDP: "A system that had all the potential for disagreements, impossible scheduling and disaster worked just fine. RATNET communities, through choice by their own committee representatives received virtually everything they wanted, and urban communities throughout the State received virtually all the live news and sports programming they desired. RATNET was and is a classic example of dedicated people working together to achieve a highly desirable goal."[37] The commercial networks also signed off on the concept as an acceptable model for continuing operation.

The legislature appropriated funds in 1978 to extend the project to include 176 villages, but Governor Jay Hammond modified the appropriation so that only 27 communities would be added, to serve a total of 40 villages and 10 regional centers.[38] However, in 1980, the legislature provided funding for a full-time, satellite-based channel for instructional and educational TV, technically similar to, but operationally separate from, the TVDP.

Planning for Rural Public Broadcasting

A study for the Alaska Public Broadcasting Commission (APBC) completed in mid-1977 made several recommendations for the expansion of radio and television coverage and content in rural Alaska. Authors Theda Pittman, Heather Hudson, and Edwin Parker noted that the activities of the two previous years had resulted in significant public investment in improving Alaska's communications facilities, including the acquisition of 120 small earth stations, the TVDP, founding of the Alaska Public Radio Network (APRN), and establishment of new public stations in Anchorage, Wrangell, Petersburg, Ketchikan, Kodiak, Barrow, and Dillingham. They recommended the following steps to build on these accomplishments:

- Provide "some form of public radio and television to every community of 25 or more that requests it and can organize to accept responsibility as a licensee"
- "Establish a statewide public broadcasting system which integrates existing stations and services with new technology under appropriate organizational arrangements"[39]
- Establish a "nonprofit users organization . . . to lease at least one transponder from RCA Alascom with time divided between commercial broadcasters and APBC"
- APBC should request state funding for its share of the transponder and coordinating activities[40]

Other recommendations included that APBC should upgrade technical quality of APRN and develop program exchanges, statewide newscasts, and public affairs programs, and should provide training for local production, program exchanges, local boards, and so forth.[41] They also recommended that local access capability should be provided for all low-power rebroadcast facilities (local radio and mini-TVs). This recommendation was based on the Canadian Broadcasting Corporation (CBC) model of providing the capability for communities to substitute some of their own content for network radio feeds, in effect, establishing community radio stations with a mix of local and network programming.[42] The authors concluded: "Public broadcasting stations have a major responsibility to provide the information Alaskans need to take greater control of their lives and to plan for their future."[43] Eventually, most communities did receive public radio from regional stations, but the public access model of allowing communities to broadcast over their local transmitters was never adopted.

Planning for Instructional TV

Following the ATS-6 satellite experiments, the state took several steps to plan for future development of statewide K-12 instructional television under the leadership of OT Director Robert Walp and Ernest Polley, director of the state Department of Education (DOE) Office of Planning and Research. Following the end of the ATS-6 educational broadcasts in 1975, OT and the state DOE received funding from HEW's National Institute of Education (NIE) to prepare "A Planning Document for Future Educational Satellite Experiments in Alaska" to build on the ATS-6 experience and design permanent K-12 telecommunications services and support. Some stations were already transmitting instructional programs. KYUK-TV and Kuskokwim Community College broadcast instructional programs in English and Yup'ik. KAKM in Anchorage broadcast two hours of instructional TV daily for the Anchorage School District.

In 1976, in response to the consent decree in the Molly Hootch case (officially known as *Tobeluk v. Lind*) that was settled by a consent decree mandating establishment of high schools in all of the 126 villages covered by the litigation, the DOE began a major reorganization of rural education, assisting the development of 21 new rural school districts to provide schools to every village with more than 25 residents. Wilke notes: "This effort was greatly affected by the lack of reliable communications and educational resources; every meeting of the newly elected school board for the Aleutian school district cost the district $20,000 in air fare."[44]

In January 1977, the DOE and OT submitted their report to NIE entitled "Planning to Meet Alaska Educational Needs through Telecommunications." The plan identified needs for technologies including audio and video conferencing, radio, computers, and television as well as resources for instruction including content, staff development, information exchange, and management and administration. NIE then funded a four-year project known as "Educational

Telecommunications for Alaska," the ETA Project. Audio and computer equipment were used in a pilot project in four school districts, focusing on 9th and 10th grade education, and staff and administrative support. The ETA project emphasized data communications and educational uses of computers. Electronic mail was introduced to support administration of the state's schools, linking all 53 school district offices (but not all schools) during the project. The ETA established a remotely accessible data base with resources relevant for Alaska known as the Alaska Knowledge Base and access to national data bases to provide educational resources for remote communities. Early personal computers (then called microcomputers) plus audio tapes and print materials were used for instructional support, particularly for the new high schools that were being established as a result of the settlement of the Molly Hootch case.[45] Eight high school courses known collectively as IST (Individualized Study by Technology) were developed but were not available in all schools because not all had access to email at that time.

The ETA project included a major effort to inform educators statewide about the potential uses of instructional TV and interactive satellite-based services.[46] As part of this "TV for Learning" initiative, DOE awarded $300,000 in grants for school district projects and productions, including support for cultural studies courses and critical TV viewing skills for children. DOE also initiated professional ITV series productions and teacher guides, including the award-winning elementary series *Home in Alaska* by Alaska Film Productions, and an elementary series on Alaska geography, followed the next year by a high school–level Alaska history and Alaska biography series. Some IST courseware was used in more than 100 schools out of the total of about 160 to 180 rural schools with fewer than 50 students in 1982.[47]

In 1977, UAF's School of Engineering ventured into education via telecommunications with a course in Computer Techniques for Engineers presented before a small class of students at Capital Community Broadcasting in downtown Juneau and transmitted via RCA's Satcom-2 satellite and microwave to UAF and UAA, as well as over the Juneau cable system. At that time, the transmission spanned three time zones, with the course offered at 10 am in Juneau but 8 am at UAA and UAF. (A single time zone for most of Alaska was not introduced until November 1, 1983.) Local faculty acted as monitors; students could send questions to instructor by email, and students could access course materials online.[48]

The next venture by the School of Engineering into education by telecommunications was a videotaped course in Arctic Engineering in the fall of 1981. Tapes of the course recorded as it was taught at UAF were later shown for engineers on the North Slope, by cable to individual living quarters for ARCO employees, and in a classroom with TV monitors for SOHIO engineers. The

UAF course had practical value, as completing the course could fulfill one of the requirements for professional engineering registration in Alaska.

In fall 1982, the School of Engineering offered four more courses taped in a retrofitted classroom at UAF to be viewed later by students at other sites, supplemented with audio conferencing and onsite instructor visits. The estimated cost was $17,000 per course plus the instructor's salary.[49] UAF also joined AMCEE (the Association for Media-based Continuing Education for Engineers), a national consortium of 22 engineering universities that produced and distributed courses via videotape.

LearnAlaska

State Senate Concurrent Resolution 35, dated April 24, 1979, directed the Legislative Council (and its Telecommunications Subcommittee) to work with the APBC and the DOE's ITV office "to conduct a feasibility study of educational TV and prepare a plan of service including recommendations that address desired level of service to identified educational audiences, technical capabilities required to provide service, estimated capital and annual costs of service, interagency management, operations, and funding; utilization support services to users, etc."[50] The study was to be coordinated with the APBC and DOE, and to produce a report by January 30, 1980.

The DOE and University of Alaska worked together closely for the next ten months. DOE identified 19 guidelines for effective ITV and developed evaluation criteria for selecting and producing K-12 ITV. DOE also surveyed successful educational telecommunications systems in 11 states and Canadian provinces, defining management, programming, and utilization models relevant for Alaska. The university explored models for audio conferencing for instruction and administration, as well as the delivery of credit courses.

In February 1980, the APBC, DOE, and UA completed "A Report on the Feasibility of Telecommunications for Instruction in the State of Alaska," which called for development of a comprehensive interactive telecommunications network to support life-long learning in Alaska. It specified three communications systems: a full-time instructional TV service linking all permanent Alaskan communities with multiple uplink sites at major rural and urban centers, a separate instructional audio conferencing network, and an expansion of the University of Alaska's computer network to community colleges. Commissioner of Education Marshall Lind and Jay Bart, president of the University of Alaska, proposed a plan that would tie 285 communities into a statewide instructional network. It would have one-way video and two-way audio so that students could talk to teachers. They wrote to Senator George Hohman (D-Bethel): "An integrated instructional telecommunications system will dramatically expand Alaskans' access to life-long learning opportunities. . . ."[51] The report recommended multiyear development of network services based on shared use by a consortium of educational users, regional and interactive services, and continued development of programming and extension of the network to all rural schools.

The funding request of $6 million for FY81 was to develop a broadcast center and begin broadcasting of an all-educational network managed jointly by DOE and UA. The APBC proposed to manage the new network. Rural legislators requested a revised plan for faster implementation and increased the funding accordingly.[52] The bill appropriating the funding, HCSSB 165, was signed into law by Governor Jay Hammond in June 1980. In his letter of transmittal, Hammond made explicit the decision to curtail expansion of entertainment TV with state funds and to support new instructional TV systems.[53] The result was two satellite-distributed channels for rural Alaska, the RATNET news and entertainment channel and the instructional channel, known as LearnAlaska.

LearnAlaska involved three agencies working together to install and operate network: the Department of Education, Department of Administration, and University of Alaska. In 1981, DOE also created the Office of Technology and Telecommunications (OTT), combining two separate units: the ETA project and the Instructional TV unit.[54] The tasks for the small staff included developing procedures and support materials for audio conferencing, conducting demonstrations of effective uses of audio conferencing, obtaining broad input from teachers on selection of video instruction series, developing and distributing ITV schedules and catalogues, producing program guides and in-service training for teachers, and obtaining ITV programming for LearnAlaska.[55] The OTT supported the establishment of the Alaska Association for Computers in Ed (AACED). Managers Jane Demmert and Jennifer Wilke adopted a "planning while you do" approach: "The rapid legislative response to the UA-DOE Feasibility Study which proposed comprehensive instructional telecommunications services for Alaska has required that we adopt a simultaneous planning and implementation strategy."[56] The result was a network that provided local and taped educational content, school curriculum materials, statewide audio conferencing, and live coverage of events such as the Indian-Eskimo Olympics and Alaska Federation of Natives (AFN) conventions.

LearnAlaska began services in fall 1981 with a broadcast schedule of 18 hours per day, seven days per week including in school K-12 programming, after school children's programming, college classes for home viewing, general interest educational programs, and some specialized and interactive programs. Among the innovative features was transmission overnight of videos from educational program libraries for which the state had obtained licenses for educational use. Rural teachers were able to request programs to supplement materials available in village schools. They could then simply set the timers on their VCRs to record the programs and use them whenever they wished in the classroom.

The instructional TV programming was available to the new village high schools built under the *Tobeluk v. Lind* consent decree (Molly Hootch decision). As of the 1982–1983 school year, 101 Native villages covered by the lawsuit had new or expanded school programs, and by January 1, 1984, 84 of the 92 high schools required under the consent decree had been finished.[57] These high schools had teachers responsible for multiple grades, no laboratories, and few resource materials. Access to televised instructional content and courses were a valuable addition to their curricula. ITV programming came from vendors in the United States and Canada; in addition, the project produced three series specific to Alaska on fisheries education, Alaska geography, and life in Alaskan communities.

LearnAlaska was supported by the computer services of the University of Alaska Computer Network (UACN), which was one of the early implementers of electronic mail. Administrators, instructors, and students within the UA system could exchange "mail" by computer. Initially, terminal access was limited for students, but there was high demand and congestion during office hours. As Demmert and Wilke pointed out: "The lack of time constraints involved with computer usage, one of the most potentially liberating dimensions of computer-based communication, is thus contradicted by the scarcity of network access."[58] The Department of Education also developed an electronic mail system serving the DOE, public school districts, resource centers, and libraries.

By 1982, about 120 sites received the ITV signal broadcast by minitransmitters after reception by satellite, and some others received the programming over local cable systems. About 160 sites were equipped with audio conferencing. Examples from the Bering Straits region show how the facilities were being used. In 1982, 26 people from six Bering Straits villages enrolled in college level courses delivered by UIATC. Perhaps of equal importance was the impact of informal (noncredit) learning on rural residents: 78 persons received education or job counseling, 6 people questioned candidates for governor, 3 parents received information about their handicapped children, and 15 people were introduced to the concepts of resource management.[59]

Teleconferencing

Audio conferencing was another feature of the state's telecommunications facilities. In spring 1982, the audio conferencing system had more than 90 sites in more than 60 communities; a year later, there were more than 150 sites, with instructional television at more than 120 sites. Using dial-up circuits and bridges, the network could support up to 16 audio conferences of 5 sites each. Local coordinators were trained to facilitate public access and to demonstrate how to use the push-to-talk microphones so that two people wouldn't talk at once and thereby block the transmission. Audio conferencing could be used for instruction, testifying in hearings, and for public meetings, and was available for other nonprofit education-related agencies that paid long-distance charges for each site.

University health educators were among the most active users of teleconferencing to deliver programs for upgrading skills of health professionals and training nurses. The University of Alaska School of Nursing began using telecommunications to deliver parts of its curriculum. In 1981, the first Statewide Nursing Program began; students could take their first year at a local community college, where they could take some courses using the teleconferencing network. Within a couple of years, the Nursing School had developed 25 courses using audio teleconferencing or video, or both technologies.[60]

In 1981, UAITC and UAF cosponsored a course on the social impact of instructional telecommunications in Alaska. Materials included printed texts and readings, with student and faculty interaction using electronic mail and audio conferencing. About 60 students participated statewide, many of whom were faculty and staff eager to gain proficiency in the new electronic mail and conferencing systems.[61] Some other faculty also used audio conferencing for guest lectures. In what became known as the "Great Skinner Coup," UAA psychology professor Robert Madigan arranged an audio conference with famed psychologist B. F. Skinner: "The students dressed up for the class, even though Skinner couldn't see them. It took a lot of very hard work. Everyone loved it."[62]

Another initiative adopted teleconferencing to increase participation in state governance. Efforts to reach out to Alaska residents began in 1972 with the first regional legislative information office in Anchorage; additional sites were opened in Fairbanks in 1975, Ketchikan in 1978, and Nome in 1979. The offices were initially open only during the legislative session and primarily limited to distribution of print materials but linked by teletype. In 1977, Legislative Resolve 93 called for establishment of a Legislative Teleconference Network (LTN) to serve Anchorage, Bethel, Fairbanks, Juneau, Ketchikan, and Nome. The LTN began operations in February 1978 and expanded from 6 to 11 sites, including Barrow, Kodiak, Kotzebue, Sitka, and Soldotna in time for the 1979 legislative session. Eventually, year-round information centers were added in Dillingham, Valdez, and Wasilla, as well Washington, DC. Five of these sites were also teleconference centers, where residents could participate in public hearings, constituent meetings, and legislative work sessions. They could also access several data bases on legislation through computer terminals.[63]

Legislative teleconference coordinator Kathleen Baltes explained the legislature's commitment to participation: "The entire network of information resources and communications opportunities is based on an elementary assumption about representative government that Alaska's legislature has made: when it is made easier for the public to increase its participation in the legislative process, it will do so. This has been the case in Alaska."[64] She used a hypothetical example of a nurse in Kotzebue who could track a bill on health care and testify from Kotzebue rather than having to fly to Juneau, which would cost at least $1000 in airfare and living expenses and leave the clinic without a suitable substitute.[65] Baltes noted that many staff in rural communities were bilingual, so that testimony could be given in local languages and translated for the record. Also, audio conferencing offered an "offnet" feature linking in other sites by telephone, so that the legislature could get testimony from an expert witness outside Alaska or from a snowbound villager. Perhaps foreseeing interactive broadband services, Baltes asked in 1982: "How long before we need to take only five steps to reach our television set to testify before the legislature?"[66]

Foresight and Future Challenges

The late 1970s and early 1980s brought public broadcasting to many rural communities, as well as investment in telecommunications services for instruction, administration, and public participation. There were significant benefits in extending access to instructional content to rural schools and adult learners and in enabling rural residents to participate in state governance. There were also cost and time savings compared to sending faculty to provide face-to-face instruction, flying to Juneau to testify in hearings, or waiting for government information to come by mail.

Some participants in these activities also began to raise questions about the role of the state and the need for more citizen involvement in telecommunications planning. One concern was whether the instructional content was relevant, and what its impact might be. Michael Metty of Northwest Community College in Nome stated that the presence of a telecommunications signal in a rural Alaskan village clearly and obviously increases access. "At the same time we have concluded that the question of access is too simple. Succinctly stated the policy question must be more clearly viewed as access for what purposes, to what information, by whom under what circumstances."[67] He thought that courses produced elsewhere (at that time, from the Open University; Dallas, Miami and Coast Community College districts; the University of Maryland; and the University of Mid-America) "probably cannot address the needs, capacities, experiences and learning strategies of our residents without major modification. To not modify means that we insist that viewers adopt the vocabulary and orientation of the producers. This is de facto cultural homogenization."[68]

In retrospect, the foresight of state planners in anticipating and demonstrating the uses of electronic mail, teleconferencing, and instructional television is impressive. Alaska was truly an innovator in harnessing telecommunications for rural development during this period. Other services that state planners envisioned would take decades to implement. The Office of Telecommunications proposed the "development of Community Information Centers which will be developed in a library-like manner to all registered users."[69] Although the state never implemented this model, it predated the concept of telecenters in developing countries and federal funding in the United States for public computing centers (PCCs) 20 years later.

LearnAlaska's Bill Bramble expressed an optimistic view of the benefits of using telecommunications to extend educational access: "As revenues decline, the capacity of the state to support traditional capital and labor intensive approaches to education will be reduced accordingly. Technology may then provide one of the few (if not the only) ways to provide high quality educational opportunities for all residents."[70] Although correctly forecasting state budget cuts because of the decline in oil prices, Bramble was wrong about the future of the networks he had helped to start, at least for the 1980s. LearnAlaska lacked a sufficiently strong constituency to advocate effectively for its survival. In 1986, the state legislature canceled the funding, killing LearnAlaska.[71]

Chapter 11

Rural Television: From RATNET to ARCS

The issue is control.

  Canadian professor Gail Valaskakis at the 1990 Chugach Conference[1]

State Funding for Rural Television

By 1986, the state government faced a stark choice about whether to support television distribution to rural Alaska, which then consisted of two channels, the LearnAlaska channel of educational and instructional programming and the RATNET channel primarily consisting of commercial programming selected by a Native committee from the commercial TV networks and PBS: "The original objective of the Television Demonstration Project, now called RATNET, was to demonstrate the technical feasibility of using earth stations for television broadcasting. Nine years later, this objective clearly has been achieved. . . . The state needs to determine whether it has an obligation to provide entertainment television, and, if not, whether it wishes to continue the subsidization of it in its present form with state revenues declining."[2] In addition to the amount that the state was paying for satellite transmission and operation of the mini-TV sites, and the question of whether the state should be subsidizing delivery of entertainment television, there was the issue of urban versus rural equity in access to educational programming: "Residents of Anchorage, Fairbanks, and Juneau must pay cable charges to receive instructional television while rural students receive state subsidized programs."[3]

  Governor Sheffield's FY86 budget proposal initially zeroed out the entire RATNET program: "In my view, it would make no sense for the state to start to pay Alascom an additional space and power charge now, when the RATNET program is in the most precarious position of its 10 year history." The attorney general concluded: "The state's dealing with Alascom are varied, complex, and costly. Nonetheless, many of those dealings are not based on written contracts or even FCC or APUC approved tariffs. Just for RATNET satellite uplinks, downlinks, and transponder costs, the state pays Alascom around $2,250,000 per year."[4] Until the end of FY1986, the state continued to provide the two channels, LearnAlaska and RATNET, on separate satellite transponders to each rural community, and each had LPTV transmitters to rebroadcast the signal. At end of FY86, the state cancelled its lease for the satellite transponder that had distributed LearnAlaska programming but continued to fund RATNET. Political pressure to retain commercial television for the bush had triumphed.

Lack of continuity in policy and plunging state revenues in 1986 led to the legislature's decision to dismantle the LearnAlaska network after four years of operation. There were no public hearings and no evaluation or other research on whether the network was meeting rural instructional or broader educational needs. Arguments in the legislature against continuing funding focused on whether entertainment TV was more popular than educational TV. Thus RATNET survived but LearnAlaska was abolished. In a study entitled A Crisis in Educational Television Programming, Richard Taylor found that production of Alaska-specific educational programming was down 98 percent, and Alaska-specific public TV production was down an estimated 25 percent.[5] And little funding was available to procure instructional content from other sources. When LearnAlaska ceased operation in July 1986, requests from schools and school districts for tapes of instructional TV programs from the State Film and Tape Library more than doubled, but its budget had also been slashed. In 1987, only $28,000 was allocated for purchase of ITV programs, compared with $150,000 in FY86.

However, the question of what role, if any, the state should play in television distribution was debated for several more years. The elimination of LearnAlaska funding saved the cost of leasing a second satellite transponder, but the RATNET transponder itself was not fully utilized. In response to requests for increased bilingual/cultural programming, the RATNET Board at their Dec 3–4, 1987 meeting agreed to schedule one hour per week for programs such as council meetings, sporting events, and school activities.[6] But that was still just one hour per week. Other options to transmit existing content were ruled out. For example, the state owned broadcast rights to less than one-tenth of the state film library's collection, leaving little content available to transmit.

Another option could have been to eliminate RATNET and substitute a feed from an Alaska public broadcasting station such as Anchorage's KAKM on the transponder. A report to the legislature estimated that this solution would save about $330,000 used to operate a tape delay center in Anchorage and $50,000 in eliminating travel for the RATNET Council. However, the report concluded: "If the RATNET system ceases to deliver commercial programming and, instead, delivers public television programming, a majority of the current communities serviced by RATNET would be receiving only an alternative with no conventional commercial programming to provide balance. It is probably an understatement to say that such a decision would be unpopular in the eyes of many people in rural Alaska. Also, the loss of the RATNET Council . . . would create the low of rural participation in program decisions."[7]

Another option was to replace the smorgasbord of RATNET programs with content available from Canada (distributed on its domestic Anik satellites) or from other U.S. distributors. Counterarguments were that the community would have to choose only one of several available channels as there was only one LPTV transmitter, and that other domestic U.S. and Canadian satellites might not cover the entire state. And there was major opposition from the Alaska Broadcasters Association, which was "on record opposing any importation of programming from outside Alaska."[8]

State Purchase of Rural TV Earth Stations

Although the elimination of funding for LearnAlaska saved the cost of leasing a satellite transponder, the state still had a large inventory of satellite facilities for television distribution, including 235 LPTV transmitters that were left inactive.[9] RATNET programs were uplinked using Alascom facilities to Alascom's Aurora satellite and downlinked to 248 earth stations around the state, of which 146 were leased from Alascom and 102 were owned by the state. The state also owned 248 LPTV transmitters to rebroadcast the RATNET signal.[10]

Efforts to reduce the costs of rural television distribution focused on reducing the cost of the transponder lease charges from Alascom and replacing the earth stations leased from Alascom with state-owned satellite stations. The state negotiated with Alascom to reduce the annual charges for RATNET services by $360,000 for 1990 and 1991. The legislature accordingly reduced operational funding for RATNET, but at the same time it appropriated $2.8 million to purchase and install 146 earth station for the network in villages where the state now purchased services from Alascom. The state calculated that it would save $375,800 in FY90 in operating costs that would have been paid under Alascom leases, nearly $$877,000 in FY 91, and more than 1 million in FY 1992—the $2.8 million capital investment would be recovered in about two and a half years. The capital investment would also make it possible once the earth stations were installed for the state to request competitive bids for the satellite transponder that carried TV to rural earth stations. The Commissioner of Administration concluded: "In previous years, Alascom and rural legislators have prevented attempts to delete RATNET from the budget, but this was the first time the

Legislature seriously discussed funding State-owned earth stations to reduce RATNET costs and still leave the network intact."[11]

Alascom contended that the Division of Telecommunications could not adequately operate and maintain 146 additional earth stations, to which the Commissioner of Administration responded:

> The Division has had little difficulty over years of servicing the 102 the State already owns because the earth stations require very little maintenance and because the transmitters associated with the earth stations require the most attention. *The State already owns all 248 transmitters, handles all the licensing and does all the maintenance and repair on them.* (italics in original)[12] In addition to concerns about losing the revenue for operating and maintaining its own earth stations, Alascom asserted that the State's real purpose for these earth stations was to expand state-owned telephone service to the RATNET sites. The Commissioner responded: 'The Department has several times . . . assured Alascom that this Administration does not intend to use those small earth stations for anything other than television.'[13]

By 1992, the state government had spent $2.4 million replacing leased earth stations with state-owned earth stations, saving about $500,000 in operating costs per year. The state then owned 224 earth stations in Alaska to receive RATNET, plus the tape delay center in Anchorage and LPTV transmitters in 244 rural communities. By that time, cable was also becoming available in rural areas. A 1992 inventory of communication facilities showed that 103 out of 244 communities receiving RATNET also received television from at least one other source, primarily cable. However, rural residents continued to rely on RATNET for Alaska programming and information and wanted more local programming.[14]

The Impact of Television in Rural Communities

Former FCC commissioner Nicholas Johnson remarked, "All television is educational television. The only question is, what is it teaching?"[15] In 1977, Theda Pittman and Heather Hudson stated that "Alaska remains a foreign country in a media sense; almost all the television programming is produced Outside, and there are not enough reliable sources of information about Alaska issues."[16] The ETA and LearnAlaska projects funded some Alaska-produced instructional television, but most TV still came from outside, although the delivery costs to reach the villages were paid by the state. Norma Forbes and Walter Lonner identified the role and responsibilities of the state in Alaska television, noting that in the United States, television has developed as a commercial medium: "the industry is constrained by few obligations to its viewers other than those of the marketplace. In Alaska, on the other hand, satellite telecommunication is funded by the state; therefore, the obligation to serve all citizens equally is inherent in the system. The audience served by telecommunication in the Arctic is extremely diverse. It includes Native grandmothers with no formal education, teachers with advanced degrees, Native leaders who regularly travel the conference circuit from Paris to Tokyo to Washington, DC, and miners who have not been away from Fairbanks for forty years. This diversity places special demands on Alaska's telecommunications decision makers and producers in their attempts to serve television's viewers."[17]

By the early 1980s, anecdotal reports on the impact of TV cited less visiting in bush communities, and a declining role for elders: "children don't come to elders as much to be told the old stories." More generally, "Television might be eroding the communal interactions and storytelling upon which the sustenance of Native cultures depends." There was also concern that "television . . . apparently accelerates the urbanization of visions and values." Douglas McConnell found that "Native children who have access to television want to move to more urbanized places than do children without television."[18] Unlike most of the country, in Alaska, the state was a major funder of television delivery and instructional content for rural areas. McConnell concluded that "the State of Alaska, as the prime financier, has a special obligation to understand the social and cultural consequences of the television it is providing."[19]

Norma Forbes of the University of Alaska shared these concerns. She and colleagues conducted a five-year study begun in 1977, examining the effects of the introduction of commercial and educational television to isolated communities in rural Alaska, hoping that the findings would be useful for "residents of villages which participated in the study, for the staff of the schools which were involved, and for Alaskan policy makers." Funded by the National Science Foundation (NSF), the studies began with baseline data from five villages about to receive their first satellite TV and five comparison villages that were not to receive TV at that time—two non-Native, two Inupiaq, two Kodiak Island Aleut, and four Athabaskan. The second phase returned to villages and added two pairs of TV and non-TV villages, including two in southeast Alaska Tlingit and Haida communities.

The studies used cognitive tests as well as interview items from studies across the Canadian North, allowing comparison of effects across Arctic North America. The researchers found that two years of television did not alter either the range or the level of Alaska children's occupational aspirations, but children in TV villages were more likely to want to live in a city than children in non-TV villages. "In the Kotzebue area, only 17 percent of the children interviewed desired to live in their home village as adults."[20] Unlike the earlier research in Wales, this study found no significant difference in English vocabulary among children in TV and non-TV villages. However, they found that the superior spatial-perceptual skills of Alaska Natives might be degraded by prolonged exposure to television. They noted that acculturation and/or TV seemed to depress spatial-perceptual skills in females but not in males, and suggested that TV might be a more powerful acculturative force for females. They concluded: "It would be ironic if the introduction of a telecommunication system designed to expand the educational opportunities of people in the Arctic should contribute to the loss of their superior cognitive skills."[21]

In addition to specific findings, Forbes's report discussed three types of influences to be considered when television is introduced into traditional societies: "that television may make people dissatisfied with the goods and services they buy and the lives they lead, television's role as a 'thief of time,' and television's influence on values." To reduce the negative impact of television, the report suggested that the content and form of television in rural Alaska need not be a carbon copy of television in the rest of the United States, and that the way television was viewed could be changed by teaching critical viewing skills.[22]

There were also anecdotes about the perceived positive benefits of TV. Then APBC director Marvin Weatherly personally carried the LPTV equipment to St. Paul and installed the station there. He remembered:

> Now these are people who—many of them—had never seen television in their life. . . . I had in my baggage a couple weeks worth of tapes and Reeve (Aleutian Airways) lost the bags. I had one tape, one tape from PBS. It was ice skating at Rockefeller Center. I installed that station at St. Paul and put that tape on. They had in the community hall . . . a television set up and the kids and the old people and everyone sat around that and kept playing the tape over and over and over. But what was interesting about it was these young kids: they had never seen buildings like . . . Rockefeller Center. . . . I left there, and later on the tapes got there. . . .
>
> I went back to St. Paul about two months later just to check everything out. Before I made that trip, I had received a call from the University of Alaska complaining that I was corrupting the values of the Natives and changing their way of life. I said, 'Isn't that up to them to decide as opposed to you?'. . . . I got off the plane, and I'm walking down the street of St. Paul . . . and this lady came up to me, grabbed me and hugged me and kissed me. And I said, 'I appreciate this but what's it all about?' She said, 'I want to thank you, Marv, for what you did.' And I said, 'What did I do?' She said, 'Before we had television here, my husband would stop

off at the bar and get drunk and the kids had nothing to do after school. They'd go down and throw rocks at the seals.' She says, 'We're a family again. They come home and we watch television together, we eat together, we are a family.' And she said, 'I can never thank you enough for that.' I said, 'That story . . . what you have told me is thanks enough.'[23]

The debate about the impact of television in rural Alaska continued into the 1990s. At the 1991 Chugach conference, Rosemary Alexander commented: "We were healthier, we were less lazy before television."[24] She said that children played outside more, and people visited each other more. Now they had to schedule community meetings "when there's not a great show on." But others pointed out that watching television also helped people to speak more grammatical English, reporting that one woman said "I watch *The Price Is Right*. It teaches me numbers. I watch *Wheel of Fortune*. It teaches me English."[25] Alexander concluded: "Many people . . . say that television has speeded up the cultural confusion in an already changing society. It definitely accentuates the differences between the haves and the have-nots. And this is basically a have-not culture."[26]

Debating the Future of Rural TV

In the early 1990s, the value of RATNET was again the subject of debate. Television viewing in rural Alaska was definitely widespread; one village survey found 97 TV sets in 62 homes, with only 3 homes without a TV. A study by the Institute of Social and Economic Research also estimated that 84 percent of rural households had videocassette recorders (VCRs), about the same percentage as in urban areas.[27] Congressman Don Young had sent promotional videocassettes to rural voters during his 1990 campaign.

RATNET was definitely popular in the bush. In 1990, Paula Rasmus-Dede recalled: "I remember when RATNET was started because I was sitting in a room with some health directors from the regions and we were supposed to be picking 15 villages to get a telephone with a health aide. One of [Governor Jay] Hammond's aides came running into the room and said, 'Oh, by the way, you will be glad to hear that the 15 villages you pick will also get television.' We had heard nothing about television prior to that meeting. And we sat in that room and fought and scratched to pick the first 15 villages to get television . . . and it became a status thing for a village to have television."[28]

But now cable systems were also spreading to the villages, where community-owned cable systems were being installed, financed by grants or by a commercial vendor such as Microcom offering the communities $70,000 loans financed over nine years. A cable subscription would cost $85 for the connection and $50 per month—about the same price as water for a village household. As one participant at the 1990 Chugach Conference commented: "When RATNET was the only television channel, you could go into the village and find two dead televisions with another television on top and the television was always on, because that was all the television we had . . . but now people have choices."[29] More choices brought demand for 24-hour sports, news, and movies: "Not that they're terribly unhappy with RATNET; they're just tired of seeing the same old thing every day."[30] However, a survey in the NANA region of northwest Alaska found that most people said they would be willing to pay $10 per month to keep RATNET, especially if it meant that they could keep their statewide news and statewide weather.[31]

Charles Northrip, then executive director of the APBC (which he characterized as "a state agency located within the Department of Administration, physically if not emotionally")[32] defended the value or perhaps inevitability of television: "why shouldn't rural Alaska enjoy all the benefits and suffer all the consequences of television as does the rest of the country? Or to put it another way, what conceit would drive us to believe that we could keep that from happening?"[33] Commercial broadcaster Augie Hiebert, long an advocate for RATNET, was concerned that the legislature was going to kill its funding. "The direction the legislature is going is to turn off something it has been demonstrated has a very great benefit to the Bush. Where is the Native constituency? Where is the Alaska Federation of Natives? Where are the 13 corporations that should have some interest in their people out there?"[34]

There was also recognition of the lack of content by and about rural Alaskans. Russell Nelson, then chair of RATNET's board said: "I always felt that RATNET was trapped in a strange sort of irony. About all they can do with the amount of money that they have is serve the most people with the least objectionable, usually entertainment programming that they could find. But this traps them into a position . . . that there is no Native language programming on there."[35] However, at the time, there were some local programs produced in Bethel, Kotzebue, and Barrow that could have been transmitted statewide. Hiebert proposed what he called the "One Percent Solution," that 1 or 2 percent of the Department of Education's budget of approximately $7 million be allocated for producing—developing and transmitting—meaningful programming for the Alaska bush.[36] The proposal was never adopted.

From RATNET to ARCS

Funding reductions in 1995 finally did force the demise of RATNET, but after a debate in the Legislature about whether publicly supported broadcast services deserved *any* funding, KYUK in Bethel agreed to assume responsibility for the rural service, which was reconstituted as ARCS (Alaska Rural Communications Service). The ARCS council was similar to its predecessor, consisting of a member from each of the 12 regional nonprofit Native associations in Alaska: one representative from the Department of Education; one representative from the University of Alaska; one member from the Alaska Public Broadcasting Commission; a representative from the Bethel Broadcasting Inc. (which was the manager station for the Council); and two public members selected at large by the governor.[37]

The Legislature appropriated $700,000 to match nearly 1 million dollars in federal funds for the Satellite Interconnection Project for digital conversion of the satellite transmissions, with the result that digital compression made it possible to carry three channels on one transponder. The three public stations in Fairbanks, Juneau, and Bethel agreed to divide programming responsibilities in order to reduce each station's expenses and overall public broadcasting system fees. KUAC operated Alaska One, uplinked from Anchorage, KTOO started Alaska Two from Juneau with news coverage of the capital including gavel-to-gavel coverage of the Legislature, and KYUK in Bethel uplinked the ARCS content. In July 1997, Alaska Three was created by the University of Alaska Southeast (UAS) and the Distance Delivery Consortium (DDC) to produce and transmit instructional content for K-12 and university level courses.[38]

ARCS has survived until the present day as a free over-the-air channel despite availability of both cable TV and direct satellite broadcasting to individual households in rural Alaska. Members of the current ARCS council point out that the network remains important to reach villagers with emergency broadcasts and other critical information. However, it faced yet another funding challenge; the FCC mandated that low-power TV stations upgrade from analog to digital signals by September 1, 2015. In 2014, the Alaska legislature provided a $5 million capital grant to upgrade the ARCS equipment for digital transmission.

Meanwhile in Northern Canada . . .

By the 1990s, Canada had taken a very different approach to providing television for its northern communities. Television for the Canadian North began in 1967 with a package somewhat similar to what RATNET offered a decade later—a "Frontier Package" of tapes taken directly from the Canadian Broadcasting Corporation (CBC) and sent by plane to Arctic settlements. There was no native content on the Frontier Package, and none originally on the feed to the north on Canada's Anik satellite launched in 1972. The CBC introduced its Accelerated

Coverage Plan (ACP) in 1975, promising television service to all communities in Canada with at least 500 people, but there was no provision for local or indigenous content.

However, the Canadian government began to support indigenous communications, and sponsored several satellite experiments somewhat similar to the Alaska experiments on NASA satellites in the 1970s. Beginning in 1980, the Inukshuk project included programming produced in six Inuit communities and transmitted on Canada's Anik B satellite for eight months, providing not only Aboriginal content but training and experience for young Inuit in television production. In addition, teleconferencing with one-way video and two-way audio allowed Inuit in the North to see each other, discuss important issues, and exchange information in their own language.

Based on this experience, the Inuit formed the Inuit Broadcasting Corporation (IBC) and received a network television license in 1981, beginning broadcasts in January 1982. However, IBC was dependent on access to timeslots allocated by the CBC's Northern Service channel. IBC soon proposed the creation of a dedicated northern channel. For several years, a consortium of Aboriginal broadcasters, territorial governments, and the CBC Northern Service held numerous meetings and discussions with federal agencies about the proposed northern channel. Meanwhile in 1990, the Canadian government adopted a Native Broadcasting Policy that required Aboriginal broadcasting in northern regions. In the same year, the Canadian parliament passed a new Broadcasting Act which entrenched Aboriginal broadcasting in legislation. Finally, federal funding was approved for Television Northern Canada, "a truly northern pan-arctic channel."[39] In June 1997, the Television Northern Canada board voted to move forward on plans to establish a national Aboriginal television network. In September 1999, the Aboriginal Peoples TV Network (APTN) began transmission, not only to the North but as a must-carry channel on cable systems throughout Canada.

Missed Opportunities?

Canadian academic and Aboriginal researcher Gail Valaskakis recounted the history of indigenous broadcasting in Canada at the 1990 Chugach Conference in Anchorage. She was astounded to learn about the lack of funding for Native production for RATNET: "It is phenomenal from my perspective that you have a network sitting there. . . . I mean if we had a network sitting there when we started this two and a half decades ago in Canada, I tell you we would have moved pretty fast, a lot faster than we did. . . . If you've got a network sitting there, then you would have to consider seriously whether or not it was worth looking at production possibilities."[40]

RATNET was saved, while LearnAlaska was shut down, based on politicians' fears about voter backlash if they discontinued subsidies to transmit commercial TV. Yet they were apparently unconcerned that RATNET did not provide Native-produced content or other programming designed for rural Alaska. Alaska government officials and legislators themselves seemed to have learned little from LearnAlaska about the importance of educational and public affairs programming for rural Alaska. State officials also did not pursue opportunities to learn from strategies pursued in northern Canada, nor to obtain rights to content produced in the Canadian North that could have been relevant for Alaska.

And unlike their counterparts in Northern Canada, Alaska Native leaders did not seem concerned about the lack of Native content on their rural TV channel. Gail Valaskakis' speech was entitled "The Issue Is Control." Yet for Alaska Natives who had struggled to gain control of their land and their schools, controlling media beyond selecting programs from national networks was not a priority.

Participation in communications planning was important but also took time and sharing of relevant information. Metty felt that local people should be involved in setting priorities, but "involvement in setting priorities by permanent rural Alaskans involves at least a six-month community organizing effort."[41] He added: "Our experience . . . points out that if useful prioritizations are to take place, the following conditions are important. The participants must have: 1) sufficient information about the capacity of the system, the options available and the needs of their constituency, 2) marked influence within the provider structure and the constituency, 3) adequate resources to insure meaningful impact, 4) enough time to accomplish the process. The absence of these almost ensures a hardware-driven rather than a user-driven system."[42]

In retrospect, the U.S. commercial broadcasting model represented what Alaska Natives and the legislature thought rural Alaskans needed to be like the lower 48. They were not exposed to the CBC model of publicly financed broadcasting including news, public affairs, sports, and other entertainment. Public broadcasting followed long after advertising-supported commercial broadcasting in the United States. And Alaska Natives lacked the mentoring and support for local production that helped Northern Canadian indigenous people start community radio stations, the Aboriginal Peoples Television Network, and the Inuit Broadcasting Corporation. Yet this different context does not appear to explain fully the absence of rural and indigenous content in Alaska television, nor the lack of any commitment to build on the legacy of LearnAlaska.

Chapter 12

Deregulation and Disruption

We changed the paradigm for the delivery of telecommunication services in Alaska, and we did it over the strenuous objections of the policy makers . . .

Ron Duncan, cofounder of GCI[1]

The Entry of GCI

The demise of the Governor's Office of Telecommunications (OT), which some observers attributed to pressure from Alascom, ironically provided part of the impetus for the founding of its competitor, GCI. Jay Hammond was re-elected governor in 1978, and proceeded to cut OT out of his budget. Bob Walp, who had been the director of OT, recalled: "I ended up down in Juneau in the governor's office with Kent Dawson and Mike Harper, and the governor said, 'Bob, my advisors say that the Office of Telecommunications has outlived its usefulness, its charter. It has done what it was put here (for). You have done an outstanding job, but there just seems to be no need for it to continue.' They said, 'What will you do?' I said, 'You know, one thing that occurs to me is to start a competitor to Alascom because I know Bill McGowan who put MCI on the map.'"[2] During the 1970s, MCI had successfully challenged AT&T to open the lower 48 market to long-distance competition.

Walp called Bill McGowan (then president of MCI) to discuss his proposal, but MCI lacked enough financing at that time to serve U.S. cities with populations of more than a million—let alone Alaska with a total population of less than half a million. However, discussions with Ron Duncan, who had sold a cable system in Fairbanks, and B. Richard (Dick) Edwards, counsel for Arctic Slope Regional Corporation (ASRC), were more promising. ASRC agreed to provide the initial financing for the new venture, which they named General

Communications Inc. (GCI). The founders incorporated GCI in 1979 in time
to bid on a proposal for the General Services Administration (GSA) to provide government communications between Fairbanks, Anchorage, Juneau, and
Seattle. They asked for an extension of the due date for bids, stating that GCI
was being established expressly to provide service to this market and that they
had a concept that would benefit the federal government.[3] However, Alascom
won the contract.

When ASRC decided that it wanted to terminate its involvement with GCI,
Ron Duncan approached John Malone, CEO of TCI (Tele-Communications,
Inc.), a major national cable television company based in Denver. Duncan
had taken courses from Malone as an engineering student at Johns Hopkins.
After extensive negotiations, GCI was acquired by WTCI (Western
TeleCommunications, Inc.), the company that operated microwave links to
bring television signals to the headends of TCI's cable systems.

From the beginning, GCI's strategy was based on taking advantage of new
technologies. As Ron Duncan explained:

> The concept that GCI was leveraged off [of was] the idea that you could
> provide a better quality of service at a lower price using the new digi-
> tal technologies. At the time we entered the market, Alascom was still
> running huge earth stations with tube amplifiers and analog service and—
> even in an environment where they didn't have an incentive to spend a
> lot of money—it would have been very expensive to provide service the
> way they were providing service. And we saw new technology . . . that was
> much smaller and much cheaper and much better quality. . . . One of the
> fundamental visions was you could do it better for less.[4]

The other advantage that GCI recognized in the late 1970s was the national
shift toward competition, as various elements of the communications networks
were liberalized, from customer premises equipment to data communications,
satellite systems, and long-distance communications. And in 1982, AT&T and
the Justice Department entered into an antitrust consent decree that required
the breakup of the Bell System. Bob Walp explained GCI's initial strategy: "We
concluded that it would be far better to do what MCI did and start off with the
commercial switched market (conventional dial-up phone service), especially
because we [could] really price our service lower than Alascom.[5]

In 1979, RCA sold Alascom to Pacific Power and Light, Inc., a utility holding company based in Portland, Oregon, which later transferred Alascom to
its publicly traded subsidiary, Pacific Telecom, Inc. (PTI). In Alaska, the carrier continued to be known as Alascom. From 1979 to 1982, GCI struggled to
get authority from the Federal Communication Commission (FCC) to offer

long-distance services, fighting many battles against the incumbent Alascom, which urged the FCC to deny GCI's application, arguing that the market could not support competition. Finally, in 1982, "the Alaskan interstate MTS-WATS (voice) market, with a limited exception prohibiting duplicative facilities to bush communities, was opened to competitive entry. . . ."[6] GCI received its authority to provide long distance services in November 1982, and chose to start service on Thanksgiving Day.[7]

Rate Integration

Rate integration was a national policy to provide subscribers throughout the United States with uniform interstate long-distance rates. The FCC listed two subelements to rate integration: first the averaged rates charged, which might not necessarily relate to the underlying costs of providing the service, and second, the revenue settlement arrangements between carriers serving the noncontiguous points and AT&T. "It is this process through which carriers settle the differences between the amount each collects from end users and the amount to which that carrier is entitled for providing its part of the service offering. These agreements in the past have been negotiated between carriers on an individual basis."[8]

Both components of rate integration, the rates charged to users and the settlements that allocated revenues among the various entities that carried the calls from origin to destination, became critical to providing affordable telecommunications services to Alaskans and to providing incentives for carriers to expand their networks and modernize their facilities. Revenue settlements began in Alaska in the 1960s when ACS began making token settlement payments to the local companies for their part in supplying toll (long-distance) services. When the local companies became aware that settlements in the lower 48 were more favorable, they began to negotiate with ACS, resulting in a payment of an amount per message plus an adjustment for higher costs in Alaska. When RCA Alascom purchased ACS, there was no consistent basis of revenue settlement, with settlements per message varying from $0.66 to $1.33, a range of almost 100 percent.[9]

Throughout the 1970s, Alascom instituted many changes to settlement arrangements with local telephone companies, but there was still no consistent pattern. By 1976, 42 percent of Alascom's toll (long-distance) revenues were being paid to local companies through settlements. The percentage that Alascom retained ranged from 76 percent for the Anchorage Telephone Utility (ATU) to minus 31 percent for Interior Telephone, one of the few companies that had conducted cost studies so that it could present its actual costs in negotiations.[10] For interstate calls, Alascom retained 66 percent of the total revenue, paying a settlement averaging $2.58 per message, but ranging from $1.71 per message to Ketchikan Public Utilities to $10.80 per message to Interior Telephone. For intrastate calls, Alascom retained 36 percent of the total revenue, paying an average of $0.87 per message, ranging from $0.37 cents to Whittier to $1.56 to Interior.

Companies in the lower 48 had adopted a settlements methodology called the "Ozark plan" in 1971, which brought many intrastate rate schedules into near parity with interstate rates. At that time, Alaska did not participate in the interstate revenue pool through the Ozark plan, which was not a problem until the issue of rate integration was raised in the FCC's domestic satellite decision. In 1977, the FCC approved adoption of the Ozark plan for Alaska, defining the amount Alaska telephone companies would receive from the interstate revenue pool under rate integration, but clearly indicated that the specific terms could be modified to reflect conditions in Alaska. "The separations procedures adopted under the FCC approved interim arrangement between AT&T and Alascom allowed the allocation of satellite related costs on a 'distance sensitive' basis for recovery from the interstate revenue pool."[11] Transition to the revised settlement process was to occur in three steps, the final one being in 1980. That step was postponed until December of 1984 so that Alascom could keep its revenues up and continue building the state network. In addition to delaying the final step, AT&T agreed to pay Alascom an extra $15 million a year.

In a report to the Alaska state legislature in 1978, William Melody stated: "Alascom did not wish to provide the local carriers with sufficient revenues via settlements to become any stronger financially than was necessary to keep them minimally viable. In this sense, Alascom was acting very much like a regulatory agency. It was determining the minimal revenue needs of the various Alaska companies and granting the allowed revenues through the revenue settlement process."[12] Although some Alaska carriers began to undertake cost studies and used them in their negotiations, most were too small to carry out cost studies. For example, in 1976, the Anchorage Telephone Utility (ATU) originated 44.5 percent of all interstate and 30 percent of intrastate messages, but of the total of 21 local exchange carriers, 11 each originated less than 1 percent of interstate and intrastate messages.[13]

Melody commented in 1978: "Small companies simply cannot afford to hire . . . experts to negotiate for them. Thus, for most companies, the settlement negotiation process is not a negotiation process at all. Most companies simply wait to see what they will be granted in revenue settlements."[14] Douglas Goldschmidt elaborated in 1979: "Joint efforts, both in collecting data and in negotiating with Alascom, have never fully materialized largely because the large companies . . . don't want to provide any subsidies to the smaller companies, and because each company believes it can better negotiate a settlement on its own. . . . This anti-collusive policy leaves the rural companies in a perpetually marginal financial state, allows Alascom to regulate the operations of all the exchange companies, and allows AT&T to regulate Alascom. This pyramiding of bargaining power leads to economic results that are tied to bargaining skills rather than to any clear relation with the economic requirements for providing telephone service efficiently."[15]

In 1972, the FCC's "Open Skies" order that authorized competition in satellite communications had also stated: "The advent of service via domestic satellite facilities should be accompanied by an integration of services, and more particularly the charges for such services between Alaska, Hawaii and Puerto Rico and the contiguous 48 states, into the domestic rate pattern." The FCC's "rate integration" plan was designed to equalize rates for calls between points within the continental United States and calls between the continental United States and offshore points including Alaska, Hawaii, and Puerto Rico. Thus, long-distance rates between the contiguous 48 states and Alaska would be significantly reduced, although the rates remained relatively high because the average mileage (average length of haul, ALOH) of Alaska interstate calls was 2152 miles,[16] whereas the ALOH of interstate calls in the lower 48 was 225 miles, comparable to the ALOH of *intrastate* calls in Alaska.[17]

Phase I of rate integration took effect in March 1976. With the reduction in rates, the revenue per Alaska interstate message declined from an average of $8.92 to $7.03. However, the FCC made no provision for changes in revenue settlement arrangements between Alaska (with Alascom acting as agent) and the lower 48 (with AT&T acting as agent). Nor were any arrangements made regarding cost separation principles and methods for Alaska traffic. Melody warned: "Thus, at the present time, Alaskans are being provided with the necessary rate reductions. . . . However, no provision has been made for the cost or revenue consequences. Unless provisions are made for changes in revenue settlements, then the price to Alaska for participating in the uniform nationwide interstate rate structure will be substantially higher rates for intrastate toll and local services."[18] This result would not only penalize Alaska residents but limit the benefits of telecommunications for Alaska's social and economic development.

In 1977, the FCC created a Federal State Joint Board to determine how to apply the Ozark plan in Alaska with the introduction of rate integration. In addition to higher costs to provide service, Alaska differed in other ways from other states. Long-distance calls were more valuable relative to local service than in other states because communities were small, and residents often needed to make long-distance calls to reach businesses and services, as well as friends and relatives. In 1978, two-thirds of all toll (long-distance) calls were within Alaska, although 60 percent of the toll revenues came from interstate calls. All local telephone companies except those that were municipally owned received more than 50 percent of their revenues from toll traffic, considerably higher than in other states.[19] Also satellite communication had become widespread in rural Alaska, and the cost of satellite use was unrelated to terrestrial distance, as all calls were routed to and from the satellite, so that the actual transmission cost of a call between communities 100 miles apart was the same as a call between communities 500 miles apart.

As a basis for settlement, AT&T and Alascom agreed on a "keep whole" calculation so that Alascom would not suffer financially as a result of phase I. Alascom's share of Alaska interstate revenues then increased to 99 percent.[20] In the negotiations for phase II, AT&T resisted providing Alascom with more than 100 percent of interstate revenues, and negotiations came to a standstill. However, the state took a strong position before the FCC, insisting that rate integration proceed as planned. AT&T and Alascom finally reached agreement, with AT&T refusing to exceed 99 percent, but making very liberal forecasts of traffic growth and revenues and guaranteeing settlements on that basis, so that Alascom would receive guaranteed minimum revenue settlementsthat could exceed 100 percent of actual revenues.[21] Thus AT&T was able to offer Alascom a higher settlement without setting a precedent that could apply to other carriers.

In effect, AT&T acted as a regulator of Alascom's profits on its interstate services through the revenue settlement process. Melody noted in 1982: "Thus, in a very real sense, Alascom and AT&T are negotiating the financial viability of the Alaskan telecommunications system and the course of its direction of its development."[22] But Alascom was only an intrastate carrier. Melody had warned in 1978: "there is a risk that more favorable revenue settlements for Alaska would accrue primarily to Alascom in terms of increased profits. The share that is passed through to local companies would depend on the Alascom revenue settlements with the local companies and the regulatory action of the APUC. Given this negotiating structure, it is quite clear that there will be no satisfactory solution to the revenue settlements issue until the regulatory agencies become involved."[23]

In 1978, RCA Alascom asked the APUC to approve an 87 percent intrastate rate increase. RCAA was trying through an intrastate rate case to bring intrastate revenues to an acceptable level so that a large decrease in interstate revenues would not financially damage the company. However, as Melody pointed out, this would require exorbitant intrastate or local rate increases, but "Survival of the Alaskan telephone industry depends upon the ability of the utilities to keep local rates low and to maintain intrastate toll rates equal to or reasonably close to interstate rates while enjoying an adequate rate of return."[24] The only other source was the interstate revenue pool which was a revenue-sharing device in which each company placed all its interstate revenues into a common pool and drew out its costs based on cost studies plus its FCC-authorized rate of return. In fact, if interstate toll revenues generated less than the local company's costs, the pooling approach would return more than 100 percent of the revenues it generated. Goldschmidt commented: "Billed revenues, however, play no part in the amount of revenues taken from the revenue pool. The pool is purely a social device devised by industry practice and regulatory precedent which allows a single nationwide rate structure, which permits simplicity in tariffs, and which doesn't economically penalize residents of high-cost predominantly rural areas, in terms of higher rate structures."[25]

Alascom estimated its revenue deficit for 1979 at $23 million. A 1981 study carried out for Alascom estimated that for 158 bush communities, the establishment of local exchange service would result in annual cost deficiencies of $15 million to $23 million.[26] While these costs would raise rates substantially if they had to be absorbed by Alaska subscribers, they were "insignificant compared to the magnitude of the costs and revenues for the nationwide telecoms system." Again, Melody warned: "Alaska will get its subsidy from the national telecommunications system . . . only if it successfully negotiates for its subsidy rights in the face of resisting economic interests of AT&T and Alascom. . . . This approach to the resolution of the outstanding issues affecting Alaskan telecommunications calls for a major assertion of policy-making both by the APUC and by the State."[27]

GCI and State Regulation

Although GCI had begun to provide services in 1982, its battles were far from over. Duncan explained, "it also became clear early on that Alascom had made a conscious decision to do everything possible to stop GCI from getting a toehold, and in retrospect [it was] a very rational decision. . . ."[28] Also, rate integration and changes in subsidies after the breakup of AT&T led to distortions in the subsidies Alascom relied on. Through rate integration, a call to or from Alaska was similar in price to a call of the same distance, time-of-day, and length within the lower 48. The actual cost of the Alaska call was still high, but the price to customers was lower because the settlement process described above resulted in AT&T subsidizing Alascom's costs.

In 1983, the state of Alaska petitioned the FCC to initiate a rulemaking to determine how to rationalize the policies of rate integration and competition in the Alaska market in light of the rapid changes in the telecommunications industry brought on by the planned divestiture of AT&T's local telephone companies from AT&T's long-distance business and by changing FCC competition policies. The state and the APUC requested the rulemaking to establish a permanent mechanism for payment of the subsidies resulting from Alaska's participation in interstate rate integration.

GCI contended that the mechanism providing rate support must allow for participation by all carriers serving routes that required support.[29] However, Alascom did not believe "that competitive providers should be compensated from such a fund since that would lead to unnecessary duplication of facilities."[30] The FCC summarized the position of the state of Alaska: "While not proposing a specific mechanism, Alaska [State and APUC] suggests that the high costs could be supported by a fund similar to the universal service fund and funded through the carrier common line element. It submits that payment from such a fund could be made to all carriers that provide service to the noncontiguous points under integrated rate structures, with payments being proportionate to traffic volumes and cost factors."[31] The FCC concluded in 1984: "The record developed in connection with Alaska's Petition is inadequate for us to make a recommendation of any rule changes at this time. However, we conclude that the public interest will be served if we begin a rulemaking proceeding to evaluate the rate integration policy and associated settlement arrangements in the light of more recent developments."[32] To resolve the competition/rate integration issues, the FCC created a Federal-State Joint Board in September 1985. Members of the Joint Board were FCC commissioners and state public utilities commissioners. Its work would continue for eight years.

Alascom described the separations procedure: "Every year, Alascom and the local telephone companies perform their own 'separation studies' that determine what part of telephone costs are attributable to the interstate network and what will be recovered from the interstate revenue pool. . . . Its basic circuits carrying most of its MTS service are assigned to the interstate jurisdiction based on 'equivalent circuit miles,' a distance-sensitive factor. Because the long-distance calls travel between Alaska and the lower 48, 88 percent of basic circuit costs are allocated for recovery through the interstate revenue pool."[33] The subsidy worked very well for Alascom. Ron Duncan explained: "it was probably around 1980 [that] Alascom went to . . . the commitment to move to fully integrated (rates), which meant that AT&T picked up the revenue requirement irrespective of what Alascom's rates were. And it didn't take Alascom long to figure out that they had . . . a blank check on AT&T's checking account, and that they had a subsidy worth $100 million bucks a year, and $100 million bucks a year is, you know, $300,000 a day, so every day you can fend off the competitor, you made another $300,000 bucks."

In 1983, the subsidy had reached $158 million per year. As legislator "Red" Boucher later pointed out: "The subsidy mechanism worked reasonably well until interstate competition started and AT&T divested its local companies. To pre-divestiture AT&T, $150 million a year was peanuts. To AT&T the competitor, such a large subsidy is a burden, and AT&T no longer wants to pay it."[34] By mid-1983, GCI was claiming that it could not compete with Alascom's subsidized rates. GCI petitioned the FCC for relief in the form of reduced connection rates with AT&T in Seattle. "In December 1984, the FCC decided to preserve the status quo until the issue of making rate integration compatible with competition was decided. The last step of rate integration scheduled for that month was postponed indefinitely. The FCC further ordered AT&T to give GCI the reduced connection rates."[35]

In late 1984, the FCC ordered AT&T to continue subsidy payments to state firms until it could study how costly deregulation would be in Alaska. In March 1985, as a side effect of an out-of-court settlement between AT&T and GCI, AT&T agreed to continue providing Alascom $1.25 million per month in transitional settlements "until all the issues involving rates in Alaska can be solved."[36] However, Alascom also sought a 15 percent interim and later 20 percent rate increase from the APUC for intrastate rates. The APUC ordered that the interim hike could not go into effect until a week after the FCC accepted the AT&T–GCI agreement. Robert Walp, then president of GCI, said his agreement with AT&T "should have the side effect of making Alascom's proposed rate increase unnecessary."[37] He added: "After the agreement . . . the future of GCI in providing a lower-cost long distance service is secure."[38]

Interstate Competition in Major Markets

The 1985 out-of-court settlement cleared the way for GCI to expand service to Juneau and Fairbanks from its base in Anchorage.[39] The *Juneau Empire* noted "The agreement guarantees GCI's future, at least for the next seven years, paving the way for the company to spend the estimated $3 million for equipment to service the state's second and third largest cities." But at the time, there was no means of equal access (1+ dialing) to a competitive provider: "callers will need to dial GCI's Juneau number, add a six-digit identification code and then dial the number they are calling. For dialing the added 13 digits the users can save money. Anchorage users save 25 to 20 percent."[40] The newspaper story added that within 18 months to two years, Juneau-Douglas Telephone planned to upgrade its software so that callers would no longer need to dial the access code.[41]

GCI had announced plans to begin service to Anchorage and Fairbanks in 1983, but it put plans on hold when AT&T proposed to hike the access fee to AT&T's network. Alascom paid nothing to access AT&T's network, while GCI's rate was set at 22 cents per minute.

The FCC ordered in 1985 that GCI would initially pay 8.5 cents per minute and gain access to a digital switch that would make GCI's system capable of carrying data as well as voice.[42]

Ron Duncan commented on the new competitive paradigm:

[Alascom] understood very correctly, I think, that it really was a battle of systems. Whether it was GCI or anyone else, if competition succeeded in the Alaska marketplace, then their whole paradigm of delivering service was doomed to fail. Because the only defense for what ultimately became the $100 million a year subsidy was, 'Well it's the only way to serve Alaska.' When we started, the argument was, 'Well, GCI only wants to serve Anchorage. The only market in the state that's ever going to be profitable is Anchorage, and we just eke out a bare little profit there and lose all this money in the Bush, and GCI never wants to serve those places, so don't let them serve Anchorage.' And then when we went into Fairbanks, it was, 'Well, the only two profitable markets are Anchorage and Fairbanks, and here's the evidence that the rest of the state's not profitable.' Then the evidence just gradually kept shifting. Every time we would go into a new market, well, that was now the profitable market. The unprofitable market was always one step away. But what they understood was that, sooner or later, people would figure out that you weren't subsidizing consumers, that you were subsidizing a competitor in a way that wasn't necessary.

Duncan continued:

. . . as soon as we came into the market, rates that for years Alascom had been trying to drive up, they immediately started slashing . . . and thus enlarging the subsidy that AT&T was paying [since AT&T paid the difference in their costs].

". . . we really concluded that the only way we were going to get any adequate relief was through the courts, because the FCC was taking much too long to deal with these issues. And there were people at the FCC who understood what was going on, and over time, more and more of them began to understand that the whole incentive structure was wrong for Alaska—that Alascom had a structure where it made more money by spending more money, and by deliberately driving up the cost of service. But it took an awfully long time to get the folks at the FCC interested in doing anything about it, and we felt that it would be faster relief to go through the courts.[43]

GCI therefore filed an antitrust suit against Alascom and AT&T in the U.S. District Court in Seattle, alleging anticompetitive and discriminatory conduct.

The GCI Antitrust Suit

Professor William Melody, who had previously consulted for the Governor's Office of Telecommunications in the 1970s, was an expert witness for GCI at the trial and endeavored to explain how Alascom was damaging GCI's ability to compete through its anti-competitive practices. Bob Walp described the testimony:

> . . . [Alascom's lawyer] was asking Bill Melody questions. . . . Bill would give these long tutorials on whatever question Mr. White asked him, and Mr. White is getting exasperated and his face is getting redder and so forth and he said, "Dr. Melody, on this next question I want you to give me a "yes" or a "no" answer and no tutorials. . . . Do you understand that?" Bill said, "Yes." White then says, "Isn't it true that in an industry such as this that a larger company because of economies of scale can deliver service at a lower per unit cost than a small company?" This is the standard monopoly argument. And Bill looks at the ceiling and he looks around the courtroom and at the jury . . . and finally he says, "In theory yes, but in practice no." Mr. White just went over and sat down and just sat there. . . . [The judge] called a recess and he called us into his chambers and suggested it was time for a settlement. He said he'd reviewed everything and there just wasn't any sense in taking anybody's time.[44]

In the settlement, GCI received $28.5 million and access to Alascom's North Pacific cable. At the time, Alascom had just announced that it was building the cable. Duncan explains:

> . . . clearly we had to have access to the cable in order to be able to continue to do our business. And it was another place where Alascom helped teach the judge a lesson because it was clear from the materials in the antitrust suit that they weren't going to be able to use more than a tiny fraction of the capacity of the cable, but they . . . resisted putting in the antitrust settlement agreement anything that would sell us a piece of the cable at cost. And it was ultimately there because the judge understood the only reason not to have [cable access by other companies] there was if you were really trying to drive GCI out of the market. And the whole name of the game of why we filed the lawsuit was because it is illegal to engage in those sorts of monopolistic activities.[45]

Before the antitrust suit was settled, GCI had completed a spin-out from WestMarc Communications, Inc., formerly WTCI (Western Tele-Communications, Inc.), and arranged its own financing. In 1987, WSMC distributed all of the outstanding shares of its common stock of GCI to its shareholders. Following the distribution, GCI became an independent publicly traded company.

Facing continuing subsidies from AT&T to Alascom, GCI returned to the FCC in 1987 and argued that Alascom should continue to be regulated as a dominant carrier until the Joint Board decision: "Full dominant carrier regulation should continue unless and until the Alaska Joint Board acts to eliminate the cost-plus incentive structure under which Alascom operates."[46]

GCI explained the ongoing effects of the subsidies:

> . . . AT&T subsidizes Alascom by permitting it to keep all of the revenues earned on Alaska service, amounting to $132 million annually, and by paying Alascom an additional $34 million in cash annually. In addition, AT&T gives Alascom $63 million worth of free distribution of its traffic in the lower-48 states. The total value of these payments ($229 million) is almost double the revenues earned by AT&T and Alascom in providing Alaska services. . . . Alascom effectively represents an analytical 'no man's land' that defies economic logic. . . . If Alascom reduces its price for MTS/WATS services, its revenues do not change because its revenues are determined not by the rates it charges, but instead by the amount of cost it assigns to its MTS/WATS service and thereby recovers from AT&T. Since Alascom's revenues do not cover its costs, any reduction in rates simply increases its subsidy payments from AT&T.[47] (MTS refers to long-distance services, and WATS refers to toll free "800" services.)

The Era of Arbitrage

Despite high rates, Alaskans depended on long-distance calls to stay in touch, particularly intrastate long-distance calls, which were generally more expensive than interstate calls, because interstate calls were subsidized through the nationwide rate averaging process. A study in 1986 found that in Anchorage, 87 percent of respondents had made long-distance calls in the previous two weeks, and 97 percent of the long-distance callers had made at least one out-of-state call. A majority (64 percent) said that most calls were to their relatives. Urban demand for long distance was quite inelastic: 36 percent of long-distance callers said they would make about the same number of calls if rates were halved, and 40 percent said they would make about the same number of calls if rates were doubled.[48]

In rural Alaska, 97 percent of residents with telephones had made long-distance calls within the previous two weeks, and 78 percent had made out-of-state calls. Forty-five percent said most calls were to relatives, 20 percent said most were for business, and 14 percent said most calls were to friends. Demand for long distance in the bush was much more price sensitive: 73 percent of long-distance users said they would make more calls if rates were cut in half, and the same percentage said they would make fewer calls if rates were doubled. The implication was that a change of rates would have more effect in rural Alaska than in Anchorage.[49]

Technology was once more creating new opportunities for competition and new policy challenges. The era of arbitrage had arrived. Although intrastate competition had not been approved, customers could save money on intrastate calls by routing their calls through lower-priced interstate carriers based in the lower 48 states. And GCI could use this approach to provide intrastate long distance. At an APUC hearing in 1986, Alascom accused GCI of arbitrage—routing intrastate calls over interstate facilities and charging for calls under an interstate tariff. "It deprives the intrastate network of revenues from the most profitable intrastate routes and thereby makes it more difficult to provide affordable service to high cost areas such as rural Alaska."[50] GCI pointed out to the APUC that arbitrage calls could be made by calling GCI's Seattle switch through a national 800 number, entering an authorization code, and completing the call back to Alaska, or by placing a call to the access numbers for MCI, Sprint, or other long-distance carriers, entering an authorization code, and calling an Alaska number. GCI noted that, using Long Distance USA in Hawaii, a customer in Juneau could call Anchorage through Hawaii at a rate of 43 cents per minute, while Alascom's rate was 85 cents per minute.[51]

In a letter to its customers dated November 25, 1986, GCI stated:

> The APUC apparently believes that these 907 calls—which both originate and terminate in Alaska but are routed through Seattle—may deprive Alascom, the in-state monopolist, of revenues. . . .
>
> These calls occur because it is less expensive in many cases to make two interstate calls—one to the Lower-48 and a second back to Alaska—than to place a single intrastate call on Alascom. In no other state are the rates so high as to make two interstate calls less expensive than one in-state call. The consumer has every incentive to use available technology to make those calls.
>
> GCI believes that the APUC does not have the authority or technical ability to stop all means of making these 907 calls. For this reason, GCI has told the APUC that the only real solution to the problem is to lower in-state rates.

GCI believes the only way to force lower in-state rates is to authorize competition for in-state calls. Accordingly, GCI has filed an application to provide in-state service at substantially lower rates than Alascom.[52]

The Alaska Consumer Advocacy Program summarized: "Intrastate competition is here already—except it came in the back door with arbitrage instead of using the front entrance as a certified utility."[53] However, the APUC was not ready to authorize intrastate competition.

The End of Negotiated Settlements

Finally, in 1988, the APUC determined that the problems inherent in negotiated settlements made it appropriate to consider replacing the settlements process with a system of access charges. Therefore, in 1989, the APUC established "a system of compensatory access charges to be paid by interexchange telephone carriers to local exchange carriers . . . where the charge would help move telephone service rates closer to a cost-of-service basis, thus making telephone service more affordable and more widely available, and would also promote competition, which in turn would enhance efficiency and innovation." The APUC concluded that access charges were a better means than settlements for compensating local exchange carriers (LECs) for the use of local network facilities in the completion of toll calls by interexchange carriers (IXCs) because they provide for regulatory review of LEC costs, relieving IXCs from the burden of monitoring such costs; negate need for lengthy rounds of negotiations; provide for predetermined, known and certain uniform payments; are prospective in application; and would help to introduce intrastate toll competition.[54]

Chapter 13

State Planning and Policy

Can we harness communication technologies so that we can turn information into an economic resource for Alaska?

State Representative "Red" Boucher[1]

I n 1981, Governor Jay Hammond stated in an Executive Order:

> As governor, I find that adequate telecommunications facilities and services at reasonable and affordable rates are essential to the conduct of government, commerce, and private life. . . . The future development of the state's private and public sectors will depend greatly on the innovative use of new telecommunications services and techniques now becoming available. It is in the interest of all people for the state to facilitate the development of both basic and advanced telecommunications services and facilities to be available to all citizens for their individual and mutual benefit.[2]

Carolyn Guess, chair of the APUC from 1981 to 1985, later called this statement "Alaska's telecommunications policy." However, as discussed in Chapter 12, the 1980s brought disruptive change to Alaska telecommunications. The collapse of oil prices in 1986 resulted in severe state government budget cuts, and the introduction of competition preoccupied state legislators and regulators.

State Planning and Policy: Three Stages

In a report to the Legislature in 1987, Larry Pearson and Doug Barry divided telecommunications activities in Alaska into three stages:

- The early or start-up stage from 1971 to 1975
- The middle or operational stage from 1976 to 1981
- The advanced or applications stage from 1982 to 1986 when government attention was weakest[3]

During the start-up stage as Professor William Melody noted, "The costs of inefficiency in the Alaska telecommunications system were compounded many times. The effects rippled throughout the state, limiting economic, political, and social development. Telecommunication provided a major barrier to the integration of Alaska into the United States."[4] Improvements in broadcasting also became a priority. Pearson and Barry commented: "Besides the need for improved telephone service, there was also a belief that a state that received its television programming often three weeks after the rest of the country, and where national and international news could be frustrated if the videotapes missed their plane in Seattle, was not very well positioned to leap into the Information Age."[5]

As described in previous chapters, the state government took an active leadership role under Governors William Egan and Jay Hammond, with the establishment of the Governor's Office of Telecommunications (OT) and key involvement by the Alaska Public Broadcasting Commission (APBC). Organizational changes mandated by Governor Hammond improved coordination of state telecommunications activities. OT tackled several long-range planning issues including extension of telecommunication services by satellite to remote areas, development of new technologies adaptable to Alaska's harsh environment, and advocacy of state's interests before the Federal Communications Commission (FCC). Governor Hammond appointed a task force that recommended that the state develop its own system of small earth stations to serve rural communities not reached by RCA. As described in Chapter 9, the Legislature appropriated $5 million to purchase 100 earth stations, and after extensive negotiations RCA agreed to maintain them.

The "operational stage" from 1976 to 1981 saw the rise of publicly funded television and teleconferencing networks. In 1976, the Legislature appropriated funds for a satellite demonstration project, and later funded transmission of television to villages, in the program that became known as RATNET—the Rural Alaska Television Network. In 1977, the Legislative Teleconferencing Network (LTN) was created, enabling legislators to hold meetings and hearings involving rural constituents. "Legislative decision-making seemed to be guided by good will. . . . As long as budgets were fat, there was apparently little need to ask if these networks were providing what people needed or wanted, or whether they could cooperate more closely to provide better, more cost-effective service."[6]

In 1978, William Melody pointed out: "Alaska is unique among the states in establishing a governor's office of telecommunications to act as a catalyst for policy planning and an advocate for the interest of the public of Alaska. . . . Continuation of this public advocacy is essential if the multitude of continuing policy decisions and negotiations at the federal and state level . . . is to reflect the interests of the state."[7] However, as Pearson and Barry noted: "This opportunity to place telecommunications high on the government agenda lasted several years, then went away."[8] In 1979, Hammond disbanded the Governor's Office of Telecommunications.

In what Pearson and Barry called the "advanced stage" from 1982 to 1986, systems planned earlier were implemented, including LearnAlaska, educational audioconferencing, and state computer networks.

> If the 1970s were the years of building infrastructure, the 1980's were characterized by the emergence of competition and public policy quandaries over how to balance the working of a competitive market with the system of financial supports that had made universal service affordable in Alaska under the technology of the time. Increasing, there was recognition at the federal market level that Alaska's market structure was not typical of that in the 'Lower 48,' and that an explicit system official support might be necessary to maintain the federal policy of universal service in rural areas like Alaska. Alaska benefitted from both general FCC policies to promote and enhance universal service, as well as policies specific to Alaska. The decade of the 1980's was also characterized by the gradual emergence of new technologies and battles over competition in the long distance market.[9]

A Planning Vacuum

By 1986, the state had spent more than $200 million in the previous 11 years on telecommunications, with a legislative history including 11 operating appropriations totaling approximately $172 million; ten capital appropriations totaling approximately $79 million; nine additional/miscellaneous appropriations totaling approximately $21 million, and 21 special/supplemental appropriation bills for telecommunications facilities or programs totaling approximately$20 million.[10] These appropriations funded the state's electronic information delivery systems, including a statewide TV network reaching 240 communities plus microwave system on Kenai Peninsula; two main audioconferencing systems— LTN linking about 70 sites and the LearnAlaska audioconferencing system serving primarily school districts and university users; four public TV stations and 15 public radio stations that received state funding; a cable TV channel in Anchorage operated by the University of Alaska, and one main computer network operated by Department of Administration (DOA), with other government departments, legislative and university computer networks connected to the DOA network.[11]

In 1986, the price of oil fell to $10 per barrel. The Legislature eliminated LearnAlaska and cut the budget of the LTN almost in half even after Governor Sheffield asked state agencies to make greater use of it to save travel costs.[12] In a report to the Legislature in 1987, Larry Pearson and Doug Barry sounded an alarm: "This is happening because of the absence of a statewide telecommunication and information policy. . . . Until recently, Alaska could count itself among a handful of states which not only identified the strategic importance of telecommunication, but invested huge sums of money to build public networks that were without equal. Now this distinction and benefits derived from it are in serious jeopardy. Alaska may be in the process of unwittingly squandering its strategic information assets."[13]

The state had taken a piecemeal approach to funding and failed to establish other communications targets after adopting the goal of providing telephone service to all communities of at least 25 residents in the 1970s. State Representative "Red" Boucher observed, "out of the 59 listed session laws, 40 were strictly appropriations, while only 19 dealt with policy issues. This seems to support the observation that money has been spent without careful thought or accountability given as to the most efficient use of that money."[14] Boucher later concluded:

In the late 1970s we had begun a series of noble experiments in telecommunication. We brought computers to schools in rural Alaska, indeed we led the nation in computers per student. We created a satellite network, across an area a fifth the size of the United States, that brought information, education and entertainment in a new form to parts of the state that had no telephone, television or radio. We began to use telecommunication to provide medical assistance to isolated villages. We created an emergency broadcasting system. . . . But the policy infrastructure that created them was short-lived. The emphasis had been placed on the delivery system, not the product; technology, not information. In a word, without a plan, without a vision . . .[15]

By Executive Order in 1981, Governor Hammond had placed all state telecommunications under the Deputy Commissioner of Administration with two divisions, for operations and services. Alex Hills, who was appointed deputy commissioner, stated that reorganization of the telecommunications functions reflected the need to streamline management of telecommunications, institute a planning function, and make assistance available to state agencies.[16] The DOA's many responsibilities included managing and coordinating all state programs in telecommunications, providing technical and consulting services to state agencies, preparing a statewide telecommunications development plan, administering the budgets of RATNET and APBC, as well as advising the governor and the legislature. Executive Order 50 alone required the DOA to prepare at least six major reports annually for the governor and legislature.

Ironically, despite Governor Hammond's policy prelude in Executive Order 50, the many operational responsibilities under the DOA resulted in a lack of policy continuity. Pearson and Barry point out that the new organization had a mixed mandate, limited staff, and no real power to force compliance from other agencies.[17] By Executive Order in 1987, Governor Steve Cowper consolidated the authority of the DOA by merging the Division of Telecommunications Services into the Division of Telecommunications Operations, thereby consolidating the general planning of state telecommunications with operations and management of the state's network. He concluded: "As Governor, I find that it would be in the best interest of efficient administration to eliminate the statutory requirements for two separate divisions and a deputy commissioner in the Department of Administration with telecommunications powers and duties. These powers and duties will be more efficiently exercised with greater flexibility given to the department."[18] However, operations took priority over planning, and no one in state government was responsible for identifying the communication needs of the state as a whole, rather than those of state government agencies.

The Special Committee on Telecommunications

In 1984, the House convened a Special Committee on Telecommunications chaired by Representative "Red" Boucher. Its original impetus was concern about the impact of the AT&T divestiture on Alaska. Boucher characterized the divestiture issue as "the most fundamental readjustment in our national economy since World War II, with repercussions that will be felt through the next generation."[19] A 1986 report by the committee identified several areas of policy concern for Alaska, including possible increases in local and long-distance rates because of fewer cross subsidies after the breakup of AT&T and "allowance of full competition in the long distance toll market, and the shifting of costs from the long distance companies to the ratepayer."[20]

The report also questioned whether intrastate competition should be allowed. If the competitor took a significant share of the revenue for traffic from Anchorage, Fairbanks, and Juneau, would the result be higher rates for the bush? Committee members were also concerned about universal service for Alaska: "Universal service in its broadest sense refers to the affordability and availability of communications and information. . . . Universal service for Alaskans could mean affordable availability of communications and information whatever technology is used."[21] Should Alaska introduce Lifeline, the new federally initiated program that would provide discounted basic telephone service and installation charges for low-income residents? Would these rates be considered discriminatory, which the APUC claimed was illegal according to their statutes? Should the legislature amend statutes, and then require the APUC to hold a rulemaking proceeding on eligibility?[22]

The Committee was also concerned about the state's own information and communication systems. The members wondered whether the state should build its own telecommunications network for state use rather than buying service from the carriers. They also found that the state's information technology resources were underutilized. "The committee's purpose therefore was not only to provide legislative oversight to the multimillion dollar investment Alaska had made in technology but to mandate that state government implement a comprehensive policy for managing the state's telecoms networks."[23]

The Committee concluded that in order to share information and to extract important information more rapidly, the state needed a telecommunications plan and an information plan. The state would then need to support these plans both at high levels of state government and through involvement of affected workers, and keep them current.[24] Boucher summarized: "Alaska's telecommunications system is suffering from neglect. . . . The state has no comprehensive plan for the use of telecommunications technology, as exemplified in the lack of any information resource management plan."[25] But it was not until the 1990s that the state would produce a plan.

The Creation of the TIC

A report prepared for the Special Committee on Telecommunications concluded:

> The state's record of providing policy direction for telecommunications is much weaker than its record providing technical support. At one time there was a policy body: The Governor's Office of Telecommunications. However, in the 1980s responsibilities for various parts of the state's telecommunication system have fragmented. . . . The absence of telecommunications policy direction has been particularly unfortunate in a time of declining state revenues. The search for economies with the telecommunication system has taken place in an environment that has provided

little opportunity for discussion of the implications for Alaskans of alternative options."[26]

Legislation passed by the 15th Alaska Legislature and signed into law by Governor Steve Cowper in June 1987 created the Telecommunications Information Council (TIC). A Cabinet-level policy-making body, the TIC's mandate was to develop a plan for the state's own use of telecommunication and information, rather than to address broader telecommunications policy issues affecting the state. It was mandated to:

- Develop a statewide telecommunication/information plan
- Establish institutional arrangements for developing and implementing improved information management in Alaska
- Establish information management policies and guidelines to implement the plan[27]

Membership could include the governor, lieutenant governor, department commissioners, representatives of the court system and Legislative Affairs Agency, and two legislators, one from the House and one from the Senate, to serve as nonvoting members. Fran Ulmer was a member as a legislator and later became chair of the TIC in 1994 as lieutenant governor.

The TIC then adopted a more specific agenda "to prepare an overall plan that will allow it to intelligently manage information; to provide direction to individual agencies in the development of information systems plans; to provide guidelines for public access to computer-based information; and to provide policy direction for broadcast telecommunication systems."[28] The TIC noted that broadcasting systems "are to be regarded . . . as information systems. Broadcast policy is now formed by government entities with no one to appeal to."[29]

The inclusion of broadcasting was a recognition of the demise of LearnAlaska and preservation of RATNET, as the state appeared to have abandoned educational broadcasting but continued to subsidize rural distribution of televised entertainment. "The state has gone from having a statewide network exclusively for educational programming to broadcasts of educational programming between the hours of 1:30 am and 5:30 am on the RATNET channel. Is this a better use of state-funded broadcast channels? . . . The challenge for the Telecommunications Information Council in the area of broadcasting is to provide a policy framework where none has existed before."[30]

The TIC's intent was to focus on management and access. It began with a harsh critique of the state's approach to technology: "technology is being managed by people who are trying to protect managers from what they fear most, rather than by people who are trying to provide managers with the best tools for doing their jobs."[31] Instead, "the Telecommunications Information Council puts the managers back into information management. In addition, it shifts the emphasis from management of a technology to management of what that technology produces."[32] They also recognized that the converging technologies of telecommunications, computing, and broadcasting "provide an additional argument for focusing at a policy level on what telecommunication does—and what we want it to do—rather than on its internal workings."[33]

Links with Siberia: "Opening Another Door"

The end of the 1980s heralded many changes, drawing Alaska into the international context of telecommunications, as well as further into the paradigm of competition, which was to be expanded by the Telecommunications Act of 1996. While the debates continued over whether competition should be allowed within Alaska, the end of the Cold War brought opportunities to provide telecommunications links across the Bering Sea to the Russian Far East.

The indigenous people of Northwest Alaska had once freely crossed the Bering Strait in walrus-skin umiaks and later motor boats to trade and visit relatives, but the border had been closed in 1948 as Cold War tensions grew. In the spring of 1973, a small group of Siberian hunters from families who had been evacuated from Big Diomede Island was allowed to return for a month of hunting and trapping. On the ice, they encountered hunters from the U.S. island of Little Diomede, three miles away, and enjoyed reunions with their relatives.[34] There was no further contact across the Bering Strait until June 13, 1988, when Alaska Airlines flew 80 dignitaries, including Governor Steve Cowper and a Native delegation, on a "Freedom Flight" from Alaska to Provideniya in Chukotka. The Freedom Flight took one and a half hours from Anchorage to Nome to pick up more passengers, and just 45 minutes through the "ice curtain" from Nome to Provideniya.

The Eskimo (Inupiat and Siberian Yup'ik) people from Northwest Alaska speak the same dialect as the Native people in Provideniya. They were delighted to converse in their own language, many crying when they met distant relatives and laughing as they shared tea and rolls together. U.S. and Soviet officials talked of Glasnost a year and a half before the fall of the Berlin Wall. The historic meeting was covered on television beamed back to North America from an Alascom transportable earth station on Alascom's Aurora I satellite.

The Alascom team had two weeks to prepare for the visit, fly the telecommunications equipment to Siberia, and get set up. Arranging the transmissions meant not only shipping and installing equipment but also obtaining permission from the FCC and from Intelsat to carry an international signal on a U.S. domestic satellite. "We'll go until we get stopped," an Alascom official decided.[35] They finally got FCC authorization and flight clearance and took off in two planes, with a 15-minute window to get through the Soviet Air Defense System. The flights were hindered by bad weather and fog and air traffic controllers who spoke little English because English-speaking Soviet controllers who had been sent to Provideniya were stuck in Anadyr. However, the plane with the technical crew and the second plane carrying equipment both landed safely in Provideniya.

Alascom decided to put the earth station in the town near the reporters, with the electronics and power supply in a van. CBS reporter Terry Drinkwater covered the trip, with live video over the satellite link. The communication system itself became a showcase. Alaska Senator Al Adams from Kotzebue demonstrated how he could go into the van and get a U.S. dial tone to call Alaska. Alascom was also able to transmit the video signal to Washington, DC, where the Soviet broadcasters could convert and retransmit the signal to Moscow from their U.S. gateway earth station. Lee Wareham, Alascom's chief engineer, recounted that a local technician told him: "If the Soviets worked like the Americans work, we could complete a five year plan in two years!"[36]

In the winter of 1989, the Bering Bridge Expedition consisting of six Soviet and six America adventurers trekked for two months across Chukotka and then the Bering strait, the International Date Line, and on to Little Diomede. The expedition was organized to promote better relations between the United States and the USSR and to encourage Alaska Natives to visit relatives in the Soviet Union. Alascom sent a portable earth station to cover the arrival of the expedition near Little Diomede. Lee Wareham recalled putting survey markers on the ice to mark the International Date Line, and actually painting the line on the ice. The Alascom earth station was set up 10 meters on the U.S. side of the line. Three Russian helicopters also landed on the ice.[37] Ultimately, the Bering Bridge Expedition led to an agreement allowing free travel across the Soviet-American border for the region's indigenous population.

In the late 1980s, scientists at the University of Alaska began several collaborative research projects with counterparts in the Far Eastern Branch of the Russian Academy of Sciences. Communication was difficult, with many attempts needed to make a telephone call over the congested circuits through North America and Europe and the outdated infrastructure across Russia. Mail from Alaska to the Russian Far East took several months. In January 1988, Lee Wareham, a senior engineer at Alascom, was a member of a U.S. delegation to Moscow for an exhibition on "Technology in everyday life." The State Department arranged meetings for him with his counterparts in the Soviet Ministry of Communications, where Wareham raised the issue of communication over the Bering Strait. The timing was propitious, as Gorbachev wanted to re-establish ties with the rest of the world. Wareham and his Russian counterparts developed a project to link St. Lawrence Island with Provideniya. In 1990, Wareham's Russian colleague got approval from the Ministry of Communications for a wireless terrestrial link from a repeater on a mountain near Provideniya to St. Lawrence Island. Alascom was to provide the equipment and get State Department approval. Both the Russian and U.S governments seemed willing to use the project as a test bed for collaboration between the two countries to "open another door" to the west.[38] The FCC approved a waiver for Alascom to use frequencies in the 450–460 MHz band to establish an international fixed public telecommunications link between St. Lawrence Island and the USSR. The FCC noted: "If granted, this will be the first direct terrestrial link between the United States and the Soviet Union. The proposal to provide up to 12 two-way voice channels has the support of the United States State Department."[39] Wareham initially proposed a microwave link with up to 120 channels, but the Russians preferred a much more limited UHF link with four telephone circuits plus one data circuit for telex. At that time, there were only 16 circuits between Moscow and the United States.

The UHF radio link connected Gambell on St. Lawrence Island with a site near Provideniya. One circuit went to Provideniya, and an additional four circuits carried traffic on a troposcatter system and then the Russian Gorizont satellite to an earth station near Magadan on the Sea of Okhotsk. All of the calls were routed manually, with an operator in Alaska contacting a Russian operator who completed the calls using a manual switchboard.[40] However, the circuits were often noisy and congested, and the university researchers also wanted to use electronic mail to avoid the inconvenience of arranging real-time communication across the international date line and five time zones. Alex Hills at the University of Alaska and Lee Wareham at Alascom began to make arrangements for a satellite-based telecommunications link between the University of Alaska Fairbanks and universities and research institutes in the Siberian coastal city of Magadan.

With the opening of international air routes over Russia in 1991, there was additional demand for reliable communications with the Russian Far East. Wareham set up a joint venture called Magalascom with Alascom and Rossvyazinform, the carrier in the Magadan region. Wareham convinced the Russians to move an Intersputnik satellite to an available slot at 145 degrees E, which would make it possible to communicate in one hop from the Diamond Ridge Alascom site near Homer as far as Novosibirsk in central Siberia. Each partner would pay in its own currency for half circuits—"a mid-air meet" at the satellite, so that currency would not have to be converted or exported. Alascom paid about $15,000 per month for half a T1 circuit. However, the aging Intersputnik satellite, which was in an inclined orbit, began to wander out of its position, resulting in signal outages. Eventually, traffic was switched to Alascom's Aurora II satellite, which could cover Provideniya and Anadyr, where the FAA had installed an earth station for monitoring air traffic over the newly opened routes.[41] Once again, Alascom needed permission from the FCC and Intelsat to carry international traffic on its domestic satellite.

UAF researchers were able to use the Aurora satellite to communicate with colleagues in Magadan. Voice and data went by microwave from Fairbanks to Alascom's Eagle River earth station and on Alascom's Aurora II satellite to a Russian earth station at Kapran, about 400 kilometers. northeast of Magadan, and then again by microwave to Magadan. The circuit was installed and tested in 1992 and became operational in July 1993, providing a voice and email link between UAF and three Russian research institutes and the International Pedagogical University in Magadan.[42] After AT&T purchased Alascom in 1995, it eventually turned down these circuits.[43]

The Exxon Valdez Oil Spill

On March 24, 1989, the tanker Exxon Valdez ran aground on a reef in Prince William Sound, spilling more than 11 million gallons of oil, the largest spill until that time in U.S. waters. Alaskans began a massive cleanup that would last for years. Satellite communications proved to be a key component of the cleanup effort at a time when satellite phones were not yet available. Alyeska Pipeline Service Company had installed a Ku-band satellite system connecting the Valdez terminal to the Anchorage office in the summer of 1988. Chuck Schumann, founder of Microcom, the Alaska company that had installed the equipment, states: "Because Alyeska had this backbone system in place, they were ready to quickly deploy a VSAT terminal to Reef Island, where the Exxon Valdez ran aground, shortly after the oil spill. The initial terminal installed on Reef Island provided for the extension of marine radio (VHF) communications over the satellite back to the Valdez terminal."[44] Previously, communications from Valdez out into Prince William Sound were extremely limited. Schumann continued:

> When Exxon set up their command post over the next few weeks, we installed a separate 3.5 meter Ku hub at their building in downtown Valdez. They used this terminal to set up links to two VSAT terminals on Knight Island in Prince William Sound and on Pearl Island in the Gulf of Alaska south of Homer. . . . Those terminals provided extensions for marine radio and aviation communications out into the sound and the gulf for cleanup coordination and flight safety. The Knight Island terminal also provided a telephone link to the National Guard fuel barge southwest

of that location. Exxon also connected directly to Houston and to their Anchorage office from the hub in Valdez.[45]

The FCC granted a waiver for use of VHF radio equipment during the cleanup. As William Harris, senior advisor to FCC Commissioner James H. Quello, pointed out later that year: "Waivers of our rules are granted almost routinely when it's apparent that they did not contemplate the unique situations and conditions faced by Alaskans."[46] Indeed, Alaska had argued several times for waivers from the FCC, and continued to do so.

Changing Geopolitical and Technological Contexts

In addition to the newly opened communication and transportation links with Siberia and the global attention to Alaska after the Exxon Valdez oil spill, other events helped to frame the context of communications planning in the 1990s. In June 1989, Chinese protesters in Tiananmen Square were brutally attacked by their own military, reminding people around the world of the fragility of freedom of speech and the specter of control of information. In addition to geopolitical developments, technological advances began to change thinking about traditional distinctions between information and communications technologies and services. Larry Pearson discussed this convergence in a report of the Alaska Legislature's House Special Committee on Telecommunications:

> Computer systems are just one means of moving information and providing access to it. As telecommunications technologies evolve, it is becoming increasingly difficult to distinguish among them. Distribution systems such as satellites and fiber optic cable move text, visual and audio information simultaneously. Computers talk and provide animated graphics. Telephones contain computers (a convergence of technologies that preoccupied the FCCs in the 1960s and '70s). Television monitors show textual information. A decoder, a telephone line and a keypad can turn a television set into a computer terminal linked to remote databases. Laser discs can substitute for computer memory and traditional libraries for high volume storage of information. These convergences of technologies provide an additional argument for focusing at a policy level on what telecommunication does—and what we want it to do—rather than on its internal workings.[47]

Converging technologies and the need for "electronic highways and byways" to extend access to information throughout the country, including its rural and remote regions, would become key themes of the next decade.

State Planning in the 1990s

In August1989, the University of Alaska Anchorage hosted the first of three annual Chugach Conferences. Participants from state government, broadcasting, the press, communications and computer industries, librarians, and lawyers met to discuss the future of communications in Alaska. They heard from national and international experts and as well as their own legislature and the APUC. In March 1993, a statewide telecommunications forum entitled "Visions of Alaska's Future" was held in Juneau. The participants of these conferences shared their concerns, insights, and recommendations for the future of communications in Alaska. Several of the speakers and participants are quoted in this book.

The author was a keynote speaker at the "Visions of Alaska's Future" conference in Juneau in 1993, where she urged the participants to set goals and develop strategies using telecommunications to help achieve them: "The first step is to get together the people who can address these issues at the statewide level and to enable them to communicate with each other so that they can develop a comprehensive statewide plan."[48] Clearly, Alaskans were starting to take these steps at the Chugach conferences and the Juneau conference. Later in the decade, they would develop the statewide plan.

The TIC languished during the Hickel administration (1990 to 1994), but was revitalized by the Knowles administration (1994 to 2002) when Lieutenant Governor Fran Ulmer took over as chair. The Knowles/Ulmer administration set the following goals in telecommunications:

- Improve public access to government information
- Maximize service to the public through voice, video and data systems
- Optimize government efficiencies
- Explore innovative and cost-effective services that meet Alaska's challenges
- Stimulate the development of private and public services[49]

The TIC decided to do an intense internal examination of where the state was and where it should go in the short, medium, and long term. Eight internal task forces were formed that examined economic development, public telecommunications, emergency communications, telemedicine, the Telecommunications Act of 1996, public transactions, education, management of information systems and data processing. After two months of meetings, these groups gave recommendations, which formed the building blocks of the plan.

In September 1996, Lieutenant Governor Fran Ulmer introduced the draft Telecommunications and Information Plan, stating:

Much of our state's history has a common theme: overcoming great barriers and daunting distances. From the Iditarod to the Alaska Highway, we are a people who have had to overcome great odds to live and prosper. Today, telecommunications and information technology holds the promise of breaking many of those barriers and bridging much of that distance. But to take full advantage of that promise, we must have a well-ordered planning process to guide us. Like the Iditarod and the Alcan, we must know where we want to go before we start our journey or build our highway.[50]

A primary focus of the plan was on strategies to increase public access to state information and services. Ulmer explained: "What we're talking about in this plan is a 'paradigm shift' in how people access needed information from their government. We are striving towards a system where people will no longer have to know the intricate structure of government in order to find out what they want to know."[51] After public hearings in the fall of 1996, the TIC passed the plan in December 1996.

The plan was the state government's portion of the "Alaska 2001" plan, which resulted from an inquiry initiated in September 1994 by the Alaska Public Utilities Commission (APUC) concerning the future of telecommunications policy in Alaska. Ulmer noted: "That document attempted to articulate the needs of Alaska's citizens; this plan attempts to articulate needs of State government."[52] The Commission's goal was to assess the role of regulation and to help formulate a strategic telecommunications policy for the benefit of all Alaskans. To help accomplish this goal, the Commission established task forces to research issues and prepare policy proposals for public presentation in four broad areas of telecommunications:

- Competition and Regulation
- Universal Service
- Government Use and Provision of Telecommunications Services
- Economic Development

The response to the Commission's call for volunteers was overwhelming, with about 200 Alaskans volunteering to participate on the four different task forces. The APUC also created an Advisory Committee composed of members representing the telecommunications industry, state government, consumers, rural residents, business, libraries, and telemedicine.[53]

Among the "Alaska 2001" recommendations were establishment of state standards for universal service, investigation of whether there should be a state universal service fund, authority from the state legislature for the APUC to adopt regulations and possible preferential rates regarding the universal service needs of schools, libraries, and rural health care providers.

They also urged the governor to establish formal communication between the state's telecommunications agencies (including the APUC) and the state's economic development agencies to ensure more consistent and coordinated public policy. The report also endorsed the administration's creation of a website concerning state government information and recommended that its content be expanded to serve as a telecommunications clearinghouse of state initiatives, case studies, and grants and other funding opportunities.[54]

New Federal Policies: The NII and the Telecommunications Act of 1996

The timing of these reports came at a major turning point in U.S. telecommunications policy, with the passage of the Telecommunications Act of 1996[55] in February of that year. In 1993, the Clinton administration had called for investment in a National Information Infrastructure (NII), a "seamless web of communication networks, computers, databases and consumer electronics that will put vast amounts of information at users' fingertips."[56] Its NII Initiative was guided by five principles: to encourage private investment, promote and protect competition, provide open access to the network, avoid creating a society of information "haves" and "have nots," and encourage flexibility and adaptability so that policies are broad enough to accommodate change.[57]

The Telecommunications Act of 1996 updated the Communications Act of 1934 for an era of convergence and competition. It preserved the themes from the 1934 Act of the public interest, just and reasonable rates, and universal service. There were many changes relevant for Alaska. The Act stated that services and prices in rural areas were to be "reasonably comparable" to those available in urban areas, rejecting the assumption that "something is better than nothing" in rural areas because minimal service was all that was either technically feasible or economically justifiable. Subsidies including the High Cost Fund for carriers serving rural and remote areas helped to keep prices down for telephone customers in rural Alaska. Several of these topics, including competition, universal service, and extension of Internet connectivity to rural Alaska appear in the Alaska 2001 report and the State Telecommunications and Information Plan.

The Telecommunications Act expanded the definition of universal service to include "advanced services" that should be universally available. "Advanced services" were not defined but were interpreted to include Internet access, with the flexibility to include future generations of services. It also specified institutions, including schools, libraries, and rural health centers rather than households as the means through which these services could be made accessible.[58] New connectivity subsidies for schools and libraries and for rural health facilities brought the Internet to rural Alaska schools and libraries and expanded telemedicine services for Native hospitals and village clinics. For the carriers, these customers became "anchor institutions," providing a guaranteed source of revenue that could provide a business case to expand Internet connectivity in rural communities. The impacts for Alaska of the "E-rate" subsidy for schools and libraries and the rural health care subsidy are analyzed in Chapters 16 and 17 on distance learning and telemedicine.

The Telecommunication Act set rules and incentives for competition to achieve goals of extending affordable access to telecommunications and "advanced services." For example, after meeting certain conditions, local exchange carriers (LECs) could provide long-distance services, and long-distance providers (interexchange carriers or IXCs) could provide local services. The act eliminated state-imposed barriers to competition and forced incumbent local exchange carriers to cooperate with their potential competitors. It required incumbents to allow competitors to interconnect with the incumbent's existing local network, so that new entrants could use the incumbent's existing network to provide competing local telephone service. It also required incumbents to provide competitors with access to elements of the incumbent's network on an unbundled basis, enabling new entrants "that have not completely built out their own networks to offer services over a combination of their own facilities and those leased from incumbents." The act also required incumbents to provide services at wholesale rates to competitors who could then resell them to customers at retail rates.[59] As discussed in Chapter 15, these provisions eventually resulted in local competition in many Alaska cities and towns.

State Universal Service Support

The APUC (now the Regulatory Commission of Alaska) established the Alaska Universal Service Fund (AUSF) which came into effect in March 1999 "to help maintain affordable telephone service throughout Alaska," supplementing the federal USF support programs. The AUSF provides support for Lifeline (reducing local charges for low-income customers), some small local exchange companies with high switching costs, and "Public Interest Pay Telephones (PIPT) . . . pay telephones in locations where they are needed for health, safety or public welfare and would not otherwise exist as a result of the operation of the competitive market."[60]

The APUC explained the rationale for the subsidy to high-cost companies. Previously, "the bulk of state universal service support was paid directly by long distance companies to local exchange companies through charges for access to local networks when originating and completing long distance calls. The purpose of these 'access charge' subsidies was to help keep local rates reasonable by subsidizing local exchange companies with high switching costs. However, the federal Telecommunications Act of 1996 requires that these subsidies be competitively neutral, and for that reason the Commission determined that it was necessary to remove them from access charges. The Commission created the Alaska Universal Service Fund in part to continue this switching support in a form that was consistent with the federal Telecommunications Act of 1996."[61]

The AUSF was calculated by dividing the total AUSF costs (including administrative costs) by total intrastate telecommunications revenue. In its first year the rate was 1.8 percent; by 2014, the rate had risen to 9.2 percent, as a result of regulations adopted by the RCA in 2011 that allow the pricing for telephone calls within Alaska to decline to approach the prices for out-of-state calls. The regulations also provided for economic support for rural Carriers of Last Resort "to maintain the essential infrastructure that connects customers to the local, wireless, and long distance networks."[62] The funds are collected through surcharges on local, instate, long distance, and wireless services. As with the federal USF fees, telecommunications companies are "permitted, but not required," to pass this charge through to their end user customers on a monthly basis. And as has been true with federal USF fees, the carriers have passed the charges through to their customers. Alaska telephone companies received a total of more than $29 million from the AUSF in 2013.[63]

The Planning Vacuum Returns

The Telecommunications Information Council lasted some 18 years. In 2002, the TIC was transferred by Governor Murkowski to the Alaska Department of Administration and abolished by the Governor's Executive Order in 2005. Since that time, there has been no entity within the state government responsible for overall planning and policy in telecommunications. This vacuum once more became evident when issues about statewide access to broadband arose several years later (see Chapter 18).

Chapter 14

Alaska's Local Telephone Companies

In a community as small as Tanana, when you saw there was a job to be done, you did it.

Paula Eller, cofounder of Yukon Telephone[1]

Local Phone Companies

In the early days, both local and long-distance services were provided by the military through the ACS network. Dial telephones and electromechanical exchanges gradually replaced operators for local calls, but operators at manual "plug boards" connected residents and businesses with other communities and the world outside Alaska. When villages were added to the long-distance network, they often had only a single payphone.

As demand for telephone service grew, local telephone utilities were established to provide local exchange service and communications to other locations through interconnection with the long-distance network. In smaller towns, a few hardy pioneers were building local phone companies: "One phone company was founded when a young technician obtained some PBXs (private branch exchanges) that had been deluged in the Fairbanks flood of 1964, cleaned them up, and began serving sites in rural Alaska. Other companies had been founded in rural locations through the efforts of similar entrepreneurs, often operating on shoestring funding and relying on 'sweat equity' to build the phone systems."[2]

With the privatization of ACS and investments by RCA in the 1970s, the expansion of the long-distance network created opportunities for new local companies. Their business model was based on a combination of local subscriber fees and a share of the toll revenues from the long-distance networks as compensation for connecting end users to the long-distance network. "Building the infrastructure cost money. The influx of money from the Native claims settlement the pipeline boom, and the increase in intrastate revenues contributed to the cash flow to make this possible. The adoption of the Ozark methodology of separations in Alaska in 1971–6 shifted large amounts of settlement dollars from interstate to intrastate services, thereby making both intrastate long distance and local service more affordable."[3] The Rural Electrification Administration (REA)—now the Rural Utilities Service (RUS)—provided low-cost loans to provide local service and later to upgrade local facilities.

Several ownership models were adopted for local telephone service, including private commercial ventures (the smallest of which were known as "mom and pop" companies), city-owned utilities in Anchorage, Fairbanks, and Ketchikan, and cooperatives. The first Native telephone cooperative was OTZ in Kotzebue. By 1995, local service was provided by 23 local exchange carriers (LECs), the largest being the Anchorage Telephone Utility (ATU) with 139,000 of state's approximately 325,000 access lines, and the smallest being Circle Telephone with about 30 lines. Among privately owned companies serving multiple locations were Telephone Utilities of the Northland serving 65 communities and United Utilities serving 57 communities.[4] As described in the next section, over the next decade several of these utilities would be bought and merged into larger companies, and many would diversify into other businesses, including cable TV, Internet services, and cellular communications.

Commercial Telephone Companies

Local entrepreneurs began to build local phone networks to connect people in their communities with each other and the outside world over the military's ACS network, later bought and extended by RCA. Some of these local companies expanded to serve several isolated communities. There has been further consolidation as the small companies were acquired or merged with larger enterprises.

Yukon Telephone

Yukon Telephone is an example of a "mom and pop" company, founded in 1960 in Tanana on the Yukon River by Cliff and Paula Eller, who literally built the company themselves. With the help of local people, they made telephone poles by cutting down and limbing trees. The poles were set in hand-dug holes and were put in place by a hoist that Cliff had constructed on the back of a flatbed truck.[5] Tanana was finally connected to the outside world with its own telephone system in May 1961. Initially, there were about 25 subscribers, growing to about 100 after ten years. Their son, Don, who now manages the company, got started at an early age: "My earliest memory of doing telephone work was sometime between the ages of 5 and 10 when my father had me run inside telephone wire under teacher housing. The housing was actually trailers which were low to the ground with lots of obstacles underneath. Crawling under the trailers would be a challenge for me today as much as it was for my father at the time, but it was great entertainment for a young child and so began my telecommunications career."[6]

In 1966, the Ellers took over the Tanana Power Company, which Paula managed for many years. Their son, Don, explained: "In order to provide phone service, stable reliable electricity is required. In the early days of Yukon Telephone, my father found that he was spending more time keeping the community generator going than working to provide phone service. He concluded that if he was going to be doing the work he might as well be paid for it, so we ended up purchasing the local electric utility in Tanana as well."[7]

In 1979, the residents of Ruby, downriver from Tanana, asked YukonTel to provide telephone service for their village. After a few months of discussion, the Ellers agreed to install and maintain the phone system. On December 6, 1980, Cliff Eller and his friend George Richardson, a missionary from Ruby, took off from Nenana to Ruby in Cliff's plane to deliver phone equipment to the exchange site and Christmas presents to Richardson's family. Near Ruby, the plane suddenly lost altitude, and Cliff had to make a crash landing. The plane hit a power line located between the town and the airstrip, fell to the ground, and caught fire. Both men escaped and were soon transported by plane to Fairbanks. Cliff Eller was later transferred to Anchorage, where he endured several weeks of painful treatments and skin grafts. He started flying his own plane again three months later.

Paula Eller served in many different capacities of the company, from being secretary/treasurer to doing the billing, bookkeeping, and also financial management. She also worked part time as a nurse, a substitute teacher, and a counselor for the high school students "In a community as small as Tanana," she said, "when you saw there was a job to be done, you did it."[8] She later became an advocate for the need for affordable, state-of-the-art telecommunication services in rural areas, as a leader in several national telecommunications organizations, serving as president of OPASTCO, board member of the U.S. Telephone Association (USTA), and board member and president of the Rural Telephone Finance Cooperative (RTFC).

Today, Yukon Telephone Co. Inc., Tanana Power Co. Inc., Supervisions Inc., and BBN Inc. are affiliated companies managed by the Eller family. Yukon Telephone provides local telephone service in the communities of Ruby, Tanana, and Whittier; Tanana Power generates and distributes electricity for the community of Tanana; Supervisions Inc. provides cable TV and high-speed Internet services for Tanana and Whittier; and BBN Inc. provides wireless high-speed Internet service to Nenana and the surrounding area.[9] Don Eller has installed a local fiber network in Tanana, making it the first community in Alaska to be served with fiber to the home.

TelAlaska

TelAlaska grew out of two of the larger "mom and pop" companies, Interior Telephone and Mukluk Telephone, and has added long distance and cable services. In 1969, the Rhyner family began providing telephone service to Fort Yukon. At age 16, Jack Rhyner salvaged telephone equipment damaged in the Fairbanks flood of 1967. Using emery boards to smooth out the gold points on the mechanical switches and high-pressure water hoses and brushes, he cleaned and repaired each component. Jack and his father, Richard, used the salvaged equipment to build the first telephone system in Fort Yukon. Before that time, Fort Yukon had only a single phone at the airport.[10] Their company, Interior Telephone, expanded to serve 11 Alaskan communities spread from north central Alaska to Dutch Harbor on the Aleutian chain.

In 1982, Jack took over management of Interior Telephone, and with his wife, Donna, developed and expanded the company they called TelAlaska. In 1992, TelAlaska purchased Mukluk Telephone Company, another pioneering enterprise, and began service to 13 additional communities, including six located along the Iditarod Trail near Nome and the island of Little Diomede Island in the Bering Sea. In 2000, TelAlaska became the new provider of telephone service to Nome, which was incorporated into Mukluk Telephone Company, and Seward and Moose Pass, which were incorporated into Interior Telephone Company.[11]

While Jack Rhyner battled cancer, the company was run by his wife, Donna Rhyner, following their corporate motto "Of Course You Can." After his death in 2008, TelAlaska was acquired by American Broadband, a company that specializes in acquiring small family-owned telephone companies. American Broadband also owns rural telecommunications companies in Nebraska, Missouri, and Louisiana. Today, TelAlaska is headquartered in Anchorage and provides local, cellular, and long-distance phone service; advanced data services; dial-up, wireless, DSL, and cable modem Internet services; and cable television service through a family of companies including Interior Telephone, Mukluk Telephone, TelAlaska Long Distance, Eyecom Cable, and TelAlaska NetWorks.[12]

GTE Alaska

GTE, the largest "independent" (non–Bell System) U.S. telephone company for many decades, provided telephone service in southeast Alaska and other regions until the year 2000. In 1999, GTE agreed to sell its Alaska operations, including approximately 21,000 customer access lines and 12 exchanges, to the Alaska Telephone Exchange Acquisition Corporation (ATEAC). ATEAC was an alliance of four Alaska local exchange telephone companies: Alaska Power & Telephone (APT), Arctic Slope Telephone Association Cooperative (ASTAC), TelAlaska, Inc., and United Companies. The transaction was part of GTE's previously announced initiative to sell approximately 1.6 million telephone lines in the United States.

On June 30, 2000, Bell Atlantic, one of the consolidated former Bell companies, merged with GTE, naming the new entity Verizon Communications. Two months later, the parties finalized the sale of GTE Alaska's assets. The exchanges were allocated to the four Alaska companies:

- Alaska Power & Telephone: Haines, Metlakatla, Petersburg, Wrangell, Klukwan
- ASTAC: Barrow
- TelAlaska: Moose Pass, Seward, Nome
- United-KUC: Bethel, McGrath, Unalakleet[13]

Phone under a tree in Wiseman near the Dalton Highway for summer visitors. Bug spray included! (Heather Hudson)

Alaska Power and Telephone (AP&T)

Alaska Power and Telephone also grew by acquiring privately owned telephone companies. Established in 1957, AP&T now provides electric and telephone service to a total of 34 communities located above the Arctic Circle, in the Wrangell Mountains, and throughout the islands of southeast Alaska. Originally called National Utilities, Alaska Telephone Company (ATC) is the largest local exchange carrier in AP&T's family of companies. Serving 1995 access lines in 1991, ATC acquired unserved territories and built new exchanges over the next eight years. In 1993, AP&T purchased Bettles Telephone, and in 1998 it acquired the Allakaket exchange from Pacific Telecom Inc., serving Bettles, Jim River Camp, and Allakaket with approximately 177 access lines. In 1994, AP&T purchased North Country Telephone, another pioneering company that served Eagle, a town with a rich history dating from the days of the Yukon Gold Rush.

In the year 2000, ATC purchased GTE Alaska exchanges in southeast Alaska, increasing ATC's facilities to 13,000 lines.[14] Today, AP&T provides local telephone service, long-distance services, Internet and broadband services, as well as operating a microwave network and providing engineering services. It also operates several hydro-electric systems in southeast Alaska.[15]

United Utilities

United Utilities was established in 1977 to provide local telephone service in 56 bush communities in the Yukon Delta region that previously had limited or no telephone service. In 1984, the company was reorganized with United Companies as a parent corporation with four affiliates—United Utilities, Manley Utility Company, United-KUC, and Unicom, Inc. The holding company was owned by Sea Lion Corporation and Togiak Natives Limited, the Alaska Native Village Corporations for Hooper Bay and Togiak. In addition to telephone services, the company operated DeltaNet, a long-haul broadband microwave network ringing the Yukon-Kuskokwim Delta in southwest Alaska.[16]

In 2008, GCI paid about $42 million for the stock of United Companies, Inc., the holding company for United Utilities, KUC, and Unicom. Today, owned by GCI, United Utilities provides telecommunications services to 58 communities, and United-KUC serves Bethel, McGrath, and Unalakleet.[17]

Alaska Communications Systems (ACS)

Alaska's telecommunications history returned to its roots with the formation of Alaska Communications Systems (ACS) in 1998, adopting the same name as the government-owned entity that had been sold to RCA in 1969. The president of the new ACS was Charles E. Robinson, who had worked with the White Alice system in the 1960s, then joined RCA Alascom, and was president of Alascom when it was owned by PTI from 1981 until its sale to AT&T in 1995, and president and CEO of PTI until it sold its remaining communications holdings to Century Tel in 1997.[18] ACS was formed by Fox Paine & Company, LLC, a San Francisco–based investment firm, and a management team led by Robinson, with a team including other members of the senior management team at PTI and other industry executives.

In 1999, ACS acquired Century Telephone Enterprises, Inc.'s Alaska properties formerly owned by PTI, and Anchorage Telephone Utility (ATU). CenturyTel had been the incumbent provider of local telephone services in Juneau, Fairbanks, and more than 70 rural communities in Alaska, and provided Internet services to customers statewide. ACS was now Alaska's largest local exchange carrier (LEC), with more than 300,000 access lines, providing service to 75 percent of Alaskans and to the state's major population centers of Anchorage, Juneau, and Fairbanks. In addition, it became one of the largest providers of cellular service with over 66,000 cellular subscribers and a long-distance provider to approximately 26,000 customers, primarily in Anchorage.[19]

The APUC approved the sale, although the staff had some concerns that it could reduce competition in the future. The staff also recommended that either the Fairbanks or Juneau operations be excluded from the sale. However, the Commission did not follow its staff's recommendation. APUC chairman Sam Cotton commented: "We didn't agree that it would negatively affect the public interest."[20] In 2000, ACS also tried to buy the Matanuska Telephone Association, but the offer was rejected when not enough MTA members voted on the sale (see below).

Facing severe competition in the local exchange market, ACS focused on wireless communications and the enterprise data market. It operates high-speed data networks and data hosting centers and offers data management services. In December 2014, ACS agreed to sell its mobile customer base and remaining wireless assets to GCI.

Adak Telephone Utility

Since 2003, a privately owned company known as Adak Eagle Enterprises (AEE) has provided services to Adak, the farthest west community in the United States, situated near the end of the Aleutian chain. Formerly the site of the Adak Naval Air Station, Adak now has a population of about 325 and provides a fueling port and crew transfer facility for foreign fishing fleets, as well as fishing and seafood processing.

AEE provides communications services through subsidiaries including Adak Telephone Utility, Adak Cablevision, Windy City Cellular (WCC), and Windy City Broadband.[21] Communication with other locations in Alaska and the rest of the world is by satellite. In 2005, the telephone utility borrowed $6 million from the Department of Agriculture's Rural Utilities Service (RUS).

Like other telephone companies serving remote regions of Alaska, the companies serving Adak receive operating subsidies because of their high operating costs and isolation. Subsidies in 2011 for AEE and WCC amounted to $15,000 per line per month, or more than $4.2 million per year for fewer than 300 fixed and wireless lines. In 2013, Adak Enterprises and Windy City Cellular petitioned the FCC for an exemption from a new rule that caps subsidies at $250 a month or $3000 a year per phone line. No other company in Alaska requested a waiver from the new cap. In July 2013, the FCC denied their petition, stating that the companies had not demonstrated that their costs were reasonable, and citing high salaries for management, costs for operating a retail store, renovations and other office expenses in Anchorage and Adak, and other assets including a fishing boat.

The telephone company threatened to file for bankruptcy, which would result in closing down operations and defaulting on its RUS loan.[22] However, GCI, which also offers cellular service in Adak, stated that if AEE or WCC were unable to continue operations, GCI would continue to provide services on Adak so that "Adak Island will not 'go dark.'" GCI further added that it was "fully capable of setting up the infrastructure necessary to continue providing service to Adak in a relatively short amount of time [120 days], for no more USF funding than currently supports service."[23]

In mid-2013, the FCC extended the subsidy at interim support levels higher than the $250 per line cap for six months. In early 2014, the FCC authorized interim support in the amount of $73,380 per month for the two companies while they completed their review of the filings.[24]

City-owned Utilities

Ketchikan (KPU)

Today, the only remaining city-owned telephone utility is in Ketchikan. KPU Telecommunications is one of three divisions of Ketchikan Public Utilities; the others are electricity and water. KPU was created as an electric utility in 1932, when the City of Ketchikan purchased the Citizen's Light and Power Company, which first started delivering power to Ketchikan in 1903. KPUTel provides telephone, Internet, and cable television services.[25]

Fairbanks

Like Ketchikan, the city of Fairbanks operated municipally owned telephone, electricity, and water systems. The mayor and city council debated whether to try to compete in a competitive environment or divest utilities that had been government-owned for more 50 years. The profitable telephone utility had been subsidizing the water and electric utilities, but by 1996, competition and the passage of the Telecommunications Act were contributing to declining profits, and the value of the telephone utility was dropping rapidly. In addition, the electric utility was threatened by its neighbor utility, Golden Valley Electric Association, whose commercial rates were nearly 35 percent lower.[26]

Eventually, in 1996, the mayor and city council of Fairbanks decided to sell. Having received an unsolicited proposal to buy the telephone utility, they passed a resolution that all the utilities be sold as a group, in order not to be left with the distressed water and sewer system. Only one proposal was submitted by the deadline, from a group composed of ten local investors known as Fairbanks Sewer & Water, Inc. (FSW), PTI Communications (telephone), and the neighboring electric cooperative, Golden Valley Electric Association. After lengthy negotiations, the parties agreed on a division of the assets and set the amount each buyer would pay to the city. On October 7, 1996, Fairbanks voters approved the package sale by a margin of 54 percent to 46 percent.[27]

Opponents of the sale then redirected their effort to regulatory agencies, challenging the application for the transfer of the sewer system to the federal Environmental Protection Agency (EPA) and both the sewer and water systems to the Alaska Public Utilities Commission (APUC). After months of delays, the City and FSW enlisted the support of the Alaska congressional delegation to encourage EPA officials to approve the transfer of the federally funded wastewater treatment facility. After five months of hearings, the APUC finally issued an order on September 29, 1997, approving the application for privatization of sewer and water, providing the final approval necessary to complete the sale of all three utilities—only a week before the contractual termination date of October 7, 1997.

However, the APUC failed to establish the rate base for the sewer and water utility that was provided for in the buyer's negotiated agreement with the city, stating that it would postpone the rate base issue until the applicant filed a full rate case, expected in about three years. This omission became critical because closing the sale contracts negotiated by the city and FSW was conditioned on the establishment of a guaranteed rate base. Failure of the Commission to decide this issue now gave the buyers an option to withdraw, which would have doomed the entire sale. FSW quickly applied to the APUC for reconsideration, but the APUC denied the request on October 2, only five days before the "drop dead" deadline when the parties could withdraw.

After lengthy meetings, all parties agreed to close, and the three buyers and city officials met at the city hall on October 6 to sign the papers allowing the city to privatize its utilities. "Ten years under study, two years under public scrutiny in a bitterly contested process, it came down to two days to open for business. On October 7, 1997, the last day allowed under a self-imposed contract deadline, the City of Fairbanks transferred its utilities to three buyers."[28] Perhaps only in Alaska would the privatization of a telephone utility depend on regulators of water and sewers.

Anchorage Telephone Utility (ATU)

ATU, which was the state's largest local phone utility, went through an even longer and more arduous privatization process. Unlike Fairbanks, the Anchorage city charter required 60 percent of the popular vote for approval. In 1989, only 52 percent of the voters approved a proposal to sell ATU to Pacific Telecom (PTI) for $412 million.

In July 1991, the Anchorage Assembly again decided to sell ATU, this time to Citizens Utilities, Inc., of Stamford, Connecticut, for $416 million (twice the book value),[29] rejecting a higher bid by Pacific Telecom of $457 million because they were concerned that regulators would think owning ATU would give PTI too much power in the Alaska marketplace.[30] Citizens then raised its offer to $450 million, apparently to overcome objections about its low bid. PTI then sued the city over allowing Citizens to negotiate another offer after it had submitted its written bid to the Assembly. The judges refused to block the sale, but found "serious questions" about the city's decision to accept the modified bid.

Former legislator and then Anchorage Assemblyman H. A. "Red" Boucher cosponsored a new ordinance calling for a dual ballot, proposing one question to ask voters to approve the sale and a second to choose a company, Citizens or PTI. PTI then offered the city $43 million more than its initial bid of $457 million. Groups organized to oppose the sale—some to block the sale to Citizens and others to keep ATU city-owned. Keep ATU Alaskan, a group opposed to sale of ATU funded largely by GCI, collected almost 22,000 signatures and got a proposition on the ballot requiring the city to keep ATU in public hands but to change the way it was managed through creation of an independent board of directors. Boucher complained: "With enemies like . . . Keep ATU Alaskan and GCI, it would be tough for Mother Theresa to get 60 percent. . . ."[31]

The *Anchorage Times* obtained an internal memo from Alascom general manager John Ayers to his Anchorage employees, asking them to circulate PTI's petition to put its Assembly-rejected offer of $457 million on the ballot, and stating that they would receive $2 for every signature they collected in their neighborhoods. Ayers said the employees' efforts contributed to getting 8,000 signatures in less than a week—they needed 5672 certified signatures to get an initiative on the ballot.[32] However, PTI said it would not turn in its signatures if the Assembly chose the two-part ballot.

Questioned about his support for PTI, Boucher said he was an old friend of PTI Chairman Charles (Chuck) Robinson, but the friendship did not influence his support for the company. "Is someone paying me under the table? No," Boucher said.[33] However, Boucher sent out a letter to Anchorage residents on September 25, 1991, on PTI letterhead that stated:

> During the period 1985 to 1990 the House Special Committee on Telecommunications, which I chaired, passed legislation establish a new direction for telecommunication in Alaska, a direction designed to carry our communication infrastructure into the next century. I understand the importance of telecommunications to Alaska's economic future.
>
> I have asked Pacific Telecom to provide me the opportunity to share with you, my constituents, my thoughts on the proposed sale of the Anchorage Telephone Utility. . . .
>
> It's time to pass the ATU baton to a company who knows how to carry Anchorage's telecommunications needs into the next century.
>
> That's why I'm writing to you today. To personally appeal for your sense of community, your sense of striving for a stronger tomorrow to be passed on to the next generation. How better than with a company committed to teletechnology, with a company dedicated to serving Alaskans and providing for their needs in the telecommunications world? . . .
>
> I urge you to vote on October 1st. Vote for the sale of ATU. Then follow up with a vote for PTI . . . for the telecommunication future of Anchorage.[34]

But Anchorage voters spurned the sale. The majority preferred selling to Citizens rather than ATU, but the approval did not get the required "supermajority" of 60 percent. And to add to the confusion, the alternative ballot item from Keep ATU Alaskan got more than 50 percent of the vote. So a majority approved selling to Citizens, but a majority also wanted not to sell ATU at all.

In 1995, the city again solicited bids to sell ATU. In early January 1996, Citizens Utilities bid $278.1 million to buy ATU, but city officials said the offer was "lower than we had hoped for" and further review was needed before the issue would be submitted for voter approval. The city imposed conditions that may have discouraged more bids, such as that ATU could not enter the long-distance market or join an alliance that would boost long-distance costs.[35] At the end of January, Anchorage Mayor Rick Mystrom rejected Citizens' bid, saying the offer was too low and "nonresponsive" to the request for proposals.[36]

ATU then sought to diversify its offerings, providing cellular services through a subsidiary, and launching long-distance service in Alaska in 1997, making it the first LEC to enter the long-distance market in the state. However, AT&T and GCI also received authorization to provide local service in Alaska, subjecting ATU to competition in the biggest local market in the state.[37]

In 1998, Mayor Rick Mystrom and the Anchorage Assembly again advocated selling ATU in a competitive bid, reasoning like their colleagues in Fairbanks that a private company was better equipped to deal with the rapid changes and increasing competition in the telecommunications industry. A city appraisal put the value of ATU at between $277 million and $298 million, more than a third less (in constant dollars) than what Citizens and PTI were willing to pay in 1991, and about what Citizens had offered in 1995. This time the voters approved the sale by a margin of 62.5 percent to 37.4 percent.[38] The buyer was Alaska Communication Systems, a newly formed company founded by former employees of Alascom and PTI.

Proceeds from the sale of ATU were used to establish a trust fund for the city. About $130 million was invested. The city initially took a yearly dividend of $9.4 million. After 2002, the city was allowed to draw as much as 5 percent of the fund's average value over the preceding five years, amounting to a dividend of about $6.5 million yearly. The fund had grown to $141 million by the end of 2007 but shrank to $96.9 with the economic downturn in 2008 after paying a $6.6 million dividend to the city general fund.[39]

Cooperatives

The Rural Electrification Administration (REA) in the U.S. Department of Agriculture was established in 1935 to extend electricity to unserved rural areas through low-cost loans to cooperatives established in many communities in rural America. Numerous electric cooperatives were formed to serve Alaska communities. After World War II, the REA also began to make loans to rural cooperatives to provide local telephone service. The new telephone cooperatives followed the model of the electric cooperatives, and some were jointly operated with the electric co-ops. Today, the REA is known as the Rural Utilities Service (RUS). It continues to provide loans and grants to Alaska companies and cooperatives to upgrade their facilities for broadband, extend middle mile networks, and to support facilities for distance education and telemedicine.

The cooperatives are operated as nonprofit enterprises. Users of a cooperative's services are eligible to become members and can thus participate in its management through an elected board of directors. The co-ops return surplus revenues to their member-owners as "capital credits" in proportion to their patronage. For example, Matanuska Telephone Association (MTA) explains its use of capital credits:

> As a member-owned cooperative, MTA assigns margins (profits) to its members, based on the amounts they were billed for telephone service during the year.
>
> These assigned margins are called capital credits.... The funds belong to individual member-owners, but are retained by the cooperative as working capital and equity. The availability of this working capital reduces the amount of money MTA must borrow, which assists in providing members with state-of-the-art services at the lowest possible cost.
>
> On an annual basis, MTA's Board of Directors reviews the financial condition of the cooperative and elects to either approve or disapprove a return of capital credits to its members.[40]

Matanuska Telephone Association (MTA)

The first and largest telephone co-op was the Matanuska Telephone Association (MTA), founded in 1953 to provide telephone service to the agricultural town of Palmer in the Matanuska Susitna valley. Today, MTA's service area covers almost 10,000 square miles, south to Eagle River, north to Anderson, east to Glacier View, and west to Skwentna.

Over the years, MTA expanded and modernized its facilities, installing digital switches, microwave, and even a fiber optic network linking the high schools in its region in 1993 (see Chapter 16). Today, its fiber network is extensive. It also offers Internet services and digital cable TV.[41]

In the year 2000, ACS made an offer to purchase MTA, just a year after acquiring ATU. Concerned about increased competition eroding the value of the utility, MTA directors in May passed a resolution to sell the co-op to ACS for $187.5 million. As much as $100 million of the proceeds would have been distributed to members. However, the member-owners were required to vote on the sale. Some questioned whether the co-op needed to be sold, and whether ACS, whose majority owner was based in San Francisco, would serve their needs as well as their own locally owned utility.

In September, MTA's members voted 9,665 to 5,490 in favor of the sale. However, although a majority of the telephone cooperative's members who voted favored the deal, not enough ballots were cast to approve the sale. Under MTA bylaws, any sale needed approval from more than half of the membership. The deal would have required at least 16,709 members to cast a ballot, but only 15,215 did so.[42] Today, MTA remains a member-owned cooperative that has served the people of the MatSu region for more than 60 years.

Copper Valley Telephone Cooperative (CVTC)

The Copper Valley Telephone Cooperative (CVTC) has served the Valdez and Copper River Basin areas for more than 50 years, currently providing services to the communities of Glennallen, Chitina, Tatitlek, Mentasta, McCarthy, and Valdez. CVTC offers landline telephone services, high-speed Internet connectivity, and wireless voice and data. It also offers special access services for business and telecommunications carriers over its fiber and microwave network.[43]

In 2010, CVTC received $5.2 million in grants and loans from the Rural Utilities Service to extend high-speed Internet connectivity via microwave to the isolated town of McCarthy located within Wrangell–St. Elias National Park and Preserve.[44] In 2014, Copper Valley Wireless received funding of $152,400 in the FCC's Tribal Mobility Fund auction to help to extend 4G mobile broadband throughout its service area.

Cordova Telephone Cooperative (CTC)

A city-owned utility originally provided telephone service in Cordova. In 1978, the voters of Cordova approved a resolution by a six to one margin providing for the transfer of assets of Cordova Public Utilities-Telephone Utility to Cordova Telephone Cooperative, a newly established member-owned telephone cooperative. They also approved reorganization of the electric utility as a cooperative. The telephone co-op's first general manager, Doug Bechtel, also served as manager of Cordova Electric Cooperative from 1978 until May of 1983.[45]

In 1978, CTC had 640 Members; by 2004, there were more than 1100 members. Although membership grew, prices remained almost unchanged, or significantly cheaper considering inflation. Residential single-line telephone service cost $10 per month in 1978, and a business line cost $18. Today, a residential line costs $13 per month, and a business line costs $21. The deposits, $60 for a residential line and $100 for a business line in 1978, have been reduced to $30 for a residential line and $50 for a business line.[46]

In addition to telephone service, CTC was one of the first to provide Internet access to the community and now offers high-speed service with DSL and fiber optic cable connection to submarine cables to the lower 48, replacing its satellite links. In 1997,[47] CTC established two subsidiaries, Cordova Long Distance (CLD) and Cordova Wireless Communications (CWC), which provides cellular coverage of the Cordova region and much of Prince William Sound.

Southwest Alaska: Nushagak and Bristol Bay

The Nushagak Telephone Cooperative (NTC) serving Dillingham and nearby communities was established in 1975 and later merged with Nushagak Electric Cooperative to become part of the current entity known as Nushagak Cooperative, Inc. Writing in 2005, David Bouker describes the struggles 30 years earlier to establish the cooperative and to obtain funding from the REA:

> NTC took over operation from Interior Telephone Company, which had been operating the system for North State Telephone Company for approximately eight months. North State Telephone Company was one of the very few utilities that had been de-certified by the Alaska Public Utilities Commission due to the very poor telephone service they had provided. A group of ladies . . . started a newspaper called the Bristol Bay Bylines; they actually began the decertification process by publicizing the quality of the telephone service and supporting local efforts to start our own telephone system. At that time there were approximately 200 telephone subscribers, with almost everyone on a party line. Some lines . . . had as many as 10 parties on a single line.
>
> The outside physical plant took essentially two forms. Part of it was attached to the Nushagak Electric Cooperative (NEC) power poles, in some notable cases by rope attaching the telephone cable to power poles. The line across the flats to Windmill Hill and out to Kanakanak was a 50 pair air-filled cable lying exposed on the ground. Approximately every thousand feet there were signs posted to warn people of "exposed telephone cable. . . ."
>
> Unfortunately, the management at NEC had difficulty communicating with the owners of North State Telephone Company. . . . Fortunately, Albert Ball, a long-time resident of the area and president of the newly formed NTC, did have a good line of open communication with the president of North State Telephone Company; his efforts saved the day in coming to an agreement to purchase the assets of NST. This agreement or transaction was formalized on a napkin provided at a breakfast meeting at the Anchorage Westward Hotel (later known as the Anchorage Hilton Hotel.) The purchase price was $140,000 plus the deposits (just under $5,000) of the existing customers.
>
> At last we were in business! But we had no money. All efforts up to this point had been financed by NEC [the electric cooperative], although we did borrow $25,000 from the local branch of the National Bank of Alaska. We then had to take over operation from Interior Telephone Company, the operators of the system after North State had been decertified eight

months previously. . . . To further complicate matters, Interior Telephone Company was owed approximately $22,000 from the NST customers for local and long distance services rendered; needless to say, NTC didn't have that kind of cash available. We had to gear up overnight, hire a technician, and perform countless other tasks pursuant to getting into the telephone business (about which we knew next to nothing!)

The biggest problem was a lack of money and other resources. . . . [REA] decided to make some rather onerous requirements of us that would either make us or break us. One of their requirements was that we had to enlist 358 members (when there were only 200 existing customers) and that we had to collect a membership fee of $10 from each one of them. The only way we could get 358 members was to get all the families in Aleknagik to sign up as well. . . .

We approached George Ilutsik concerning the problem. He was advised that if the people of Aleknagik signed up for telephone service, we would extend the service to Aleknagik and over to the North Shore and Island, delivering the service at the same price the residents of Dillingham would be paying. George took us at our word and he delivered every family in Aleknagik as members, which enabled us to meet the REA requirements. What made this unusual was that the economy in the mid-seventies was not very good. In fact, President Nixon proclaimed this area an economic disaster area in 1974. . . .[48]

Today, NTC provides telephone, Internet, and cable TV services and serves Dillingham and the communities of Aleknagik, Clarks Point, Ekuk, Manokatak, and Portage Creek.

Bristol Bay Telephone Cooperative serves King Salmon and several communities in the region with telephone, Internet via DSL, cable television, and cellular services. The Bristol Bay Cellular Partnership (BBCP) provides cellular service to the Bristol Bay region, including King Salmon/Naknek/South Naknek, Clarks Point, Dillingham, Egegik, Ekwok, Igiugig, Koliganek, Levelock, Manokotak, New Stuyahok, and Pilot Point.[49]

OTZ Telephone (Kotzebue)

The first Native-owned telephone cooperative in Alaska was OTZ, established in Kotzebue in 1975: "Tired of unreliable and expensive phone service, the residents decided they wanted their *own* telephone company. They met at Walsh's Store in Kotzebue to discuss incorporating and buying out the existing provider." They formed the first board of directors, including James Gregg, Alex Hills, Albert Adams, Nellie Ward, and Thomas Sheldon. Then with the help of both Native corporations in Kotzebue (NANA and KIC), OTZ Telephone Cooperative, Inc. was born "to provide locally owned telephone service at affordable rates."[50] (OTZ is the airport designation for Kotzebue; the public radio station is KOTZ.)

The first years were dedicated to installing local infrastructure and telephones in the villages of the NANA region. Following the passage of the Telecommunications Act of 1996, OTZ Telecommunications was created to provide long-distance service to members. In 1996, OTZ also began to provide Internet service, and it now provides DSL connectivity in Kotzebue and the NANA villages. OTZ was also the first communications company to bring cell service to Kotzebue and now provides mobile service for the villages in the surrounding region. It has hired and trained local people for both technical and administrative positions.

ASTAC (Arctic Slope)

The Arctic Slope Telephone Association Cooperative (ASTAC) serves communities in a remote roadless region of more than 90,000 square miles. ASTAC serves eight communities in the North Slope Borough (Barrow, Anaktuvuk Pass, Atkasuk, Kaktovik, Nuigsut, Point Hope, Point Lay, and Wainwright) and the petroleum industry exploration and production complex at Deadhorse-Prudhoe Bay.

ASTAC was founded in 1977, when "Arctic Slope residents determined that the continued development of their traditional communities; the business success of the Alaska Native Claims Settlement Act (ANCSA) corporations; the Arctic Slope Regional Corporation (ASRC) and the eight Village Corporations; and the delivery of public services by the North Slope Borough (NSB) home-rule municipality, were all dependent on the availability of, at least, basic telephone service."[51] Today, ASTAC provides local and long-distance service, Internet, wireless, and data services through satellite connections linking the North Slope communities.

Chapter 15

The Phone Wars

The battles were fought in the courts, the Legislature, the
regulatory arena, in the press and in advertising.
Sean Reid, "Can Anyone Figure out Alaska's Phone Wars?"[1]

The Beginning of Intrastate Competition

By the end of the 1980s, Alaska interstate long-distance rates had decreased considerably, and Alascom had upgraded its facilities in urban areas where it faced interstate competition. GCI had long sought entry into the intrastate long-distance market, but the APUC remained concerned about the impact of competition on service prices and quality, particularly for the bush. In 1989, APUC Commissioner Carolyn Guess said the biggest issue that year was competition: "Should the commission allow competition to Alascom which has been the only long distance carrier since the military sold the network to RCA? . . . [a]s we plan for the year 2000 in the area of telecommunications services—let us, who are the 'haves,' not become the 'have mores' at the expense of those today who 'have less.'"[2]

Finally in 1990, after a successful GCI-led initiative to allow competition via ballot, the Alaska state legislature found that facilities-based competition for long-distance services was in the public interest and passed legislation directing the APUC to adopt regulations governing competitive intrastate long-distance service. The Commission did so, and GCI began offering intrastate long-distance service on May 15, 1991, on its own facilities, where it provided interstate service and through resale of others' services where it had no facilities. Yet the APUC remained concerned that competition could be detrimental in the bush. GCI and future intrastate competitors were not allowed to build facilities in the more rural regions of Alaska because the Commission found, by a 3 to 2 vote, that those areas would not support competing carriers.[3] However, GCI was allowed to build facilities in the more populous areas of Alaska that included about 90 percent of the state's population.

The Road to Competition

Although GCI's founders had recognized the significance of the federal policy shift to competition, Ron Duncan said that they failed "to fully appreciate how much time and effort could be spent before that national policy was successfully extended to the Alaska marketplace, and certainly . . . how resistant Alaska public policy was going to be to the transition to competition."[4] GCI discovered that

> Alaska's policy makers had to go through that same education process that the federal policy makers went through. . . . I mean, if you look at the time MCI started until the time they really got the federal regulatory mechanism firmly behind the competitive paradigm, it was ten or 15 years. And I think if you take . . . 1979 as a start date it was almost the end of the 80's before . . . we got intrastate authority [to compete]. And I think that really represented the turning point in the regulatory perspective in Alaska. . . .We had to cram competition down the unwilling throats of almost the entire public policy spectrum in Alaska for more than ten years. I think the only real exceptions to that would have been the congressional delegation and that was vital to our success on the federal level, but [there was] . . . much more resistance than we ever would have anticipated—from the state public policy apparatus.[5]

The road to competition thus involved state regulators as well as the Federal Communications Commission (FCC) and the courts. The APUC was struggling throughout the 1980s with the apparent incompatibility between competition and universal, affordable telephone service. Susan Knowles, a commissioner on the APUC from 1975 to 1993, commented: "when you have a commission that is essentially maturing with the industry it is regulating, because if you haven't addressed an issue—no matter how cut and dried it is—it is new. You've got to do a lot of research. . . . You are really starting with a very modest knowledge base, and you have to catch up very quickly in the course of a proceeding. So there were a lot of those issues."[6]

Knowles continued:

> Technology drove policy, and the state regulators were not as comfortable with some of the policy changes as the federal regulators were, I think,

primarily because we were the ones who would be out in the communities listening to real people telling us about real problems with their service, etc. And those of us who had large rural populations always had a high level of anxiety about the fact that we knew it cost more to provide service to Savoonga then it did to downtown Anchorage. There's no escaping that, and we didn't want to evolve to a system where service became unaffordable in Savoonga because it was totally based on what it would cost there. So we had the state regulators (who) were in the trenches, and they had these unique concerns based on the populations they were serving. . . .

The regulatory process itself also posed challenges, as described by Commissioner Knowles:

. . . I think there's another challenge too because the regulatory process . . . it wasn't a negotiated process. It was an adjudicatory process, and sometimes that's a very painful way to get information and get to the best outcome. And I can't think of any commissioner who didn't at one time say to him or herself, "If I could just get these two people in a room and just tell them this is the way it is going to be and you're going to sort it out and this will be it," it probably would have been a very quick and happy outcome, but that wasn't the nature of the process we were allowed to use. We weren't allowed to talk to these people in the course of a proceeding. And I don't find fault with it because it was a time-proven procedure, and it also assured the public and the parties that it was absolutely ethical and above-board, but it wasn't always the most efficient.[7]

The Alaska telephone companies likely endorsed that view, particularly the small local companies without resources to retain lawyers or carry out studies. Don Eller, manager of YukonTel in Tanana, which was founded by his father, commented: "one of the best lines ever spoken at the APUC . . . was blurted out by my father in the middle of a rate case. Dismayed and disgruntled at the costs involved with lawyers and expert witnesses for the rate case, my father enthusiastically said, 'My customers cannot afford your due process.'"[8]

The Joint Board Decides at Last

The Federal State Joint Board continued to wrestle with the problem of how to continue the earlier policy of rate integration in a competitive era. In January 1989, it issued a supplemental order inviting comments on its five stated objectives:

- Continued rate integration
- A market structure that permits competitive entry
- Preservation of universal service in Alaska
- A requirement that any plan adopted be revenue neutral with respect to federal and state jurisdictions
- A market structure that is as efficient as possible[9]

Alascom continued to plead its case before the legislature and at the APUC: "But why should you in the Legislature care? Because Alascom would have no choice but to make up the lost revenue elsewhere, and that would mean higher rates for intrastate service everywhere outside the urban triangle of Anchorage, Fairbanks, and Juneau. Remember, we are the carrier of last resort. We must provide service everywhere, and, in exchange, we are entitled, by law, to an authorized rate of return on our investment." GCI proposed a surcharge on its calls of between 5 and 13.5 cents per call on its network as a contribution to offset the impact on phone rates outside the urban triangle. Alascom countered: "Our analysis shows such a surcharge would be totally inadequate for keeping those rates stable, and the APUC has not disagreed."[10] An expert witness for Alascom told the APUC: "Alaska is simply too small a market, and costs too much to serve, for competitive entry to work. . . . It is doubtful that any competitor would even want to install facilities to serve 90 percent or more of the routes, since the traffic volumes and potential revenues are so small."[11]

However, in March 1993, AT&T proposed to the FCC to terminate its Joint Service Agreement (JSA) with Alascom and enter into competition for interstate and intrastate Alaska services. It stated that it would develop its own network to serve Alaska, including a new submarine optical fiber cable to connect Alaska directly with the lower 48. AT&T would apply to the APUC to provide "intrastate switched services, with basic intrastate MTS rates equal to or less than the lowest current Alascom or GCI rates for comparable services."[12] AT&T's proposals included a "Consumer Benefit Plan" to "guarantee permanent and vigorous competition for Alaska telecommunications services, and at the same time offer customers lower prices combined with higher service quality and more service options than have every been available before."[13]

The Joint Board activity continued into 1993, weighing several alternative market structures.

After nearly a decade, the state had filed approximately 35 sets of comments, reply comments or other major pleadings and documents amounting to more than 1100 pages of material. By then, the state concluded that the tide had turned against the JSA. The Washington, DC, office of the governor concluded that "various changes have occurred which make continuation of the joint services arrangement between Alascom and AT&T not sustainable in the long run. Those include:

- All relevant carriers—including Alascom, GCI and the ATA—have at various times publicly stated that the joint services arrangement should end.
- Technological improvements have made telecommunications networks more cost efficient.
- New Federal and State policies have placed greater emphasis on competition.
- Telecommunications markets themselves have become more competitive in Alaska and elsewhere."[14]

The Joint Board sought to address these changes and presented a plan for transition to a new market structure. On October 26, 1993, the Joint Board made its Final Recommended Decision, rejecting the market structure plans previously advanced by Alascom and AT&T, and instead proposing to end the AT&T/Alascom Joint Services Agreement subject to adoption and implementation of certain transition mechanisms. It noted that the AT&T/Alascom subsidy arrangement gave "Alascom little reason to control costs" and concluded "There are few, if any, advocates for the current market structure's preservation."[15] The FCC adopted these provisions in its Memorandum Opinion and Order released May 19, 1994, and set the date for termination of the Joint Services Agreement effective January 1, 1996.[16]

Why did it take eight years for the federal and state regulators to propose and agree on recommendations for a new market structure? Again, APUC Commissioner Susan Knowles who represented Alaska on the Joint Board after Commission Marv Weatherly retired, offered insights:

> . . . you know it languished for quite awhile in part because it seemed to be somewhat of an intractable problem. Alascom and GCI were the principal protagonists, but the problem was that Alaska had just so recently . . . gotten rate integration whereby our interstate long distance rates were developed on the same basis as everybody else with a little support thrown in, that I think everybody was a little bit reluctant to go on from there. And Alascom argued very persuasively for a very long time that it had earned this position and paid its dues and any change in this would be the undoing of affordable . . . intrastate long distance service for Alaska.
>   And . . . the thinking evolves over time. So you had the recognition of the inevitability of competition. It was there. We had GCI, and there were other carriers who also expressed an interest in coming into the market. So you had that reality. On the other hand, you had a long-standing system

that had served the people of Alaska well and a sense of apprehension about whether changing the rules would have an adverse effect on the state. So this tension existed for quite a period of time, and I think the log jam was finally broken when . . . Alascom realized the inevitability of the change.

. . . a lot of people were frustrated anyway with how long the (joint) board had gone on and not gotten to the bottom of things, although when I would hear that, I would say, 'Alright, what's your idea? You think it's so easy. Let's hear what your solution is.' And of course, most of the people who were the naysayers didn't really have the answer either, which was why those of us who worked on this, my predecessor as well as myself, were not as surprised maybe as the outside world with why it took so long.[17]

Consolidation and Local Competition

The FCC action to terminate the JSA led to a buyout of Alascom by AT&T in 1995 for $365 million.[18] The regulatory order would have led to head-on competition between AT&T and Alascom, whose revenue was expected to decline as a result; absorbing Alascom made more financial sense for AT&T than competing with it. The new company, known as AT&T Alascom, continued to exist as a separate entity, as required in the APUC's regulatory approval of the purchase.

In 1995, GCI obtained permission from the APUC and the FCC to demonstrate its advanced satellite communication technology in rural Alaska, using DAMA (demand assignment multiple access) to significantly reduce satellite delay while improving quality and efficiency. DAMA allows circuits to be assigned to communities as needed, instead of on a permanent dedicated basis. Thus any circuit not being otherwise used at the time could be assigned to any location needing it. The result is that each community has access to more circuits as needed, but the total number of circuits in use can be substantially less with DAMA than required by assigning fixed circuits to each location, thereby reducing costs to the carrier. In 1996, GCI challenged Alascom's assertion that "Alaska is simply too small a market, and costs too much to serve, for competitive entry to work," when it constructed earth stations using DAMA in 56 rural locations.

The next battle in the "phone wars" was over opening local service to competition. The Telecommunications Act of 1996 opened the door to local competition, eliminating state-imposed barriers to competition and forcing incumbent local exchange carriers (ILECs) to cooperate with their potential competitors, known as competitive local exchange carriers (CLECs). The act required incumbents to allow competitors to interconnect with the incumbent's existing local network to provide competing local telephone service. The unbundled access provision of the act required incumbents to provide competitors with access to elements of the incumbent's network on an unbundled basis, that is, leasing only those parts of the network that they needed. The unbundling provision permitted new entrants "that have not completely built out their own networks to offer services over a combination of their own facilities and those leased from incumbents." The act also included a resale provision, requiring an incumbent to sell to competitors, at wholesale rates, any telecommunications services it sells to its customers at retail rates. Competitors could thereby resell to customers at retail prices the services they purchase from the incumbent at wholesale.[19]

In 1996, GCI purchased the cable companies serving Anchorage, Fairbanks, and Juneau, thus acquiring a network passing 76 percent of the state's households. It also completed the first phase its Metropolitan Area Network in Anchorage, a fiber optic network that would provide high-speed data connectivity and serve as the distribution backbone for local telephone service. GCI initiated facilities-based local phone service in Anchorage in 1997. The next year, GCI completed the first phase of a similar network in Fairbanks.

However, the road to local competition in Alaska required more than technology. In the interest of promoting universal service, the Telecommunications Act exempted rural telephone companies from the duty to compete. In April 1997, GCI requested interconnection with three rural telephone companies, PTI Communications of Alaska, Inc., Telephone Utilities of Alaska, Inc., and Telephone Utilities of the Northland, Inc., all of which had been purchased by ACS (see Chapter 14). The APUC held public hearings in December 1997 to determine whether to terminate ACS's rural exemptions, and issued an order in January 1998 continuing these exemptions, stating that the evidence in the record did not support an affirmative finding that ACS would not suffer an undue economic burden if the exemptions were terminated, and that support mechanisms had not yet been reformed to accommodate competition in local service.

GCI appealed the APUC's decision to the Alaska Superior Court, which found that the APUC had erroneously placed the burden of proof on GCI, and remanded the case to the APUC for another hearing. The APUC held a second hearing in June of 1999, and then granted GCI's petition to terminate ACS's rural exemptions, finding that adequate mechanisms were in place to preserve universal service.

In 1999, the Alaska Legislature replaced the APUC with the Regulatory Commission of Alaska (RCA), giving it broad authority to regulate utilities and pipeline carriers throughout the state. The new agency with the old acronym and the same staff but different commissioners came into existence on July 1, 1999. Apparently there was no institutional memory of the regulatory battles of the 1970s when RCA was the monopoly telecommunications carrier.

ACS petitioned the APUC's successor, the RCA, for review of the decision to terminate ACS's rural exemptions, but after reviewing the record, the RCA affirmed the APUC's termination of the rural exemptions. ACS appealed to the superior court, which affirmed the RCA's decision; ACS then appealed to the Alaska Supreme Court.

In 2001, the U.S. Supreme Court let stand a new interpretation of federal regulations for phone competition, placing the burden of proof on the competitor to show that competitive entry would not jeopardize universal and affordable service.[20] The Alaska Supreme Court therefore found in 2003 that the RCA had erred in allocating the burden of proof to ACS, reversed the lower court decision, and remanded the case to the RCA for additional proceedings with GCI shouldering the burden of proof.[21]

Meanwhile, following the lower court decision, GCI began offering competitive local telephone service in Fairbanks in June 2001 and in Juneau the following March. By 2003, GCI had approximately 10,350 access lines in Fairbanks, or a 22 percent market share, and 6,500 access lines in Juneau, or a 30 percent market share.[22]

In 2004, GCI reached a settlement with ACS concerning local competition after Governor Frank Murkowski's office had facilitated the negotiations. The terms of the settlement included:

- ACS relinquishes all claims to exemptions from full local telephone competition in Fairbanks and Juneau.
- New rates for unbundled loops would be introduced in Fairbanks and Juneau beginning January 1, 2005.
- Interconnection agreements between ACS and GCI for Fairbanks and Juneau for Fairbanks and Juneau would be extended until January 1, 2008.[23]

GCI's next target was the Matanuska Telephone Association (MTA). GCI already offered cable television and Internet services in the MTA area, as well as providing an option for long-distance telephone service. The RCA had ruled that GCI could compete in MTA's service area by building its own network (facilities-based competition), but GCI wanted to lease components of MTA's existing network. MTA's regulatory director stated: "The bottom line, if they are allowed to use our network at this deep-discounted rate, we will be out of business in five years. . . . We are looking for more long-term, sustainable competition."[24]

In February 2007, GCI launched competitive local telephone service over its own network in Eagle River, followed by Chugiak and Peters Creek, and all communities throughout the Matanuska Susitna Borough by the end of the year.[24] It also began to provide local service in Ketchikan and Sitka in southeast Alaska and in Soldotna and Kenai on the Kenai Peninsula.

Chapter 16

Distance Education and eLearning: From Satellites to the Internet

In the industrial age, children went to school. In the information age, school will come to children.

Douglas Barry, Alaska Department of Education, 1988[1]

Satellite-Delivered Distance Education

In the 1980s, satellites became a major vehicle for delivering courses for academic credit as well as for continuing education to a wide variety of learners across the United States. Some universities, such as Stanford, Colorado State, and Southern Methodist, were already offering courses, primarily in engineering, to industry sites in their regions via over-the-air broadcasting on dedicated frequencies or microwave transmissions. Via satellite, they could reach workplaces nationwide. In 1984, the National Technological University (NTU) was established as a virtual university, with master's level engineering courses uplinked from member institutions. By 1990, it had 28 uplink sites and 285 downlink sites at more than 60 organizations with about 5200 students.[2]

Businesses and professional associations had also discovered the utility of distributing courses from universities or their own training centers to work sites and satellite conference facilities around the country. Technological companies had found that they could "teach several hundred people for the cost of one tuition for technical programming."[3] The National University Teleconference Network (NUTN) formed a consortium of 204 colleges and universities "to bring professors to business." Corporations such as IBM, Hewlett Packard, and JC Penney operated their own satellite networks. Professional associations such as the American Hospital Association, American Law Network, and American Management Association offered continuing education to their members via satellite.

At the high school level, satellites also began to transmit advanced placement courses and other classes in elective subjects such as foreign languages to high schools that could not offer these subjects. This model was attractive for the small high schools in Alaska where teachers might cover several grades, and students could not take advanced science and math or other specialized courses. Traditional correspondence courses with materials sent by mail had been available, but there was growing interest in additional forms of delivery. An Alaska school superintendent responding to a distance education survey commented: "(Satellite) delivery would be very beneficial in our small high school program where only one or two teachers deliver all subjects."[4]

By the late 1980s, there were several satellite-based sources of distance education, primarily offering high school level content. One of the largest was TI-IN, a network based in Texas, that transmitted on three satellite channels, offering foreign languages, high school math, physics, astronomy, and social science courses to schools in 19 states. The Satellite Educational Resources Consortium (SERC), based in Columbia, South Carolina, was a partnership between the state departments of education and public television entities in 23 states that offered live, interactive high school credit courses via satellite and also provided a host of teacher in-service programs, student science seminars, and teacher workshops. Oklahoma State University offered high school German, math, and physics, with one-way video and two-way audio so that students could interact with instructors. These educational satellite networks received grants from the U.S. Department of Education's STAR Schools program, which was designed to encourage improved instruction in mathematics, science, foreign languages, and other subjects, and to serve underserved populations, including Native Americans.[5]

In the western U.S., the Educational Service District in Spokane, Washington, offered pre-calculus, Spanish, Japanese, and senior advanced English, also with one-way video and two-way audio. The Spokane school district emphasized the need for a classroom coordinator at the receiving school with multiple functions including motivating and supervising students, as well as administrative functions—preparing equipment and supplies, recording attendance and grades, and sending and receiving student papers tests and assignments.[6]

In Alaska, Hoonah tried out the service from Spokane in fall 2008. However, many challenges prevented most Alaska schools from taking advantage of satellite-delivered courses. In some cases, the satellite footprint did not cover Alaska, or most of Alaska. Satellite reception even where there was coverage remained problematic. The Alaska Department of Education recommended purchasing equipment that provided both C-band and Ku-band reception, also suggested community needs assessment. (Ku-band is a higher set of frequencies that had been introduced on some domestic satellites.) "Local government might be interested in contributing to the costs of the system if programs could be brought down to meet their needs."[7]

Differences in time zones were an added challenge for interactive courses; some TI-IN classes were scheduled at 4 am Alaska time. Also, the state had difficulty in obtaining rights or licenses to transmit some courses on the RATNET channel. And funding remained a problem in an era of low oil prices and state budget cuts. In 1988, the State Department of Education cautioned: "Adequate funding to purchase necessary equipment and train personnel is imperative. Recent budget cuts have made it impossible for local funds to be targeted for distance delivery courses."[8]

Alaska educators were also developing courses using new technologies. The Alaska Department of Education prepared a correspondence course on "The Alaska Studies Connection" using mixture of media including print, audio and video tapes, computer activities, audio conferencing, email, and telephone. The one-year course covered Alaska geography, natural resources, history, and current affairs.[9] The UAA School of Education developed distance education courses for teachers of learning-disabled Alaska Native students with components including mailed videotapes, audio conferencing, and computer-based exercises.[10]

The University of Alaska steadily increased its offering of distance education courses. UAA's "Livenet" system provided one-way video and two-way audio; UAS used "Livenet" to offer its MPA program to military sites throughout the state.[11] The public administration courses first went live in spring 1991 to Fort Wainwright near Fairbanks, Fort Greely at Delta Junction, Fort Richardson in Anchorage, the U.S. Coast Guard Station in Kodiak, and the Adak Naval Air Station in the Aleutians. The student in Adak said the fax machine was on 24 hours a day to send and receive coursework, and commented that with only one-way video, the professor "doesn't have the foggiest idea what I look like."[12]

By the early 1990s, the Alaska Department of Education was providing instructional TV services to a cooperative that included 54 of the state's 500 schools—using satellite, cable, and public TV stations to deliver video products that the teachers could record to use at their convenience, much as LearnAlaska had done a decade before. And some school districts were also exploring new distance education technologies. The Northwest Arctic School District based in Kotzebue experimented with an "electronic chalkboard" for a pre-algebra course taught from Anchorage. Kotzebue claimed that the "Electronic Chalkboard" had dramatically reduced the dropout and failure rate in its pre-algebra courses. Schools in the Bethel region formed the Distance Delivery Consortium (DDC) composed of the Lower Yukon, Lower Kuskokwim, and Yupiat school districts with monthly delivered satellite programs uplinked from KYUK in Bethel.

Curriculum Sharing Using Telecommunications

The distance education services followed what could be described as the "distant expert" model. An instructor at one location taught students or other participants at sites that did not have the requisite local teaching expertise, whether it was in high school physics, a foreign language, graduate engineering, or professional knowledge and skills for career advancement.

A second approach is the "curriculum sharing" model, typically linking a group of schools so that they can share instructional resources, with specialized teachers from each school offering courses to students at the other schools. In 1993, four high schools in the Matanuska-Susitna Valley were linked with the first educational optical fiber network in Alaska. Interactive video instruction from the various schools was offered over the network in a project supported by the MatSu Borough Schools and the Matanuska Telephone Cooperative (MTA).[13]

Meanwhile, the North Slope Borough School District (NSBSD) decided to blaze its own trail in distance education, primarily based on curriculum sharing among its schools. With funds generated by taxes on oil production on borough lands, the school district built a network linking ten schools in seven villages plus Barrow for both instruction and administration. In 1992, the NSBSD served 1700 students, of whom more that 86 percent were Inupiat, in the nation's largest school district, covering about 88,000 square miles (about the size of Minnesota). Village schools offered pre-kindergarten to 12th grade, with the smallest school being Point Lay with 42 students and the largest village school in Point Hope with 221 students. Barrow had both elementary and high schools.

In 1992, the NSBSD installed a fiber optic network linking the school district's buildings in Barrow. The NSBSD also implemented distance education delivery, leasing satellite capacity from Alascom to provide a private network for two-way real-time compressed video plus audio and a computer network linking village schools to each other and a production studio at the high school in Barrow. Martin Cary, then North Slope School Coordinator explained: "There may be ten high school students in a school. A teacher must be able to teach all subjects. It's very difficult to do. . . ." With the satellite network "we're now able to make the larger resources in our Barrow high school available to the smaller schools. We can put the best math teacher in the system online, for example. Last year we had an outstanding art teacher in Point Hope . . . who taught a class consisting of students all across the North Slope villages."[14]

In the morning, the school district offered courses such as algebra, geometry, advanced math, art, Alaska studies, Native land claims, health, personal finance, and writing. From mid-afternoon on, the system was available for individual student consultation, teacher and teacher-aide training, and consultations between faculty: "The North Slope Borough's 1700 school age children . . . don't simply participate in two-way video lessons and classes. Through the district's satellite-backed wide area network, they take tests, deliver their homework for review and evaluation, question and communicate with their teachers and fellow students, online and in real time."[15]

The video conferencing network was also available for activities such as nonprofit board meetings, budget hearings, elders conferences, "Meet the Candidates" sessions, and Kivgiq (the Messenger Feast), a midwinter festival in Barrow with participants invited from across the North to celebrate the successful whaling season. High school students operated the distance learning control center at the high school, learning video production skills.[16] Some students went on to media carriers. Student Telaina Kempf commented about her high school video production experience: "Once you learn and get good at it, you crave it. . . . You don't want to quit."[17]

A wide area network (WAN) connected everyone in the district for administrative and educational purposes, providing access to electronic mail, computer applications, and data processing services districtwide. The WAN was used for accounting, payroll, and student records.[18] A student network was used for exchanging files among students, sending out homework, and receiving tests.[19] As Cary pointed out: "NSBSD students don't simply participate in two-way video lessons and classes. Through the district's satellite-backed wide area network, they take tests, deliver their home work for review and evaluation, question and communicate with their teachers and fellow students, online and in real time."[20]

The North Slope model attracted interest from around the world: "School systems in remote areas of the world are viewing the North Slope Borough School District (NSBSD) as a model to emulate. School officials as far away as Australia have journeyed to Alaska to see how this district conducts its operations."[21] They were fascinated not only with the technology but also the remoteness, and the Inupiat culture in Barrow. As one visitor wrote: "Hanging in the front yards of their modest homes you could see everything from the catch of the day to racks of seal meat drying in the sun, to reindeer and polar bear hides stretched on rooftops"[22]

By 1994, the borough had decided to expand the network to provide computer communications to tie all the North Slope communities together in a network called AuroraNet. The mayor was concerned about the decline in oil revenues for the borough and wanted to make borough operations as efficient as possible: "One of the ways we are planning to do this is by use of the AuroraNet system being developed through our new Network Support Services Division. This system of satellite and computer communications will reach all villages and tie all divisions of the borough into a tighter, more efficient operation. Not only that, but some of these services, such as local access to the Internet, will also be available on a limited basis to the general population of the North Slope."[23]

AuroraNet was designed to allow employees all over the North Slope to feel an integral part of the North Slope workforce. Borough-wide email was introduced, as well as centralized data storage and backup, computer virus protection, and technical support for users anywhere on the system. A centralized information server provided access to job vacancies, borough staff information, and general information. The borough expected that the network would save time and money through standardization of applications and a standard interface, training for staff at their work sites, centralized technical support, and consolidated software control and licensing.[24]

Information Superhighways for Alaska?

Several officials from Barrow attended a technology conference in April 1994, where a highlight was a video appearance by Vice President Al Gore, who was promoting the concept of an "Information Superhighway" that would provide interactive connectivity to all Americans as part of a National Information Infrastructure (NII). The *Anchorage Daily News* commented: " No roads lead to Barrow, but distant talk of new computer-assisted communications has people here wondering how long it will be before this town on the shores of the Arctic Ocean is the northernmost terminus of the 'information superhighway.'"[25] As the reporter pointed out: "In some ways, the North Slope is ahead of the game. The school district already delivers regular classes to smaller village schools through a two-way video system." The North Slope was already anticipating how the Internet and broadband could deliver educational materials to schools and students and enable sharing of information among students as well as communities. The borough's planners also realized that connecting employees and centralizing some administrative functions could save both time and money.

Some thought that the information highway might quickly reach the North Slope: "Suddenly, North Slope communities that had been relatively isolated might become electronic neighbors with urban America through video phone calls, interactive shopping, and virtual reality games. . . ." But Martin Cary, by then Manager of Network Support Systems for the North Slope Borough and later to join GCI, concluded: "I think, realistically, it going to be quite some time before all of those services are delivered to us."[26]

However, several policies designed to help create a national "Information Superhighway" were included in the Telecommunications Act of 1996 (see also Chapter 13). One that was particularly significant for Alaska was a subsidy to connect schools and libraries to the Internet, known as the E-rate.

The E-Rate Program

In 1993, Reed Hundt, the newly appointed chairman of the FCC in the Clinton administration, remarked: "There are thousands of buildings in this country with millions of people in them who have no telephones, no cable television and no reasonable prospect of broadband services. They're called schools."[27] One of the goals of the Telecommunications Act was to provide connectivity for those schools.

The Telecommunications Act of 1996 expanded the definition of universal service to include schools, libraries, and rural health care facilities, and access to "advanced services." The goal was to provide opportunities for students and community residents to take advantage of these "advanced services" even if they were not yet available in their homes, that is, to help to bridge what became called the "digital divide." West Virginia Senator Jay Rockefeller explained the strategy to get these provisions in the Act: "The telecommunications companies wanted more competition and the ability to expand. In exchange, we insisted on a strong, continued commitment by the telecommunications companies to 'preserve and advance' universal service including access to advanced telecommunication services for schools, rural health care providers and libraries."[28]

The E-rate (short for "education rate") was created by Section 254(h) of the Telecommunications Act of 1996 to provide discounts on a wide variety of telecommunications, Internet access, and internal connections products and services. All public and private nonprofit elementary and secondary schools are eligible (except those with an endowment of more than $50 million). Libraries are also eligible, subject to conditions that they meet the definition in the Library Services and Technology Act and have a budget completely separate from a K–16 school.

The Federal Communications Commission (FCC) sets the overall policy for the program, which is administered by a nonprofit entity, the Universal Services Administrative Company (USAC). Funds come from telecommunications carriers, which generally pass through these costs to customers through itemized charges on their telephone bills. USAC makes payments from this central fund to support the Schools and Libraries program and Rural Health Care program as well other Universal Service programs including Low-Income to subsidize subscribers of local telephone service and High Cost to subsidize carriers in expensive-to-serve regions (because of difficult terrain and/or small or scattered populations).

Schools may apply for all "commercially available telecommunications services" ranging from basic telephone services to T-1 (1.544 mbps) and wireless connections, Internet access including email services, and internal networking equipment. Approved costs are billed directly to USAC, up to the limit of the subsidy. Schools and libraries are responsible for the remainder and must demonstrate that they can cover their portion of the costs.

An important element of the E-rate program is that it is designed to be incentive-based. Subsidies are not awarded directly to the communications carrier but to the user; it is the school or library which is eligible to receive the discount. An Alaska telecommunications official noted that the USF created a competitive environment in Alaska that is vendor neutral, puts the power of choice in the hands of the consumer, and offers a subsidy program that attracts long-term capital investment.[29] (In contrast, other countries generally subsidize the carrier directly to install facilities for schools or to provide services at a reduced price. Their approach creates no incentives for new entrants, nor for keeping construction and operating costs down. And quality of service may also suffer if the providers perceive serving schools as an onerous requirement rather than an opportunity.)

The applicable discount rate is based on a school's economic need and whether it is located in an urban or rural area. The proxy for economic need is the percentage of students who are eligible for free or reduced-priced lunches under the National School Lunch Program. The libraries' discount rate is based on the school district or districts in which they are located. Support for telecommunications services and Internet access is provided to all eligible applicants regardless of their level of need. In 2014, most rural Alaska schools and libraries were eligible for an 84 percent discount.

To apply for the subsidy, each school district must first prepare a technology plan stating how it will use, manage, and pay for the facilities being requested. Early on, the state of Alaska set up an office to assist schools with their applications. Della Matthis, the state's first coordinator for school and library E-rate activities, said she had never seen a federal program get to the classroom level so quickly. In 1999, she cited examples of how the E-rate had enabled small school districts such as Kuspuk in the Kuskokwim Delta and the Eagle Public Library, 66 miles up a gravel road just below the Arctic Circle, to get online.[30] In 2006, she commented on the importance of upgrading to broadband for Alaska's schools:

School with satellite antenna for Internet access in Manley Hot Springs.
(Heather Hudson)

In education, broadband access for schools is important both to enable multiple users to be online, and to allow for data-rich applications such as multimedia web access and video conferencing. An Alaskan analysis of school bandwidth requirements states: "Dial-up connectivity does not allow for efficient data flow and usually will not allow such services as e-mail for group use. Normally, a school with dial-up will only transmit information, not being able to rely on downloading or browsing." It notes that "Less than T-1 connectivity allows Internet use for data transfer, web searches, e-mail and web posting. Under normal circumstances, information flows at speeds allowing for group use, but may be overwhelmed. Video services can be used with some loss of picture and sound quality, but usually will require that other traffic, such as Internet use, be shut down."[31]

Alaska's E-Rate Success

As described in earlier chapters, Alaska has been committed to using telecommunications for rural development since the early 1970s when experiments using NASA satellites first demonstrated that reliable communications could improve rural health care and enhance rural education. When the Internet arrived, Alaskans were also quick to see its advantages for an isolated population.

By 1999, Alaska ranked first among the 50 states for Online Population and first for Technology in Schools (a weighted measure of the percentage of classrooms wired for the Internet, teachers with technology training, and schools with more than 50 percent of teachers having schoolbased email accounts); Alaska also ranked third in Digital Government, a measure of the utilization of digital technologies in state governments.[32]

The Alaska State Library and the University of Alaska libraries also developed SLED—the Statewide Library Electronic Doorway. SLED offered databases from various vendors, and later a portal of sources of information on Alaska and links to Internet sites. The state's director of libraries, archives, and museums discovered how small the world could be when a patron at a library asked for information about Russian monasteries, a task that would have taken researchers weeks to complete before the libraries came online—to request information from distant libraries by mail, photocopy relevant materials, and then mail them to the patron. She commented in 1999: "With the Internet, the citizen in rural Alaska can be as tied in to what's happening in the world as a person in Anchorage or New York or Seattle or any place else."[33]

It is perhaps not surprising, then, that Alaska took full advantage of the E-rate program. In fact, Alaska has been highly successful in obtaining E-rate support. From the inception of the program in 1998 through 2013, Alaska received more than $295 million in E-rate funds. In 2013, Alaska received almost 2 percent of the total funds available with $42.2 million, although its population was only 0.23 percent of the U.S. total.[34] With this amount, Alaska ranked 16th among the states and the District of Columbia in total E-rate funding, while 47th in population.

Behind the Scenes: The Mentor and Carrier Involvement

But *how* did Alaska manage to secure such significant funding? As noted earlier, the E-rate program requires an understanding of telecommunications and information technologies, has complex application procedures, and requires that each school complete a technology plan before being certified as eligible for the discounts. Many schools and school districts far larger than those in Alaska have found these requirements daunting. One explanation for Alaska's success is the assignment of a state librarian as state E-rate coordinator to help the schools and libraries prepare applications and navigate the E-rate labyrinth. She provides advice, explains the requirements, and assists in completing the forms and tracking their progress.[35] Initially, this was a half-time position, although it soon became much more than a half-time job. Today, Alaska has a full-time E-rate coordinator in the Alaska State Library.

Another unusual feature in Alaska is the participation of carriers in helping the schools to acquire funding. When the E-rate program was established,
GCI set up a project office and website for schools, and offered a package of
services including connectivity via leased line or VSAT, an onsite school server,
and services including email, web access, and technical support.[36] The value-
added "one stop shop" GCI offered for schools has also turned into a new
business opportunity. Not only does GCI serve several schools in Alaska that
have subscribed to this SchoolAccess service, but it has rolled out the service for
rural New Mexico and Arizona schools, although GCI is not a carrier in their
states.[37] Other carriers and local companies serving Alaska have also partnered
with schools and libraries to obtain discounted Internet access, including ACS,
MTA, AT&T, and smaller Alaska providers.[38] The E-rate initiative has proved
to be a win-win opportunity for the companies as well as the schools, as the
revenue from the "anchor tenants" provides a source of funding to serve small,
isolated communities.

The Alaska Waiver

After receiving the E-rate subsidy, some Alaska school districts began to ask
whether they could make their Internet access available to other sites in the
community because no other Internet service was available. However, FCC
rules require that any school, school district, or library buying discounted telecommunications services must certify that those services are to be used only
for educational purposes. A major concern of the FCC has been to ensure that
subsidized E-rate services do not compete with commercial services. Alaska
was able to demonstrate that no local commercial ISPs, or ISPs offering local
services, were available in these communities.

To address this matter, in January 2001, the State of Alaska filed a petition
with the FCC for a waiver of the rule requiring the "educational purposes only"
requirement so that, under certain circumstances, schools or libraries could
make the telecommunications service used to access the Internet available to be
used by an AfterSchool/Internet Service Provider (AS/ISP) which would then
provide local or toll-free Internet access to the community. In December 2001,
the FCC granted the State's petition, subject to several conditions, including:

- There is no local or toll-free Internet access available in the community or
  communities to be served by the use of the school district's telecommunications service
- The school (or library) does not purchase more discounted telecommunications services for Internet access than it needs for educational use

- The AS/ISPs can offer service only during hours when the school (or library) is closed
- The school districts (or libraries) must make the telecommunications services to be shared available to AS/ISPs in a competitively neutral manner[39]

This waiver provided an opportunity for Alaskan residents of many remote rural communities to access the Internet with a local (or toll-free) phone call for the first time. The E-rate coordinator's "other related duties" included drafting a plain language guide for schools and libraries about how to proceed.[40] In fact, the philosophy of the Alaska waiver from 2001 was reflected in the National Broadband Plan of 2010, which encouraged schools to provide broadband access to their communities. In 2011, the FCC's Community Use Order for E-rate recipients allowed communities to use school bandwidth after school hours under specified program rules, formally approving the approach that Alaska schools had advocated a decade earlier.

Distance Learning Today

Today, telecommunications, and particularly broadband, provide rural Alaskans with new educational opportunities in the classroom, at libraries and community centers, and at home. Teachers can take their classes on virtual field trips and conduct virtual science labs. Students can learn more about Alaska with materials from Glacier Bay National Park and the Alaska Sea Life Center. And students can produce their own multimedia presentations to share with students across the country and around the world. For example, KC3 Kids Creating Community Content challenges high school and middle school students "to research, develop, and present community content through various technologies and connect with classes world-wide through videoconferencing."[41] In 2013, the students of Nuiqsut Trapper School on the North Slope won first place in their division with a presentation on "Hunting the Bowhead Whale."[42]

Numerous schools now provide laptops for use in school, and in many cases, students with permission from parents may take the laptops home. The laptops can be used not only for homework assignments but by other family members, as a means of expanding digital literacy. In households with Internet connections, residents may take online courses from the University of Alaska or other sites. Mobile devices are becoming increasingly popular. Some schools now supply tablets, and an increasing number of students also have ebook readers and smartphones.

Some Alaska educators and Native organizations are developing content for these new devices to help teach and preserve indigenous languages. NANA and the North Slope Borough have collaborated with Rosetta Stone to produce CDs for instruction in Inupiaq dialects spoken by local residents. NANA provided free CD sets to its shareholders.[43] The Kashunamiut School District in southwest Alaska is translating a popular children's interactive book series in cultural diversity into Cup'ik, a regional dialect of Yup'ik. Students will be able to hear English and Cup'ik narrations and listen to pronunciations of names of illustrated objects. Students will be able to download the stories to tablets and ebook readers. Robert Whicker, director of the Alaska Association of School Boards Consortium for Digital Learning, stated: "With that much exposure on the Web, a rare language like Cup'ik—now spoken by only a few people in a remote Alaska village—has a chance to go mainstream."[44] The Cook Inlet Tribal Council (CITC) is funding a venture to develop educational video games with indigenous content.[45]

In 2014, the Alaska Department of Education & Early Development awarded grants to four school districts to strengthen instruction through digital teaching and learning both within their districts and across district boundaries through partnership with other Alaska school districts. The Digital Teaching Initiative is designed "to deliver high-quality interactive distance courses to middle and high school students; increase student access to a diverse array of courses; empower teachers to reach beyond their own classrooms; train teachers; and expand school districts' infrastructure, technology, and staffing." Alaska Education Commissioner Mike Hanley stated: "This is an opportunity to strengthen 21st century best practices in Alaska and provide greater access to high-quality teachers and content for more of our Alaskan students."[46]

Community residents may also go online at public libraries to find educational materials and take continuing education courses. In the fishing community of Craig in southeast Alaska, people come in from their fishing boats to download books from the Alaska State Library onto ereaders. In the tiny community of Lake Michumina in central Alaska, access to online equipment manuals is especially popular. Local librarian Shawna Hytry explained: "If your chainsaw is broken you can't cut your firewood to heat your house for the winter. . . . If your snow machine or four-wheeler is broken you'll have to haul water by foot with a wagon."[47] Libraries with videoconferencing facilities and connectivity through the Online with Libraries (OWL) program have also made it possible for residents to complete required training for certification in food handling (needed for local food service jobs) and training for parents of children with medical disabilities, and to participate in presentations on Medicare, cancer prevention, museum exhibits from Anchorage and Barrow, and numerous other topics.[48]

However, many rural communities still lack broadband connectivity, as discussed in Chapter 18.

Chapter 17

Telemedicine in Alaska

We had no clinic. We went from house to house taking care of the sick . . . Our tools consisted of a thermometer, a stethoscope, and a blood pressure cuff . . . We had no phones, no radios, but used the school's radio to report our patients. There was no nonsense about confidentiality.

Community health aide Paula Ayunerak[1]

Telemedicine and Telehealth

Applications of telecommunications in support of health care are referred to as "telemedicine," although some practitioners and researchers prefer to use that term for teleconsultations, and the term "telehealth" to refer to applications for medical education and administration, or more generally "to encompass a broader definition of remote healthcare that does not always involve clinical services."[2]

Information and communication technologies can be used to support a range of health services including the following:

- Emergencies: to summon immediate medical assistance; to communicate with emergency vehicles and staff
- Consultation: typically between primary health care providers and district level physicians, or between district physicians and specialists
- Remote diagnosis: such as transmission of medical data and images, interpretation of data by distant specialists
- Patient monitoring: transmission of patient data from home or rural clinic, often coupled with follow-up through local medical staff

- Training and continuing education: of health care workers, paraprofessionals, physicians, etc.
- Public health education: of target populations including expectant mothers, mothers of young children, groups susceptible to contagious diseases, etc.
- Administration: ordering and delivery of medications and supplies; coordination of logistics such as field visits of medical staff; accessing and updating of patient medical records; transmission of billing data, etc.
- Data collection: collection of public health information such as epidemiological data on outbreaks of diseases
- Research and information sharing: such as access to medical data bases and libraries and consultation with distant experts and peers

Alaska telemedicine and telehealth services today include all of these functions, and Alaskan health providers have been pioneers in implementing many of them, particularly to serve the residents of remote communities.

Alaska ranks 48th out of the 50 states in the ratio of doctors to residents. Some 65 percent of physicians are located in Anchorage, leaving 59 percent of the state's residents in underserved areas. The majority of physicians are located in Anchorage. There are also shortages in many medical specialties, as 49 percent of all physicians in Alaska are primary care physicians, compared to a U.S. average of 28 percent. At the village level, health care is delivered by community health aides who are usually local residents (primarily women) who receive basic medical training and provide first line care for the villagers. Health aides are supervised by medical staff at regional hospitals. More than 575 health aides in 200 villages provide nearly one-half million patient encounters per year.[3]

Today, Native health care is delivered through Alaska Native health corporations, which in turn receive funding from the Public Health Service, the federal agency responsible for providing health services to Native Americans. Community health aides are still the frontline providers of village health care, but their capabilities are supported and extended through telemedicine facilities, including communication of voice, still images such as photographs and x-rays, electronic medical records, and video. The networks use terrestrial links such as microwave to reach villages where available, and satellite communications to reach other village clinics and offshore facilities (such as Diomede, St. Lawrence Island, the Pribilof Islands, and the Aleutian chain).

The Beginning of Telemedicine in Alaska

In the early days of the Alaska Territory, missionaries provided some medical care in larger villages, but the only medical care along the coast came from annual visits of a Coast Guard cutter. In 1931, health care and education for Alaska Natives became the responsibility of the Bureau of Indian Affairs, which built hospitals in Barrow, Bethel, Kanakanak, Kotzebue, Mountain Village, Tanana, and Unalaska. The hospitals were equipped with two-way HF radios, and teachers sent to Native villages were equipped with a medical kit and HF radio.[4]

The health aide system was established in 1954, when a U.S. government report stated that "the indigenous peoples of Native Alaska are the victims of sickness, crippling conditions, and premature death in a degree exceeded in few parts of the world."[5] The program began with training of sanitation aides who returned to their villages to instruct others in maintaining safe drinking water and proper trash disposal. In 1955, the U.S. Public Health Service (PHS) assumed health care responsibility for Alaska Natives. In 1956, a physician based in Bethel made the case for expanding the program to train community aides as firstline health workers: "It is not a question of whether the villagers shall be treated by completely qualified medical personnel or persons with less than full qualifications, but a question of whether they shall be treated by persons with limited qualifications or go untreated altogether."[6]

An agreement between the Department of the Interior and the PHS to share frequencies on HF radios allowed schools, village clinics, and Native hospitals to communicate with each other. In 1964, the Alaska Indian Health Service (IHS) began to train Community Health Aides/Practitioners (CHA/ Ps) who worked out of their homes or clinics in some village schools.[7] They were allowed to use the school radio to contact the regional hospital. Health aide Paula Ayunerak described those early days: "We had no clinic. We went from house to house taking care of the sick . . . Our tools consisted of a thermometer, a stethoscope, and a blood pressure cuff . . . We had no phones, no radios, but used the school's [two-way] radio to report our patients. There was no nonsense about confidentiality."[8]

Described as "a marriage between necessity and innovation," the Community Health Aide Program (CHAP) continued to expand. In 1968, the IHS received funds to hire 185 health aides for 157 communities to be supported by daily "doctor calls" from regional hospitals over HF radio. However, HF radio was often unreliable because of atmospheric disturbances and infrequent maintenance, and health aides were sometimes reluctant to bother teachers to use the radio.[9]

In 1972, villages in central Alaska began to communicate with the Tanana regional hospital and the Anchorage Native Medical Center (ANMC) using a single channel on NASA's ATS-1 satellite. As discussed in Chapter 7, the experiment showed that reliable communications could indeed improve patient care and save time and sometimes even lives, and that health aides also learned from each other's experiences heard in consultations over the shared audio chan-nel.[10] In 1975, the state of Alaska authorized an expenditure of $5 million for the purchase of 100 small satellite earth stations for villages, providing reliable voice communication through RCA's first commercial satellite. Each village had a public payphone and a special phone at the clinic with a dedicated audio channel for health communications linking the villages with their regional hospital and with ANMC.

Expansion of Telemedicine

Once the earth stations were installed, dial telephone services were added in many villages, initially from a few public phones and eventually with local telephone exchanges. Eventually, the IHS dropped it shared dedicated audio conferencing circuits because of cost, and relied on dialed long-distance con-nections between the clinics and regional hospitals. The IHS soon added facsimile machines in the villages so that health aides could fax the patient encounter forms with patient data to the regional hospitals, where the doctors could review them and provide a diagnosis and treatment plan to discuss with the health aides by telephone.

The North Slope Borough installed slowscan television in its villages and Barrow Hospital as well as at ANMC in the late 1980s; the low-bandwidth video was useful for some orthopedic consultations and instruction but was eventually discontinued. In the mid-1990s, Norton Sound Health Corporation and AT&T Alascom experimented with the use of a commercial videophone for consultations, demonstrating the potential of store-and-forward video and off-the-shelf equipment.[11]

The Alaska Telemedicine Project, established in 1994, developed and tested prototypes adopted in later telemedicine programs. Founding members were the University of Alaska Anchorage, the Anchorage Providence Health Systems, and AT&T Alascom. Other members of the consortium included Native health corporations, other health care providers, government agencies, and other tele-communications carriers and technology companies, with assistance from Tripler Army Base in Hawaii. Providence Hospital provided teleradiology ser-vices through links between its medical center in Anchorage and private health care providers in the towns of Cordova, Dutch Harbor, Kodiak, Valdez, Homer and Seward. Additional services included teledermatology, dietetics and nutri-tion, ENT (ear, nose and throat) consults, and other services.[12] Difficulties encountered included limited availability of radiologists and inadequate tech-nical support for the public sector locations.[13]

In 1996, a consortium of health care providers, the University of Alaska, and telecommunications companies received funding from the National Library of Medicine for the Alaska Telemedicine Testbed Project (ATTP) to link clinics and hospitals in western Alaska with store-and-forward email. This was the same year the Telecommunications and Information Council (TIC), chaired by Lieutenant Governor Fran Ulmer, completed several reports including a "Task Force Report on Telemedicine," with findings and recommendations included in the Telecommunications and Information Plan adopted in December 1996 (see Chapter 13). Also, the Telecommunications Act of 1996 included provisions that "a telecommunications carrier shall . . . provide telecommunications services which are necessary for the provision of health care services in a state . . . at rates that are reasonably comparable to rates charged for similar services in urban areas of that state."[14] (See the following sections.)

The ATTP was designed to coordinate the replication of scaled, tested approaches to telemedicine and health care informatics to villages in rural Alaska served by several Native health corporations: Maniilaq Association based in Kotzebue, Norton Sound Health Corporation based in Nome, YukonKuskokwim Health Corporation based in Bethel, and Bristol Bay Area Health Corporation based in Dillingham. The ATTP developed, deployed, and evaluated the use of narrow bandwidth technology for otolaryngology and dermatology.[15]

Health clinic in the village of Wales, Alaska. (Heather Hudson)

The evaluation found that both patients and providers perceived telemedicine to be as good as or better than transportation-based models of health care delivery, and that these applications did not increase patient length of stay for health care providers in rural Alaska.[16] The ATTP also completed a cost/benefit study to analyze the benefits of telemedicine and telehealth and identify the cost per transaction of each telemedicine encounter. The cost analyses showed that the average cost per encounter was approximately $40, and that costs were falling.[17]

Meanwhile, based on initial successful consultations using the ATTP, the original plans were expanded and carried forward to Senator Ted Stevens, with a request to include all federal healthcare facilities in the state. By the end of 1998, funding was assured for AFHCAN—the Alaska Federal Health Care Access Network.

The AFHCAN Network

> Imagine turning on your computer, clicking onto e-mail, and finding a message from a Community Health Aide . . . at the St. George clinic [in the Pribilof Islands]. The e-mail message has an attachment that contains heart sounds, which you can actually listen to through your computer or it could contain a picture of a damaged finger or an inflamed eardrum. As a physician based in Anchorage, you would be able to help the CHA down at St. George quickly assess the health problems of a patient thousands of miles away.[18]

Today AFHCAN provides those services and more for all federally funded health services in the state, some 248 sites including military installations, Alaska Native health facilities, regional hospitals, small village clinics, and state of Alaska public health nursing stations, affecting more than 212,000 beneficiaries, the majority of whom are in rural Alaska Native villages.[19] AFHCAN provides telemedicine facilities through a partnership with the Alaska Native Tribal Health Consortium (ANTHC), the Department of Veterans Affairs, Department of Defense, U.S. Coast Guard, and the Alaska Department of Public Health Nurses In 2011, AFHCAN served 37 Tribal organizations, six U.S. Army sites, three U.S. Air Force bases, and 27 state of Alaska Public Health Nursing locations. Some 44 partner organizations are involved.[20]

AFHCAN's goal is to improve access to health care for federal beneficiaries in Alaska through sustainable telehealth systems. Project planners took several steps to design the project to be sustainable. For example, they took particular care to understand the needs of the users (aides and physicians who would use the system) and customers (those who would pay for its ongoing operation such as the regional health corporations). They noted that 67 percent of the sites have community health aides and thus made sure that equipment and training were designed for the aides and facilities in village clinics.

Further, they designed the system to address priority medical problems. The clinical committee established for the project stated it should focus on primary care (i.e., treating people in village clinics and similar installations) rather than secondary care (at regional hospitals) or tertiary care (e.g., at the Alaska Native Medical Center). An example of equipment included to address the priority problem of otitis media (middle ear infection, which can cause deafness in children) is an electronic otoscope. They also designed the telemedicine system to be scalable, to adapt to expanded requirements, new applications, and more users.

Facilities for village health clinics include a specially designed telemedicine cart with a personal computer and peripherals including a digital camera, electronic otoscope (for ear infections), and ECG, as well as printers, scanners, routers, wireless networks, and customized furniture.[21] The project technical staff chose suitable off-the-shelf equipment wherever possible (such as a rugged and simple-to-use digital camera). Where standard equipment was not suitable, they worked with vendors to make modifications (such as on the equipment

Health aide with AFHCAN telemedicine cart in the Wales clinic. (Heather Hudson)

cart, which was designed to move easily within the clinic). In some cases, wireless networking was used to avoid attaching long cables to movable carts.

Thus, instead of relying only on verbal descriptions from health aides or sending x-rays by plane to Anchorage, doctors at regional hospitals can now use the AFHCAN network. One common application is for diagnosis of otitis media, a common middle ear infection among children that can cause deafness if not treated promptly. Health aides can use the electronic otoscope connected to the computer to transmit images of the ear canal. They can also take pictures of wounds, sprains, dermatological lesions, and so forth, using the digital camera, and transmit the photo as an attachment to an email message to the doctor. The equipment can also be used to send digitized x-rays. Initially, the technicians had to digitize the x-ray film: "In the past," stated the information manager of the Maniilaq health center in Kotzebue, in 2000, "there was a big delay in the process. There would be times when the bone would set before a diagnosis could be made. Now we digitize the film, and it's in Anchorage the same day."[22] Today, film is obsolete, and digital x-rays are transmitted over the network to be read by radiologists in Anchorage, elsewhere in the U.S., or overseas.

The availability of computers in village clinics also makes it possible to use electronic training and reference materials. The Community Health Aide Manual is now available on a CD. Training materials for operating the telemedicine equipment are online,[23] and the network can also be used for continuing education of the health aides and other health workers. For example, ANMC has developed online certification courses in Understanding Telehealth and the Role of the Telehealth Coordinator, Becoming a Certified Telehealth Coordinator, and Becoming a Certified Telehealth Program Manager.[24] Pictures from telemedicine consultations such as images of the inner ear can be included in a multimedia curriculum that could be used to educate a broader audience on ear disease.[25]

The Benefits of Rural Telemedicine

ANTHC and its partners at regional Native hospitals have collected data on several applications of telemedicine and their impact on patient care and savings in costs of health care delivery. By 2011, AFHCAN had an 11-year operational history with more than 125,000 cases. A review of these cases showed that telemedicine saved both time and money. One telemedicine application is consultation on rural cases by specialists at ANMC in Anchorage. The analysis showed that 20 percent of all specialty consultations are turned around in 60 minutes, 50 to 60 percent are turned around in the same day, and a total of 70 to 80 percent are turned around within 24 hours. Some 80 percent of all telemedicine consults prevented patient travel from villages to regional hospitals or regional hospitals to Anchorage, although one or two cases "caused" travel, as patients with conditions requiring treatment by physicians were identified that would otherwise have been missed.[26] These findings are remarkably similar to data from the ATS-1 satellite telemedicine experiment 40 years ago[27] (see Chapter 7).

Preventing patient travel results in significant cost savings because medical evacuations are very expensive, and even scheduled bush flights from villages to regional hospitals typically cost several hundred dollars. In addition, evacuated patients under 18 years of age need to be accompanied by an escort such as a parent or guardian. A review of teleconsults for Medicaid cases from 2003 to 2009 found that travel was avoided for 75 percent of cases, resulting in net savings to Medicaid of more than $2.8 million. These savings were dramatic; for every dollar spent by Medicaid on reimbursement, $10.54 was saved on travel costs.[28] In addition, the teleconsultations that resulted in avoiding travel prevented an estimated 4,777 lost days at work and 1,444 lost days at school for the patients in this study.[29]

Many consultations involve "store-and-forward" telemedicine in which images such as pictures of the inner ear taken by the health aide using an electronic otoscope and digital photographs of skin lesions and other visible symptoms can be forwarded for consultation by physicians at the regional hospitals or specialists in Anchorage. The authors of a study on otology using store-and-forward telemedicine point out that "store-and-forward telemedicine has several characteristics that are highly attractive to the busy consultant. It is asynchronous, so that the referring and consulting providers do not need to be available simultaneously. It does not require the scheduling and bandwidth associated with videoconferencing. Store-and-forward cases can be reviewed during the unavoidable 'downtimes' in a physician's schedule, enhancing physician productivity."[30]

Audiology consultations are particularly prevalent in rural Alaska because of the high incidence of middle ear infections (otitis media) that can impair hearing. The Audiology Department at the Norton Sound Health Corporation in Nome completed more than 3000 store-and-forward direct audiology consultations with the Ear, Nose, and Throat (ENT) Department on ANMC in Anchorage. Using 16 years of data, they conducted a retrospective analysis of ENT specialty clinic wait times on all new patient referrals made by the Norton Sound Health Corporation providers before (1992–2001) and after the initiation of telemedicine (2002–2007). Prior to the use of telemedicine, 47 percent of new patient referrals had to wait five months or longer for an in-person ENT appointment. By 2007, only 3 percent had to wait that long, with the rest able to receive specialist consultations over the AFHCAN network.[31] By eliminating the need for a face-to-face encounter with the specialist, the use of telemedicine reduced the number of referrals from the ENT specialty clinics by more than 95 percent. The cost savings were significant as round-trip airfare between Nome and surrounding villages cost $300 to $400 per patient. Annual cost savings alone in 2007 from the prevention of in-person appointments, reached $250,000. The authors also note that "the reduction of the ENT clinic backlog may have made room for other provider referrals."[32]

A separate study by researchers at ANMC reviewed 45 cases recommended for elective major ear surgery by telemedicine evaluation matched with 45 surgeries from the standard evaluation. It found that telemedicine evaluation accurately predicted the need for surgery 89 percent of the time, and in-person evaluation predicted the surgery 84 percent of the time. Thus, the authors concluded that store-and-forward telemedicine was as effective as in-person evaluation for planning elective major ear surgery.[33]

Universal Service Funds for Rural Health Services

Although the AFHCAN equipment was designed for locations with only a dial-up data line (or equivalent), most sites now have subsidized broadband connectivity. The Telecommunications Act of 1996 expanded the original goal of universal service of extending reasonably priced telephone services to rural and other underserved areas to include support for telecommunications services for schools, libraries, and rural health care providers. In section 254 of the act, Congress sought to provide rural health care providers "an affordable rate for the services necessary for telemedicine and the instruction relating to such services." Specifically, Congress directed telecommunications carriers "to provide telecommunications services which are necessary to health care provision in a State, including instruction relating to such services, to any public or nonprofit health care provider that serves persons who reside in rural areas of that State, at rates that are reasonably comparable to rates charged for similar services in urban areas of that State."[34]

The Federal Communications Commission (FCC) sets the overall policy for the program, which is administered by the Universal Services Administrative Company (USAC), the same organization that administers E-rate funds. The Rural Health Care Division of USAC distributes up to $400 million annually so that rural health care providers pay no more than their urban counterparts pay for the same or similar connectivity. As with the E-rate program, funds are collected from itemized surcharges on customer telephone bills.

To qualify for universal service support, a health care provider (HCP) must be a public or not-for-profit organization located in a rural area. The HCP may seek support for eligible services, which include mileage-related charges, various types of connectivity from leased telephone lines to broadband circuits, and one-time installation charges.[35] All telecommunications carriers may participate, including interexchange carriers (IXCs), wireless carriers, and competitive local exchange carriers (CLECs). Each eligible HCP requests bids for telecommunications services to be used for provision of health care through postings on the USAC website. Requests for bids must be posted on the USAC website for 28 days before the HCP can enter into an agreement to purchase services from a carrier. The HCP must consider all bids received and select the most cost-effective method to meets its health care communication needs.[36]

Alaska health care providers and telecommunications carriers were among the first to apply and have been very successful in receiving support from the Rural Health Care fund. Commitments for Alaska funding reached $36.8 million in 2013, and $342.8 million out of a total of $653 million allocated by the fund from 1998 to 2013. Alaska has received the highest total funding of any state, as well as the highest amount per capita.[37] The discount to the health care providers now amounts to some 97 percent, largely because of the growing disparity between rural prices for broadband connectivity and prices in Anchorage, where there is now considerable competition.

In 2006, the FCC authorized a two-year pilot program for construction of dedicated broadband networks to connect health providers in a state or region, and to support the cost of connecting these networks to Internet2 (a high-speed network for research and education). The pilot program had several objectives: "If successful, increasing broadband connectivity among health care providers at the national, state and local levels would also provide vital links for disaster preparedness and emergency response and would likely facilitate the President's goal of implementing electronic medical records nationwide."[38] In November 2007, the FCC announced allocation of more than $417 million for construction of statewide or regional broadband telehealth networks under the Health Care Pilot Program that provided funding for up to 85 percent of eligible costs for the construction or implementation of statewide and/or regional broadband networks.[39] By 2014, there were 50 active projects involving hundreds of health care providers (HCPs).[40]

Alaska received $10.4 million to finance the design and development of a statewide broadband network (the Alaska eHealth Network, or AeHN). Comprised primarily of rural health care practitioners, the AeHN is intended to "unify and increase the capacity of disparate healthcare networks throughout Alaska in order to connect with urban health centers and access services in the lower 48 states." The coordinated network is intended to facilitate exchange of critical health information between health providers as well as to support telemedicine services and applications. It will also be used to support a Health Information Exchange (HIE), so that medical records can be shared and accessed by multiple providers. (The AeHN also received funding from the state of Alaska to provide management services for State HIE services.) The AeHN involves more than 70 stakeholders including public sector and private sector health care providers and tribal entities.[41] The Alaska Native Tribal Health Consortium (ANTHC) was designated as interim project manager, to be replaced by a public-private partnership that will manage the AeHN in the long-term.

Lessons from Alaska's Telemedicine Experience[42]

Several lessons emerge from Alaska's telemedicine experience that are relevant not only to Alaska, but to other remote and indigenous regions:

- **Saving Time:** Telemedicine links between a community health aide and doctor at a regional hospital can enable patients to be seen quickly who would otherwise have to wait for a visiting doctor or for arrangements to be sent to a regional clinic.
- **Saving Money:** The AFHCAN system saves money as well as time. As noted, these savings have been dramatic; between 2003 and 2009, for every dollar spent by Medicaid on reimbursement, $10.54 was saved on travel costs.[43] Earlier analysis of a pilot network similar to AFHCAN found that an evacuation by plane can cost from $10,000 to $25,000. The package of computer, peripheral equipment, and training is estimated to cost $22,000, so that if it saved two evacuations, it would pay for itself.[44] However, it should also be noted that a telemedicine consultation may sometimes cause patient travel because a serious problem is identified that would have been missed by the health aide.
- **Improving Quality:** Catching patients early may prevent deterioration of patients' conditions; such consultations may also be valuable for preventive care.
- **Designing for Sustainability:** The AFHCAN planners selected or adapted equipment that is rugged to withstand field conditions such as power and temperature fluctuations and cramped space, and is easy to use (taking into consideration the likelihood of high staff turnover and need for retraining). They attempted to minimize capital and operating costs by choosing low cost but highly reliable equipment, and transmitting data primarily in store-and-forward mode.

Chapter 18

The Growth of Mobile and Broadband

*We can manage our operations and payroll online, and our
employees can stay in touch with their families.*

Seafood processor using broadband in southwest Alaska.[1]

The Growth of Mobile Communications

Wireless technologies have been important for communications in Alaska since
the completion of the WAMCATS telephone network in 1904, which included
the first commercial use of radio links, across Norton Sound. (See Chapter 2.)
Until the 1970s, much of rural communication was by high-frequency (HF)
radio, with the two-way radios often located in the village school or clinic. As
discussed in Chapter 7, HF radio could cover long distances but was notori-
ously unreliable. VHF radio, with shorter range but higher reliability, could be
used where there was line-of-sight visibility between antennas. For example,
beginning in 1970, RCA installed a single VHF mobile telephone in each of
several villages on the Seward Peninsula. From 1971 to 1975, RCA Alascom
installed the same wireless technology, Improved Mobile Telephone System
(IMTS), with one telephone channel shared among several villages. The village satellite telephone system was installed starting in 1975, bringing reliable
telephone service to communities that could not be reached economically with
terrestrial technologies.

Beginning in the 1970s, citizens band (CB) radios used by truckers in the
lower 48 became a popular means of communicating within villages and over
short distances on the land or rivers. Like the HF and VHF radios, everyone
tuned to a frequency could hear the conversation, so that while they lacked privacy, the CB radios provided an inexpensive way to share information. Some

villagers used them as a form of radio broadcasting to share the news. Cam Milroy reminisces:

> The CB radio was the way that people who lived in the country communicated. It was also a lifeline. In 1980 there were no cell phones. Every morning we would hear from our neighbors. . . . GOOD MORNING would blast from the speaker and we would hear a daily scripture read from Carrie Uhl. Her voice was easy to listen to and she would bring joy to everyone. . . . Carrie would bring the morning to life and everyone would follow up with conversation and current events. It was also a way for people to help each other if it was needed. . . . It was common for the boats that traveled the river to have a CB. But I was the first to mount one on a snowmachine. It was tucked away in the rear seat compartment and the antenna mounted to the rear bumper bar. . . . If I came upon a small herd of caribou I would pass the information along so other [families] could gather meat before they left the area. A few times the CB came in handy to call for help for people broken down out on the trail. In the evening after the chores had been finished, we would brew a big pot of tea and folks would chat on the CB and say goodnight.[2]

Early commercial mobile phone service first introduced in the 1960s had extremely limited capacity. For example, there were only 12 two-way channels for the Bell system and 11 for competitors in all of New York City.[3] Cellular telephony, an innovation developed at Bell Labs, increased capacity dramatically by reusing channels in multiple cells. "Cellphones" could transmit on relatively low power, and the signal was automatically switched from one frequency to another as they moved across the cells. However, the phones were the size of bricks and were typically installed in vehicles, and the service and prices were targeted primarily at businesses like construction and professionals such as lawyers who could improve productivity—and revenue—if they could communicate anywhere and anytime. In 1982, only .1 percent of registered U.S. motor vehicles were equipped with mobile phones. The business press commented: "Bell thinks the prices will deter all but the richest non-business customers."[4]

In July 1982, *Fortune* magazine announced a "gold rush,"[5] not in Alaska, but in Washington, DC, where the FCC had announced a competition for licenses for cellular telephony. The FCC created distinct service areas for cellular telephony and initially authorized two cellular operators in each of the top 30 markets, licensing one affiliated with the incumbent landline carrier and one other operator. Competitors included telephone companies such as AT&T and GTE, competing long-distance carriers such as MCI and Western Union, paging companies, data networks, some venture capitalists, and entrepreneurs from sectors including real estate and railroads.

The FCC planned to open up other local markets beginning in September 1982. By 1985, only Anchorage had been identified as an Alaskan cellular market by the FCC, ranking low on the list at number 187. The Alaska Consumer Advocacy Program noted: "The FCC's bidding process has been swamped with entries and certification is slow."[6] Some Alaska entrepreneurs including Mead Treadwell established Tundra Telephone, Inc. to apply for a cellular license for Anchorage, and eventually other Alaska locations. The market estimates at the time predicted penetration of about 1.5 percent of the population, which Tundra Telephone thought was very conservative given cellular's appeal. With foresight that was more than a decade ahead of its time, Treadwell and his colleagues thought that cellular had much greater potential in Alaska: "In Alaska, the eventual promise of cellular service is to serve developing areas with primary telephone service before wires are laid, and to hold onto that primary market service market in competition with traditional wireline carriers. Further, in Anchorage today, a certain percentage of demand will come for primary telephone service: if cellular can do all that a regular telephone can and move around, why have conventional wireline service at all?"[7] However, Tundra Telephone's investors did not succeed in starting a cellular network in Alaska. Early analog mobile services were provided by the Anchorage Telephone Utility (ATU) later bought by ACS, and by MTA using AMPS (Advanced Mobile Phone Service) developed by Bell Labs.

Alaska's market structure was largely determined by actions at the federal level. The Federal Communications Commission (FCC) and later the APUC declined to regulate cellular services, reasoning that the duopoly in each market was effectively competitive and that cellular was not a basic telephone service. In adopting this policy, the APUC followed a precedent it had set earlier for radio paging systems, which were also treated as unregulated utilities.[8] With the introduction of more competitive providers in 1994–1995, the FCC preemptively deregulated wireless services, and as a result the APUC no longer regulated these services.

In 1993, the FCC adopted rules for licensing of PCS or "personal communication services." The FCC began auctioning PCS licenses in mid-1994 for what would now be considered modern cellular or mobile services, enabling individuals to communicate "any place, any time," although with bandwidth sufficient just for voice and text. Announcing the auction, FCC Chairman Reed Hundt stated: "What you will see emerging . . . is the first chapter of the American telephone industry in the 21st Century; this may represent the greatest one-time private sector investment in any single industry in the nation's peacetime history."[9]

In Alaska, GCI was the high bidder, paying $1.65 million for one of two 30 MHz blocks of spectrum with statewide coverage. GCI constructed the network in 1996 and began service in 1997. The other successful bid went to American Portable Telecom and was eventually transferred to Alaska DigiTel. GCI purchased a noncontrolling 80 percent interest in DigiTel in 2006 and bought the rest of the company in 2008; it was assigned to Alaska Wireless Network (AWN) (see the following sections).[10]

Beginning in the early 2000s, Alaska carriers began to upgrade their mobile networks to carry data, moving from second generation 2G voice and text services to 2.5 G and later 3G, which could provide Internet access. As discussed in Chapter 14, in 1999, former Alascom executive Chuck Robinson formed ACS to purchase the Anchorage Telephone Utility (ATU). In the package with ATU came MACTel, a cellular phone service, as well as ATU long distance. By 2001, ACS was the largest cellular provider in Alaska, covering 85 percent of the population, including business, residential, and tourist corridors.[11] In 2004, ACS constructed a CDMA2000 network with coverage including Anchorage, Fairbanks, Juneau, and other communities throughout the state. Using the CDMA network, ACS launched wireless data service in Anchorage, Eagle River, and the Mat-Su valley with data rates ranging from 50 to 80 kbps to 156 kbps, and peak rates of 2.4 mbps in strong signal areas.

GCI continued to update its network, and in 2011 launched 4G services in Anchorage, with plans to expand to other major centers. In 2012, GCI and Alaska Communications (formerly ACS) agreed to pool their mobile assets including licenses, network equipment, and backhaul and transport facilities to form the Alaska Wireless Network (AWN). Alaska Communications was to own one-third and GCI to own two-thirds of AWN. Combining their network assets resulted in AWN having the most extensive coverage in Alaska, covering more than 95 percent of Alaska's population.[12]

The other major mobile carrier in Alaska was AT&T. From 2009 through 2011, AT&T invested more than $650 million to improve its wireless networks, building new cell sites, deploying fiber optic connections to cell sites, and adding more radio carriers to its cell sites to offer faster mobile broadband with 4G speeds.[13] Other local carriers such as MTA, Copper Valley, Bristol Bay, OTZ, and ASTAC also offered cellular service, with roaming outside their coverage areas on the larger statewide networks.

In 2013, Verizon entered the Alaska market, offering 4G data service in Anchorage, Fairbanks, North Pole, Juneau, and much of the Matanuska-Susitna Borough. In 2010, Verizon had filed an application with the FCC to acquire the Long Term Evolution (LTE)-suitable 700 MHz C Block spectrum license covering Alaska from Triad, a designated entity that had purchased the license during the FCC's 700MHz spectrum auction in 2008.[14] Verizon invested about $100

"Can you talk to my brother in the village?" A Native fan calls home about meeting famed dog musher George Attla at the Iditarod race start in Anchorage. (Heather Hudson)

million over two years to build the network and planned to offer voice service throughout the state by the end of 2014. Verizon partnered with the Matanuska Telephone Association (MTA) to expand the Mat-Su service, Copper Valley Telecom to expand into Cordova, Valdez, and the Prince William Sound, and Ketchikan Public Utilities to serve Ketchikan. These local providers would own and operate their own networks but sell Verizon products and services using Verizon spectrum. Some industry analysts saw the Alaska Communications/ GCI AWN venture as a preemptive strike against Verizon's impending entry into the Alaska mobile market.[15] In December 2014, GCI announced that it would purchase Alaska Communications' wireless operations for $300 million.

Mobile communications satellites were introduced in the late 1990s, relaying signals from handheld satellite phones to multiple satellites in lower orbits than the geostationary satellites that Alaska has relied upon for rural communications. Both Iridium and Globalstar launched constellations of satellites that could be used for mobile voice communications with brick-sized portable telephones (reminiscent of early cellular phones). These satellite phones have been used by hunters, mountain climbers, geologists, and others in remote areas of Alaska, but their limited bandwidth and relatively high prices have made them less attractive for rural Alaskan residents, although the current range of cellular coverage is limited, leaving many remote regions without mobile service.

Backbone Networks: Satellites and Submarine Fiber

Satellites

Broadcaster Augie Hiebert enthused in the 1970s: "Satellite technology was invented for Alaska."[16] The Comsat earth station installed near Talkeetna in 1970 brought television to Alaska, as well as a link for international telephony. The experiments on NASA Applied Technology Satellites (ATS) between 1971 and 1975 demonstrated how satellites could provide reliable communication *within* Alaska with particular benefits for rural health care delivery and education, including programming for preschoolers, students in village schools, and rural adults.

As discussed in earlier chapters, in 1975 the Alaska Legislature appropriated funds to the Office of the Governor to initiate purchase and installation of 100 satellite earth stations for establishment of statewide satellite communications network. RCA's SATCOM 1 was launched in December 1975, and in 1976, the state of Alaska and RCA Alascom entered into agreement for procurement and operation of small earth stations that would provide telephone service to remote communities. In 1978, with funding from the Legislature, the state began transmission of television via satellite to rural Alaska villages in what became known as the TVDP (Television Demonstration Project). Later, television service was extended to more communities with the establishment of RATNET (the Rural Alaska Television Network), eventually known as ARCS (the Alaska Rural Communications System).

Alascom's own first satellite, Aurora I, was launched in October 1982 at 143 degrees west, about the longitude of Cordova. Aurora I was the only communications satellite devoted exclusively to use by a single state. Nine years later, Aurora II was launched as a replacement for the aging Aurora I, which was almost out of station-keeping fuel after nine years of service. Aurora II was put into orbit slightly farther east at 139 degrees west, so that its transponders could be leased to customers that wanted to reach the east coast, in addition to serving the Alaska market.[17] In 1995, AT&T purchased Alascom, and continued to operate its satellite services, launching Aurora III in the year 2000. In 2002, AT&T Alaska used Aurora III to provide interactive instructional video in Alaska, interconnecting six schools within the Aleutians East Borough School District.

Since 1970, the FCC had not allowed competitive services in rural Alaska, reasoning that the market was too small to sustain competition. In 1995, GCI obtained permission from the APUC and the FCC to demonstrate its advanced satellite communication technology in rural Alaska. Under a temporary waiver, GCI provided intrastate telephone service in competition with Alascom on an experimental basis by installing earth stations in rural communities. In 1996, GCI constructed earth stations in 56 rural locations, and entered into a lease/purchase agreement with Hughes Communications to secure sufficient satellite coverage for the entire state. However, it was not until 2003 that the FCC decided to discontinue the Alaska bush earth station policy, which had precluded installing or operating more than one satellite earth station in any Alaska bush community for competitive carriage of interstate Message Telephone Service (MTS) communications, (ordinary interstate, interexchange toll telephone calls), finally ending the policy adopted in 1970 that barred rural competition.

Both Alascom and GCI began to deploy DAMA (Demand Assigned Multiple Access) technology that allowed much more efficient use of satellite circuits by assigning them as needed for each call, rather than preassigning circuits for each community. In 1999, GCI acquired capacity on the Hughes Galaxy XR satellite, which it forecast would meet its needs for the following 12 years. Nine years later, in 2008, GCI successfully transitioned all of its satellite traffic to the Hughes Galaxy 18 satellite, which it estimated would be able to provide Alaska rural telecommunication services for the next 14 years.[18]

Optical Fiber

By the 1990s, the growing Alaska economy and competition in interstate telecommunications were spurring increased traffic, so that existing satellite and cable capacity between Alaska and the lower 48 was not likely to meet the growing demand. The projected demand for interstate services, particularly data, resulted in more investment in high-capacity submarine cables rather than satellite transponders to link Alaska with the lower 48 and the rest of the world.

In 1991, the North Pacific cable between Japan and the United States was completed, with a spur funded by Alascom from Pacific City, Oregon, to Seward, providing 420 megabits or the equivalent of 6000 64-kbps voice channels between Alaska and the rest of the country, as well as a link with Asia. GCI purchased capacity on the North Pacific Cable, but in 1998, GCI also began construction of its own Alaska United Fiber Optic Cable System, a 2331-mile cable network connecting Anchorage, Fairbanks, and Juneau to Seattle, linking Alaska's major population centers with the lower 48 states. Beginning service in 1999, the $125 million network brought significant increased capacity for Alaska's growing Internet, data, and video markets. GCI sold $19.5 million worth of capacity on the network to ACS, which used the cable as long-distance backbone for its local-access, long-distance, data, and cellular phone service, including services from its recent acquisitions, ATU and PTI Communications.

ACS later purchased another $19.5 million worth of capacity, and GCI also leased capacity to AT&T Alascom, which at that time controlled more than 50 percent of the state's long-distance market.

In 2001, GCI acquired the 800-mile fiber optic cable that followed the Trans Alaska pipeline, extending GCI fiber optic facilities from the major energy-producing region of Alaska to the lower 48 states. By 2003, growing Internet, data, and video use had generated enough demand for an additional cable, as GCI announced construction of a $50 million fiber optic cable connecting Seward, Alaska and Warrenton, Oregon. The cable had a capacity of 640 gigabits per second access speed and was planned to complement and provide backup for GCI's existing fiber optic cable between Whittier and Seattle.

In 2006, the Kodiak-Kenai Cable Company completed a 600-mile submarine fiber optic cable to link Kodiak Island and the Kenai Peninsula with the rest of Alaska via high-speed, broadband fiber optic technology. The $36 million cable system, owned by Old Harbor Native Corporation and Ouzinkie Native Corporation, was funded by the Alaska Aerospace Development Corporation, with additional funds from equity investors and bank loans. The fiber optic cable provides service to the Coast Guard base at Kodiak, the Alaska Aerospace Development Corporation launch complex at Narrow Cape, and to some 60,000 residents of Kodiak Island, the Kenai Peninsula, and Seward, through capacity leased to Alaska carriers such as GCI and AT&T Alascom.[19]

In 2008, GCI announced construction of the Southeast Alaska Fiber Optic System to connect Ketchikan, Petersburg, Wrangell, Angoon, and Sitka with Juneau and the Alaska United Fiber Optic system. In 2009, Alaska Communications Systems Group (ACS) completed construction of AKORN (the Alaska-Oregon Network) submarine cable, intended to provide diverse connectivity between Alaska and national and global networks. AKORN's four fiber pairs and advanced electronics more than tripled Alaska's existing interstate bandwidth capacity and followed a different route to the lower 48, going from Anchorage to Nikiski, then overland to Homer, from which the submarine cable connected to Florence, Oregon. AKORN was designed to complement ACS's Northstar cable system between Anchorage and Nedonna Beach, Oregon. The two cables were laid from their separate landing sites to network peering points in Portland and Seattle over diverse paths, creating a fiber ring providing alternate routes in the event of a network disruption.[20]

By the end of the first decade of the new millennium, capacity linking Alaska to the outside world was not a problem. However, with growing demand for broadband, Alaskan providers were looking for alternatives to geostationary satellites to serve the more than 150 rural communities dependent on satellite communications.

Rural Broadband

As Internet services proliferated in the new century, demand grew throughout Alaska for more capacity to access bandwidth-intensive services. In urban and suburban areas, phone companies upgraded their copper networks to offer "last mile" connections via DSL (Digital Subscriber Line), while cable companies offered Internet access over their cable networks. Microwave, and increasingly optical fiber, linked the Internet service providers (ISPs) to the high-capacity optical fiber backbone networks described above.

The problem in rural Alaska was not the "last mile," as some communities already had local DSL over telephone lines, and villages were often compact enough to be covered by fixed wireless networks, but the lack of sufficient bandwidth to and from these communities, the so-called "middle mile." Some rural businesses and entrepreneurs who required Internet access installed their own satellite terminals (VSATs), but costs of installation and usage were too high for most residents. Because of the E-rate subsidies for schools and libraries (discussed in Chapter 16), most communities had some form of higher speed connections in their schools and public libraries, but residents in communities without separate libraries had to use school facilities that were often available at very limited times if at all, and generally there was no access in the summer when the schools were not in session.

Most villages were still served by satellite. The carriers serving the bush could lease additional satellite capacity, which they did to serve many of the schools, libraries, and rural health facilities covered by federal subsidies, but they argued that the satellite capacity was limited and too costly to meet consumer demand. Also, the half-second delay to transmit to and from geostationary satellites was frustrating for Internet users. Pressure mounted to expand connectivity over terrestrial networks and to find means to fund the required capital investments.

Two events triggered policies that provided some of the funding. In response to the "Great Recession," Congress passed the American Recovery and Reinvestment Act (Recovery Act or ARRA) in January 2009, which included $7.2 billion "to begin the process of significantly expanding the reach and quality of broadband services."[21] Also, one of the requirements included in the act was that the FCC was to prepare a national broadband plan, to be completed within a year of passage of the act. The policies adopted to implement the plan included changes in subsidy schemes as well as new funding for rural broadband, with some programs specifically targeting remote and tribal regions.

The Recovery Act allocated $2.5 billion for rural infrastructure projects to the Rural Utilities Service (RUS), which administered these funds through the Broadband Infrastructure Program (BIP). Alaska received more than $117 million for BIP rural infrastructure projects. The largest project, TERRA (Terrestrial for Every Region of Rural Alaska), was to provide terrestrial connectivity through a hybrid optical fiber and microwave network to 65 communities, most of which were predominantly Native villages in Bristol Bay and the Yukon-Kuskokwim regions.[22] Another project, SABRE (Southwest Alaska Broadband Rural Expansion), was intended to provide wireless fourth generation (4G) broadband service to southwest Alaska through a partnership between a telecommunications company and a subsidiary of Sea Lion Corporation, the Alaska Native Village Corporation for Hooper Bay. This project floundered and appeared to duplicate some of the middle-mile broadband expansion funded under TERRA. It was eventually cancelled. Copper Valley Telephone Cooperative received $5.2 million in grants and loans to extend high-speed Internet connectivity via microwave to McCarthy, a small community located within Wrangell–St. Elias National Park and Preserve.[23] Another funded project offered free satellite equipment and installation plus discounted service to rural residents nationwide, including Alaskans who did not have other options to access broadband.[24]

The National Telecommunications and Information Administration (NTIA) in the Department of Commerce established the Broadband Telecommunications Opportunities Program (BTOP) to administer its $4.7 billion allocated under the Recovery Act. BTOP funded two Alaska projects in addition to state broadband planning and mapping. OWL (Online with Libraries) upgraded connectivity for 65 rural libraries, most of which are in indigenous communities. Facilities include videoconferencing and web conferencing, so that the libraries can serve as public computing centers. OWL was also intended to provide training and support in digital literacy to benefit community residents without broadband at home. Beneficiaries were intended to be remote library users where computer home ownership and Internet subscriptions are lowest, K–12 students to obtain homework help, adults undertaking university and vocational courses, and public service agencies serving the rural communities.[25] Examples of applications at OWL sites are discussed in Chapter 16.

The University of Alaska Fairbanks (UAF) received $5.4 million for a project known as "Bridging the e-Skills Gap in Alaska," to provide computer skills and broadband awareness training to promote broadband adoption, particularly targeting Alaska native villages. The project brought together 21 partner organizations throughout rural Alaska to increase technology literacy. It promoted broadband awareness and training activities and partnered with the Alaska Vocational Technical Education Center to train village Internet agents, who were to become resources for public computer centers and users in rural

Alaska. It also partnered with several other organizations, including the Alaska Native Tribal Health Consortium, which developed a Telehealth Coordinator Certification Program for medical professionals.[26]

All states including Alaska received funds from NTIA for broadband mapping and planning as part of the activity required for implementation of the National Broadband Plan. Alaska received $6.4 million, administered by ConnectAlaska, a subsidiary of the nonprofit organization Connected Nation. (The original recipient was intended to be the Denali Commission, which withdrew when its legal counsel concluded it could not receive the funds under the terms of the ARRA.) These funds were to be used to create a comprehensive broadband map for Alaska and to provide data on broadband availability, technology, speed, and infrastructure. Connect Alaska, in collaboration with broadband providers, was to update these maps on a routine basis to reflect real-time broadband availability. The maps were be accessible to the general public via an interactive portal.[27]

The state was also to establish an Alaska broadband task force to develop a state broadband plan. Broadband planning funds were also to be used for support to the task force for collecting information coordinating activities across relevant agencies and organizations, undertaking research and benchmarking activities on broadband needs, usage, and barriers in local communities, and measuring the impact of broadband implementation over time, providing technical assistance to communities and Alaska Regional Development Organizations (ARDORs) in identifying needs and planning for broadband usage, and assessing the availability and quality of e-government applications.[28]

The State Broadband Task Force

The Alaska State Broadband Task Force was established in February 2011, with members appointed by the governor representing the Alaska communications industry; public service users of communications including schools and higher education, libraries, health care, public safety and the military; state agencies; the Legislature; business users; and consumers. Its purpose was to develop a plan "to accelerate deployment, availability and adoption of affordable broadband throughout the state." The task force completed a draft state broadband plan in August 2013 and a final report in fall 2014. The report included a variety of options for funding build-out of broadband in rural areas, as well as recommendations including:

- Adopt an objective of symmetrical 100 Mbps service to home and businesses
- Establish an Office of Broadband Policy within the state government
- Encourage each community to implement its own last-mile solution

- Incentivize 24-hour Internet access at community centers/meeting places
- Establish public-private partnerships with industry innovators and entrepreneurs
- Streamline state e-government systems and foster improved user access, ease of use, application development, and deployment
- Create training programs for knowledge workers, technicians, and web-based industries
- Establish and fund the Alaska Center for e-learning and e-commerce (AkCee)
- Create incentives for organizations to provide digital literacy programs that facilitate broadband adoption[29]

Recommendations were to be presented to the state legislature in 2015. There was effectively no Native participation in the task force, and minimal outreach to Alaskans, particularly people in rural areas lacking broadband.

Federal Rural and Tribal Broadband Initiatives

The Connect America Fund established as part of the implementation of the National Broadband Plan is focused on supporting and expanding fixed and mobile broadband availability, and is intended to replace the former High Cost Fund that provides support to carriers serving high-cost areas, which include most of Alaska. The largest source of universal service funds for Alaska, the High Cost Fund provided more than $181 million to Alaska telecommunications carriers in 2013, and a total of more than $2.1 billion from 1998 to 2013, approximately 4 percent of the total High Cost disbursements during that period.[30]

In 2011, the FCC released the Universal Service Fund and Intercarrier Compensation Transformation Order,[31] including comprehensive reforms to modernize the High Cost Program and accelerate the build-out of robust broadband networks across the country. Several Alaska carriers have challenged the data, models, and assumptions used by the FCC to calculate their costs and determine subsidies under the new program. As of late 2014, the Alaska subsidies were still under review.

In 2010, the FCC established an Office of Native Affairs and Policy (ONAP) to promote deployment and adoption of communications services throughout tribal lands and native communities. ONAP is "charged with bringing the benefits of a modern communications infrastructure to all Native communities by . . . ensuring robust government-to-government consultation with federally-recognized Tribal governments and other Native organizations; working with Commissioners, Bureaus, and Offices . . . to develop and implement policies for assisting Native communities; and ensuring that Native concerns and voices are considered in all relevant Commission proceedings and initiatives."[32] ONAP established a National Native Broadband Task Force, initially including two representatives from Alaska, to advise on broadband plans and policies for tribal and other indigenous regions.

The FCC recognizes a direct government-to-government relationship with tribes. It considers tribal lands to include Alaska Native regions established pursuant to the Alaska Native Claims Settlement Act and Alaska Native villages, thereby making most Alaska carriers eligible for funding under several programs established to extend broadband on tribal land, as well as programs for rural and remote regions.

The FCC determined that mobile broadband is increasingly important as a means of accessing the Internet and other services, and should be available throughout the country, and therefore established a universal service support mechanism dedicated expressly to mobile services. The Connect America Mobility Fund allocates $300 million for mobile voice and broadband in high-cost areas, plus $500 million per year ongoing support. The licenses are awarded by reverse auction, so that the carrier requesting the lowest subsidy wins the bid. The first auction, held in 2012, ended with 33 carriers winning bids to share $300 million available nationwide for one-time support. GCI received $3.23 million to upgrade mobile coverage in some rural regions of Alaska.[33]

A special allocation under the Connect America Mobility Fund is to provide $50 million capital plus up to $100 million per year for tribal areas to support the build-out of current and next-generation mobile networks in areas where these networks are currently unavailable. In 2014, the FCC held a reverse auction for subsidies under Phase I of the Tribal Mobility Fund to distribute $50 million in one-time support for mobile service providers serving tribal lands lacking 3G or 4G service. GCI received $41.4 million to upgrade wireless service for 48 communities, while Copper Valley Wireless received $152,000 to serve an estimated population of 127 in the Ahtna region.[34] There were no competing bids for the regions won by GCI and Copper Valley. Phase II of the Mobility Fund will offer $500 million annually for ongoing support of mobile services, with up to $100 million of this amount designated annually and exclusively for support to tribal lands.[35]

Within the Connect America Fund, the FCC created a Remote Areas Fund with a budget of at least $100 million annually "to ensure that even Americans living in the most remote areas of the where the cost of providing terrestrial broadband service is extremely high, can obtain service."[36] As a step toward implementation of the Remote Areas Fund, in July 2014, the FCC announced funding of $100 million for rural broadband experiments in high-cost areas that would include most of Alaska. However, no Alaska carriers bid for subsidies in this pilot phase, although several had submitted expressions of interest.

The Lifeline program subsidizes the price of voice services for low-income residents, including those living on tribal lands. The FCC allocated $25 million for the Lifeline Broadband Adoption Pilot Program in order to gather data on broadband adoption and deployment among low-income consumers. It directed that at least one pilot application providing service on tribal lands be accepted; ultimately, applications from two tribally owned eligible carriers were accepted into the pilot program, but there were no applications from Alaska.[37]

It is important to note that these programs were open only to licensed Eligible Telecommunications Carriers (ETCs). The FCC sought to encourage tribal entities to become certified as ETCs and provided training for them on eligibility, opportunities for joint ventures, and the mechanics of the auction and other processes. In addition, tribally owned or controlled ETCs are eligible for 25 percent bidding credits in mobile reverse auctions.[38] As of late 2014, there were ten tribally owned ETCs in the United States but no Native-owned or operated ETCs in Alaska, although there were some Native cooperatives providing local telephone service. The eligible incumbent Alaska carriers have declined to participate in these programs, citing concern about future revenues as a result of changes in the high-cost funding support in the new Connect America Fund.

Engagement with Tribal Governments

In 2012, the FCC implemented a requirement that communications providers receiving subsidies to serve tribal lands must "meaningfully engage" with the tribal governments of these lands. While the engagement mechanisms are flexible, the discussions must include: "(1) a needs assessment and deployment planning with a focus on Tribal community anchor institutions; (2) feasibility and sustainability planning; (3) marketing services in a culturally sensitive manner; (4) rights of way processes, land use permitting, facilities siting, environmental and cultural preservation review processes; and (5) compliance with Tribal business and licensing requirements."[39] "Tribal government" may require interpretation in Alaska, as there are 229 recognized tribes, as well as Village Corporations and regional Native Corporations established under the Alaska Native Claims Settlement Act.

The FCC also established a rulemaking proceeding designed to improve tribal access to spectrum and to promote greater utilization of spectrum over tribal lands. In 2014, the Commission continued to seek comments on processes to provide tribes with new opportunities to gain access to spectrum, including the establishment of a tribal priority for spectrum use, a formal negotiation process for secondary markets agreements, a build-or-divest process, and other proposals including use of white spaces (unlicensed spectrum) on tribal lands. It also sought to foster the establishment of tribally owned wireless carriers.

These are not only new initiatives, but some of them are without precedent. For example, the author knows of no other regulator in the world that has required carriers to consult with indigenous governments as a condition of receiving subsidies to serve their communities. However, implementation of such initiatives requires much more than regulations or funding set-asides. Representatives of tribal communities must have sufficient knowledge of telecommunications issues to "meaningfully engage" with their service providers. Tribal governments and other entities must be aware of the new opportunities and have the requisite understanding of policy, technology, and FCC procedures to be able to participate. Tribal entities that are interested in providing communications services must meet requirements for ETC certification. Existing carriers must evaluate incentives to apply for new funding programs to serve regions where long-term commercial viability appears unlikely. Additional outreach and involvement of both consumers and providers of rural broadband will be necessary for Alaska to benefit from these opportunities.

Other Broadband Programs

Another federal infrastructure program involving Alaska is FirstNet (First Responder Network Authority), which is intended to provide emergency responders with the first nationwide, high-speed network dedicated to public safety using a single platform for public safety communications linking all types of first responders across jurisdictions. It is intended to provide voice, data, and video, with shared applications and access to data bases. FirstNet plans to use Long-Term Evolution (LTE) wireless technology, and to support the integration of Land Mobile Radio (LMR) networks, with implementation to be partially funded by FCC spectrum auctions.[40] The Alaska Department of Public Safety was awarded a $2 million planning grant to analyze the steps needed to integrate the many different communications systems currently used for public safety in Alaska.

Additional broadband funding for rural Alaska was provided by the Rural Utilities Service (RUS) through the state regulator, the Regulatory Commission of Alaska. The RCA's Rural Alaska Broadband Internet Access Grant Program was intended to facilitate long-term affordable broadband Internet services in rural Alaska communities where these services do not currently exist. Requirements for eligible communities included a population of less than 20,000 and a "not-employed rate" of more than 19.5 percent. Broadband speed was defined as 768 kbps (the old FCC definition) with access referring to individual households. Funding was available to telecommunications carriers, which could receive up to 75 percent of construction costs, and had to commit to keeping rates comparable to those in Anchorage, Fairbanks, and Juneau for at least two years.[41] However, the federal RUS support was not renewed.

Benefits of Internet and Broadband

With the advent of the Internet era in the late 1990s, Alaska was once more a communications pioneer in distance education for rural residents, expanded telemedicine services, and online access to state government services ranging from hunting and fishing licenses to applications for annual Permanent Fund disbursements. As broadband became available, rural Alaskans began to go online to access services such as electronic banking and shopping, to promote rural businesses from handicrafts to tourism, and to manage rural nonprofit activities and commercial businesses such as fisheries, aviation, and retail enterprises.

In general, benefits of rural telecommunications can be classified in terms of:

- *Efficiency*, such as managing operations of rural businesses, reducing travel, and filing online reports and business data
- *Effectiveness*, improving the quality of services provided such as in health care and education
- *Equity*, reducing the distance barriers between rural and urban communities by providing access to information, entertainment, education, and other services not otherwise available in remote communities
- *Reach*, enabling Alaskans to extend their range electronically to market Native crafts, tourism, and other local assets[42]

The previous chapters provide examples of many of these applications and benefits, which now also apply to broadband. In 2012, the University of Alaska Anchorage's Institute of Social and Economic Research (ISER) conducted a study of households in southwest Alaska, where broadband was soon to be introduced. Internet use was already quite widespread in remote communities, and two-thirds of users were online almost every day. Thus, many people in the region were already "Internet-savvy," but most were dissatisfied with slow speeds and uneven quality of service and wanted faster and more reliable connections.

Concerning likely uses of broadband, they ranked personal communications and entertainment highest (social networking, downloading music and video, and playing online games). However, 48 percent also said they expected to use broadband for education, 45 percent said they would use Skype or similar services for video conferencing, and 39 percent said they would use broadband for work or telecommuting.[43]

Those households who had invested in satellite service with individual satellite terminals may demonstrate how early adopters of broadband will use the service, as they chose to upgrade to higher speed Internet service than was currently available from local carriers. Some 88 percent of satellite users accessed government services online, 87 percent accessed financial services, while 68 percent used the Internet for education, and 62 percent for work or telecommuting. These early adopters of the fastest connections available provide some indication that future broadband users are likely to take advantage of broadband for work, education, and public and private sector services not available in their communities.

Community access remains important for Internet users. Outside the home, they may access the Internet at work, at school, at libraries, and tribal offices. About 60 percent of respondents thought members of their household would access broadband elsewhere in the community, even if they subscribed at home.

Cellphone penetration was high, with 87 percent of households having at least one cellphone and 60 percent of households having a smartphone. Some residents took their smartphones to school, where they could use the WiFi connection. More than 50 percent also had a tablet or e-reader, indicating strong demand for mobile broadband.

Respondents from Native organizations commented that broadband could save them time in accessing online information and software compared to time required using current Internet services and would be beneficial in applying for grants and filing reports with funders and helping tribal members applying for jobs. Some also noted opportunities to offer training in villages and to help local entrepreneurs develop websites to sell crafts and other products.

The tourism industry also wanted more bandwidth to support their operations and build their businesses. Fishing lodges and other wilderness tourism businesses rely on telephone and email to respond to potential customers, and websites and travel agencies to attract business. Similarly, businesses in hub communities use online services to attract customers and manage their operations.

The seafood processing industry, a major employer in southwest Alaska, wanted faster connectivity to run their back office operations, such as uploading catch information, payroll and other accounting data, and using other software for their business. They also employ thousands of seasonal workers who want to use the Internet to keep in touch with family and friends and to access entertainment. Broadband wireless connectivity to boats and processing vessels could be used to keep crews up to date on operations as well as to provide personal broadband access for crews and seasonal employees. These applications for logistics and back-office communications as well as for personal use by employees are also likely to apply to other key industries in rural Alaska such as mining and oil and gas.

For workers in these sectors, their jobs may be enhanced and skills improved by access to broadband. For many of these entities, economic benefits may include cost savings in terms of increased efficiency or travel substitution, as noted in Chapter 17 for telemedicine. For others, there may be increased revenue and possibly new jobs such as from additional grant funding received by Native organizations, more business for tour operators and lodges, and so on.

New sources of income could come from the opportunities for IT workers and trainers and environmental monitoring as communities collect and transmit data on climate and wildlife, electrical power management, entrepreneurial activities such as selling crafts online, and telework for distant clients. For example, using Skype, an employee at a tourist lodge was able to provide English language tutoring to students in Brazil. The villages of Tuntutuliak and Kongiganak in southwest Alaska use "smart grids" to provide data from wind turbines, diesel generators, and home and office meters to control and monitor their energy production and consumption. Local managers also use broadband to get technical advice and to access web-based applications. The Oomingmak Musk Ox Producers' Cooperative markets knitwear made from qiviut, the soft underwool from the Arctic musk ox. Owned by approximately 250 Native Alaskan women who knit each item by hand, the cooperative sells its knitwear via its website to customers from many countries.

Affordability

However, investments in broadband and other information and communications technologies (ICTs) may be necessary but not sufficient for economic development. Rural residents may lack skills to use these tools, or start-up funding for new enterprises. Costs of electric power and transportation may make it difficult to compete with other suppliers. And relatively high broadband prices may limit use. In southwest Alaska villages with terrestrial broadband (TERRA) service, 2 mbps download with upload of 256 kbps and a cap of 5 GB cost $65 per month in early 2015. The maximum speed available was 6 mbps at prices ranging from $165 to $315 per month, depending on the usage caps. The minimum speed available in Anchorage was 10 mbps without caps.[44] The median household income in many of these villages is less than 60 percent of the statewide average. Employment is seasonal, and most families depend partially on subsistence hunting and fishing.

Some small businesses and nonprofit organizations in southwest Alaska state that they cannot take full advantage of broadband applications "in the cloud" because of the charges they face if they exceed bandwidth caps. For example, in 2014 the Bristol Bay Native Association (BBNA) paid more than $76,500 per year for broadband for its facilities in Dillingham and Bristol Bay region villages, including supplemental use of satellite services in addition to terrestrial broadband. Because of the cost of overages, BBNA stated that it could not participate in webinars nor in online training offered from the lower 48 states, and was unable to use online tools to remotely train and support staff in its villages.[45] Adult students taking distance education courses report that they cannot participate fully until the end of the month, or have to limit use by other family members to avoid overage charges.[46] To date, the issue of broadband affordability has not been addressed, as the middle-mile network is not subject to federal or state regulation.

Chapter 19

Past and Future Connections

. . . telecommunications has helped to bring about a greater sense of statewide unity.

Native leader Willie Hensley[1]

Major Themes

Several themes emerge from this review of the expansion and applications of telecommunications in Alaska. The Alaska experience also yields lessons relevant for future Alaska communications planning and policy, and for other remote areas across the North, and in other rural and developing regions.

Innovation in Technology

A recurrent theme has been adoption of innovative technological solutions to cope with Alaska's climate, terrain, and isolation. From the first commercial wireless telegraph in the United States across Norton Sound, to small earth satellite stations for Alaska villages, low-power village TV transmitters, and telemedicine carts designed for village health clinics, technological innovation has been key to delivering communications services in Alaska.

Technological innovation also was critical to the introduction of competition in Alaska telecommunications. Carriers that estimated costs based on technology and practices used in the lower 48 concluded that competition in Alaska's small market was impossible. However, as Ron Duncan, cofounder of GCI, explained: "Alaska can't be a technology leader in the sense of creating whole new technologies to meet its service needs because the market just isn't big enough. The best we can do … is to take existing technology that has been developed for mass markets elsewhere and modify it around the fringes so it better meets the needs of the Alaska market. [GCI's] bush earth station product was done very much that way. We took a small earth station product that was designed for a different purpose and by paying a couple million dollars to the vendor to develop some customized software, we managed to repackage it in a way that lowered the cost of rural service by an order of magnitude."[2]

Innovative Uses and Applications

From the earliest days, Alaskans have also been innovative in how they use communications technologies. They sent messages via radio broadcasts and two-way radio to isolated villages, used a shared audio channel on a satellite to implement telemedicine, licensed rights to educational programs for teachers to download overnight, enabled rural Alaskans to testify in state hearings via teleconference, marketed Alaska crafts and other products online, and developed software to preserve and teach indigenous languages.

Alaskans continue to develop innovative applications. The partners of the Alaska Native Tribal Health Consortium are digitizing patient records so that they can be accessed at any location in their health care system. They are also adopting new applications such as conversion of doctors' dictated notes to text and remote monitoring of patients with chronic conditions in their homes, even in the villages. The Cook Inlet Tribal Council has built a fabrication laboratory (fab lab) where students learn to use high tech design programs, 3D printing, and other electronic and programming tools.[3]

Alaska Natives are also using the Internet to preserve their culture and history. A Native language map first produced at the University of Alaska Fairbanks in the 1970s has now been updated to included traditional and modern place names, with links to a GIS database. The map data is intended to be available online so that others may add demographic, scientific, or historical information about locations on the map.[4] A group of Native leaders and volunteers used the web to preserve and share materials about the Alaska Native Claims Settlement Act (ANCSA) during the 40th anniversary year of its signing in 2011.[5] The Alaska Native Cultural Center also provides historical and cultural materials on its website.[6] The Inuit Circumpolar Conference includes Inuit in Alaska, the Canadian Arctic, and Siberia and uses a variety of media to share cultural materials and to address shared issues such as climate change, ocean resources, and natural resource exploitation.[7]

Benefiting from Federal Subsidies

In the 1970s and 1980s, state communications officials and consultants helped Alaska carriers gain access to a national revenue pool that could subsidize costs of providing service in Alaska, thereby helping to keep communications affordable for Alaska residents. Since the passage of the 1996 Telecommunications Act, Alaska has benefited significantly from federal universal service funds. In

2013, Alaska received a total of $278.7 million in universal service funds, ranking sixth among the states in total amount received, following Texas, California, Oklahoma, Florida, and New York, states with 5 times to more than 50 times Alaska's population. In the 15 years from 1998 to 2013, Alaska received a total of almost $3 billion from these funds, or 3 percent of the total awarded, despite having only 0.23 percent of the U.S. population. Over that period, Alaska schools and libraries received more than $295 million. (See Table 19-1.)

TABLE 19-1   Universal Service Funds disbursements to Alaska 1998–2013[8]

	$ MILLIONS	% OF TOTAL U.S. FUNDS
High Cost Fund	2,113, 835	3.8%
Low Income (Lifeline and Linkup)	185,620	1.2%
Schools and Libraries (E-rate)	295,435	1.1%
Rural Health Care	342,765	52.5%
Total	$2,937,655	3%
	POPULATION	PERCENTAGE OF U.S. POPULATION
ALASKA	735,000	.23%

As discussed in the chapters on distance learning and telemedicine, Alaskans were very proactive in applying for these subsidies. GCI took an early interest in the E-rate program, and set up an office to help schools and libraries obtain funding. The state government also assigned a state librarian to work half time on assisting schools and libraries with the applications process. She devoted much additional volunteer time to E-rate matters and became an effective advocate for Alaska's schools and libraries.

Although the Rural Health Care fund is relatively small, Alaska's share is significant, as Alaska's rural health care providers have received the largest absolute amount and more than 50 percent of the total funds distributed, as well as the highest funding per capita. With decades of experience in rural telemedicine, it may not seem surprising that Alaska's rural health care providers were successful in obtaining USF support. They too were proactive in applying for funding and were among the first successful applicants. For both the E-rate and rural health care programs, the carriers that won competitive bids to provide the services as well as their customers benefited, but the communities also stood to benefit, as the schools, libraries, and rural clinics became anchor tenants that generated predictable revenues, helping to make a business case to extend services to the villages where these institutions were located.

The largest subsidy to Alaska is from the High Cost fund, which provides support to rural local exchange carriers (LECs) so that prices for local voice services can remain affordable for their subscribers. In 2013, the high-cost sub-sidy amounted to more than $181 million for Alaska carriers, and from 1998 to 2013, the total amount Alaska carriers received was more than $2.1 billion. However, the subsidy models are changing under the recently implemented Connect America Fund, which is designed to help achieve the goals of the National Broadband Plan released in 2010. While there will continue to be subsi-dies for rural telecommunications, more emphasis will be placed on broadband.

With subsidies come benefits, but also dependency. Alaska rural carriers and their customers, including households and businesses, schools and librar-ies, and rural health services, have become dependent on these subsidies. The rural LECs have come to depend on the high-cost subsidies as a key compo-nent of their business models. Videoconferencing for rural telehealth and for libraries and students would not be affordable without the rural health care and E-rate subsidies. Although it appears unlikely that these subsidies will disappear, they are definitely likely to change, as is already happening with the transition to the Connect America Fund. Nationwide USF disbursements totaled $8.3 billion in 2013. This amount is unlikely to increase, as there is resis-tance by subscribers who ultimately pay for these subsidies, which are passed through by the carriers as surcharges on their telecommunications bills.

Competition and Consolidation

The telecommunications industry was long considered to be a natural monopoly; it appeared that facilities and services could be provided most cost-effectively by one integrated provider. Technological innovations beginning in the 1960s gradually eroded this paradigm, resulting in competition ranging from customer premises equipment (such as telephones and fax machines) to satellite systems. Competition in services began with early data communications, and expanded to long-distance communications and eventually to local services. In Alaska, the rationale for monopoly was based primarily on the high capital and operat-ing costs of reaching customers scattered over a vast land area and the limited revenues from a small population. However, entrepreneurs who introduced inno-vations in satellite technology, and later in digital and mobile communications, demonstrated that competition could succeed even in rural Alaska.

Much of this competition has been facilities-based, with separate satellite earth stations and cellular antenna towers installed by each carrier. Yet a new era of consolidation appears to be emerging, as high-capacity backbone net-works consisting of optical fiber and microwave extend broadband to some regions of rural Alaska. Resale of this capacity would offer a means of offering competitive services over these networks, whereby the facilities owner would lease wholesale capacity on its network to retail service providers. However, potential competitors claim that wholesale rates are too high for them to compete with the facilities owner, which itself offers retail services. And as noted in the previous chapter, rural residents, small businesses, and nonprofits complain that while bandwidth is now available, it is unaffordable, so they cannot take full advantage of all the services they need.

Another example of the impact of the lack of rural competition is in the growing amount of rural health care connectivity subsidies. The FCC's rural health care subsidy is based on the difference between the price charged for connectivity at the rural medical sites and the price charged at the major urban center in the state, which in Alaska's case is Anchorage. Competition has increased significantly in Anchorage, with a growing number of providers, and decreasing prices. However, service in rural Alaska remains a de facto monopoly, with the carrier that serves the rural health providers having no incentive to lower prices once it has won the bid to provide the service. Thus, the gap between urban and rural rates has widened, and the subsidy the rural health care providers now receive from USAC amounts to more than 97 percent of their communications charges.

Special Treatment for Alaska

Alaskans have frequently sought—and received—special treatment from regulators. After more than four decades of advocacy by Alaskans at the FCC, William Harris, senior advisor to FCC Commissioner James H. Quello, pointed out in 1989, "Waivers of our rules are granted almost routinely when it's apparent that they did not contemplate the unique situations and conditions faced by Alaskans."[9] Broadcaster Augie Hiebert was able to get a license for a high-powered FM station in Prudhoe Bay to rebroadcast an AM station from Anchorage because "Things are different in Alaska." Michael Porcaro, when executive director of the APBC, managed to get village transmitters affiliated with all four television networks so that the villages could receive the RATNET channel with combined network programming. Schools receiving E-rate support got a waiver to extend their WiFi signal to the community because there was no other local provider. Recently, Alaska carriers have sought waivers from some FCC requirements mandated under implementation of the Connect America Fund.

Public and Private Sector Advocacy

Larry Pearson concluded in 1987: "Attempts at making information and telecommunications policy have been most successful when interest from the governor and the legislature is highest."[10] The waivers from the FCC and many other regulatory and policy victories for Alaskans resulted from engagement of key officials in both the public and private sectors. Governors Egan and Hammond were committed to getting affordable satellite communications to rural Alaska. Senators Bartlett, Gravel, and Stevens also played significant roles in Alaska telecommunications, including setting the terms for the sale for ACS, arranging Alaska participation in NASA satellite experiments, and obtaining federal funding to support Alaska communications. Proactive officials in state government agencies included Bob Walp, Marvin Weatherly, Bob Arnold, Alex Hills, and Charles Northrip. Legislators Fred Brown and "Red" Boucher chaired committees concerned with Alaska communications. Fran Ulmer, as legislator and later lieutenant governor, helped to establish and then chaired the Telecommunications Information Council (TIC). Dr. Martha Wilson of the Alaska Area Native Health Service was a tireless advocate for rural telemedicine. In recent times, Senator Mark Begich championed the expansion of broadband in rural Alaska.

Private sector and academic advocates have also helped to expand and improve Alaska communications. Broadcaster Augie Hiebert made his first trip to Washington, DC, in 1946. In the 1970s, he led initiatives to get satellite television to rural Alaska. Hiebert also organized "Alaska Day" at the FCC so that Alaska broadcasters could showcase their activities and explain their needs for regulatory waivers and exceptions. Professors Edwin Parker, Bruce Lusignan, and William Melody were both consultants to the state and advocates for technologies and policies that would bring affordable satellite communications to rural Alaska. Professors Glenn Stanley and Robert Merritt of the University of Alaska's Geophysical Institute advised the state legislature. Larry Pearson of the University of Alaska Anchorage also advised the legislature and organized the Chugach conferences and other fora on state communications issues. The author of this book has carried out research on Alaska telemedicine, rural communications, and Internet adoption and has advised and advocated for Alaska consumers concerning the need for universally available and affordable broadband.

Lessons from the Alaska Experience

In summary, although Alaska differs not only in climate but also in access to economic resources from many other developing regions, the Alaska experience offers valuable insights about how telecommunications can be harnessed for rural development. Among the lessons are:

- Creative approaches to telecommunications network design, operations, and maintenance can significantly reduce the cost of rural telecommunications

- Targeted subsidies can have a significant impact in providing affordable access to rural and disadvantaged populations, but they may also lead to dependency
- Affordability remains a concern where competition is limited or nonexistent, and especially where income levels are low
- Access to telecommunications can enable rural citizens to participate in government and policy making
- Informed advocacy in national regulatory and policy fora can bring significant benefits
- State government can shape the environment for development in distinctive ways
- Information and telecommunications systems are strategic assets that require ongoing public policy attention
- Information and telecommunications policy involves much more than building infrastructure; it requires assessment of user needs, training, and improvement of public access to information[11,12]

Missed Opportunities and Possible Remedies

Despite Alaska's many successes and innovations in extending telecommunications facilities and applying communications services, there have been several missed opportunities.

Lack of State Planning

Governor Egan established an Office of Telecommunications (OT) in 1972, which continued to operate through the Hammond administration until it was abolished by Governor Hammond in 1979. OT became a focal point for identifying goals and developing strategies to improve access to telecommunications, particularly in rural Alaska. In 1981, Governor Hammond placed responsibility for telecommunications operations and services in the Department of Administration. In 1987, Governor Cowper merged services with operations to improve efficiency, but the result was that operations took priority over planning, and the Department of Administration's mandate for statewide (as opposed to internal state government) communications planning apparently disappeared. The Telecommunications Information Council (TIC) established by "Red" Boucher and later chaired by Lieutenant Governor Fran Ulmer once again provided a focal point for setting goals and determining priorities until its abolition in 2005 by Governor Murkowski.

In 1987, Larry Pearson and Doug Barry sounded an alarm, highlighting "the absence of a statewide telecommunication and information policy . . . ," and concluding "Alaska may be in the process of unwittingly squandering its strategic information assets."[13] Today, despite the growing importance of broadband and numerous federal initiatives to support and expand rural communications services, the state now has no entity responsible for statewide telecommunications planning and policy. The Department of Commerce, Community, and Economic Development (DCCED) oversaw production of a state telecommunications plan, with participation by some other state agencies and by members of the Alaska telecommunications industry, and some user representatives. However, the plan was produced as a federal requirement for implementation of the National Broadband Plan rather than an Alaska-generated initiative and has not been viewed as an Alaska priority. The state clearly needs to reengage in telecommunications planning and policy, with responsibility at a senior level. It also needs to involve the many stakeholders including the public as users and beneficiaries of telecommunications services. The TIC could serve as a model for a new participatory planning organization.

Conduit versus Content

FCC Chairman Tom Wheeler stated in Anchorage in 2014: "It's not about the tracks, but what goes over the tracks."[14] Larry Pearson pointed out that in the 1970s, "Alaska was interested then in building systems. How these systems would proceed to identify and serve important human needs was discussed occasionally, but never came close to achieving the interest and support accorded the systems themselves."[15] However, some projects were funded to develop Alaska content. Perhaps the most significant was LearnAlaska, which was highly innovative, not only in developing programs by and for rural Alaskans but in distributing content from numerous sources that teachers could download and tape for use in their classrooms. Yet not only was LearnAlaska discontinued, but its approach and lessons learned seem to have vanished from Alaska's institutional memory.

Early broadcasting initiatives emphasized distribution of network programming to village residents through satellite feeds and low-power mini-TV transmitters. Legislative funding was continued (to this day) for what began as RATNET (the Rural Alaska Television Network) and is now called ARCS (Alaska Rural Communications Service). ARCS continues to bring a mixture of commercial and public television programs to viewers at 235 rural locations. The programs are selected by a council with representatives from regional Native associations, public broadcasters, government agencies, and the general public. In 1991, Canadian Native communications researcher Gail Valaskakis was astounded to learn that there was a channel set aside to distribute TV programming to rural Alaska, but no Alaska content. Yet since the beginning,

RATNET, followed by the ARCS Council, seemed satisfied simply to pick from a menu of available national programming rather than to demand that Alaska-produced programming be funded and distributed, and that northern and indigenous programming from other sources such as the Canadian Broadcasting Corporation and Canada's Aboriginal Peoples Television Network (APTN) be included.

Public radio in Alaska has developed a strong base of rural content produced both in Anchorage and in other public and community radio stations around the state. Also, Koahnic Broadcast Corporation (KBC), a nonprofit Alaska Native-operated media center located in Anchorage, produces Native news and cultural programming and operates radio station KNBA. ("Koahnic" is an Athabaskan word in the Ahtna dialect meaning "live air.")[16] There is still little television programming for or about rural Alaskans or Alaska Natives. With delivery now possible by both broadcast media and the Internet, the LearnAlaska model should be revived and updated to provide educational content relevant for rural Alaska.

Native Engagement

In 1990, when it appeared that the legislature was going to discontinue funding for RATNET, broadcaster Augie Hiebert asked: "Where is the Native constituency? Where is the Alaska Federation of Natives? Where are the 13 corporations that should have some interest in their people out there?"[17] The same questions could be asked today. Few Alaska Natives are engaged in articulating their needs as consumers of communications content and services. Yet as the majority of residents in much of rural Alaska, access to affordable communications services and relevant content should be their particular concern. There was no Native participation in the State Broadband Task Force. No Native telecommunications carriers have applied for recent federal broadband subsidies to serve rural and tribal regions.

However, as Michael Metty pointed out in 1982, participation by rural Alaskans involves organizing and preparation: "Our experience ... points out that if useful prioritizations are to take place, the following conditions are important. The participants must have: 1) sufficient information about the capacity of the system, the options available and the needs of their constituency, 2) marked influence within the provider structure and the constituency, 3) adequate resources to insure meaningful impact, and 4) enough time to accomplish the process."[18]

Metty's points are still pertinent. One possible step would be for Alaska Natives to establish a Native communications organization that can represent their interests and advocate for their needs at both the state and federal levels.

It could help tribal organizations take advantage of the FCC's recent mandate that carriers receiving subsidies to serve tribal areas must engage with tribes. It could also help Alaska Native organizations to become communications providers or to partner with other providers to obtain federal funding available for extending broadband in remote and tribal areas.

Consumer Involvement

In the 1970s a consumer advocacy organization challenged some Alaska television licenses, and AKPIRG, the Alaska Public Interest Research Group, included communications in the issues it researched and publicized. In the 1980s, there was a Telecommunications Users Advisory Consortium. Today, consumers and other users are virtually silent about the availability, affordability, and quality of their communications services. There is no organization representing communications consumers in Alaska, ranging from individual residents to small businesses, nonprofit organizations, and social and educational services. Such an organization could bring user needs to the attention of industry, regulators, and policy makers and could provide information in public hearings and filings at the state and federal level.

Telecommunications and Alaska's Future: Taking up the Challenge

When Wernher von Braun visited the ATS-6 satellite installation in Tanana in 1975, a young student in the village school asked what they could do with the satellite technology he had explained to them. He replied, "What do want to do with it?" That response is still germaine. Alaskans have the opportunity to determine how to use broadband and other new information and communication technologies for the social and economic development of the state, and to bridge the gaps between urban and rural Alaskans.

Alaska is entering a critical new era as it strives to diversify its economy, to play a growing role in Arctic development and policy, and to tackle the challenges of climate change. To prepare for Alaska's future, the State Commission on Research (SCOR) report emphasized the need to educate more Alaskans about science and technology, and to prepare students for careers that require expertise in science, technology, engineering, and mathematics (STEM).[19] Communications technologies can extend access to STEM education and other content to students throughout Alaska and to adults who want to update their skills. These opportunities may be particularly important in rural Alaska.

Alaskans have demonstrated many times that they can seize opportunities to achieve communications objectives. The goal of bringing telephone service to rural Alaska villages in the 1970s may have seemed an impossible dream to those who assumed that if the only long-distance carrier in the state refused, then it couldn't be done. Edwin Parker points out: "Having a clear goal is essential, but not sufficient to bring about the necessary changes. What is also needed is a vision of how the technology, the economics and the politics can be brought together to achieve the goal. Beyond that vision, it takes perseverance over an extended period to make that vision widely shared and to find a way around all of the barriers (and there are always many) in the way of implementing the vision."[20]

Communications plans and policies for Alaska must take into consideration the needs and aspirations of *all* Alaskans. Although Alaska's rural population is relatively small, it is also young. For example, the median age in the Wade Hampton Census Area of western Alaska is about 22 years of age, and in the Kotzebue and Bethel regions it is less than 26 years.[21] Rural Alaska youth will grow up using computers, smartphones, and other mobile devices, but they will also need employment opportunities if they are to remain in their communities as adults. Broadcaster and communications advocate Augie Hiebert's call to action in 1970 remains relevant today: "Someone must speak for the isolated villages; someone must be concerned about opening the world educationally. . . . Someone with policy making powers and judicial wisdom must care about Alaskans outside the mainstream of advantages we take for granted."[22]

Abbreviations and Acronyms

2G	Second generation mobile network (for voice and text)
3G	Third generation high-speed mobile network
4G	Fourth generation high-speed mobile network
ABA	Alaska Broadcasting Association
ACS	Alaska Communications System
AEBC	Alaska Educational Broadcasting Commission
AFHCAN	Alaska Federal Health Care Access Network
AFN	Alaska Federation of Natives
ALOH	average length of haul (distance of long-distance telephone connection)
AMCEE	Association for Media-based Continuing Education for Engineers
ANMC	Alaska Native Medical Center
ANTHC	Alaska Native Tribal Health Consortium
APBC	Alaska Public Broadcasting Commission (formerly AEBC)
APSC	Alaska Public Service Commission
APUC	Alaska Public Utilities Commission
ARCS	Alaska Rural Communications System
ARDOR	Alaska Regional Development Organization
ARRA	American Recovery and Reinvestment Act of 2009
ATA	Alaska Telephone Association
ATS	Applied Technology Satellite
AWN	Alaska Wireless Network
BC Tel	British Columbia Telephone, formerly owned by GTE (now part of Telus)
BMEWS	Ballistic Missile Early Warning System
C-band	Portion of the electromagnetic spectrum in the 4 to 6 GHz range used for satellite communications
CAA	Civil Aviation Administration, predecessor of FAA

CAP	Civil Air Patrol
CBC	Canadian Broadcasting Corporation
CDMA	Code Division Multiple Access
Comsat	Communication Satellite Corporation
CPB	Corporation for Public Broadcasting
DAMA	Demand Assignment Multiple Access. A way of sharing a channel's capacity by assigning frenquencies on demand to an idle channel or unused time slot.
DEW Line	Distant Early Warning Line
DDD	Direct Distance Dialing
DOA	Alaska Department of Administration
DOE	Alaska Department of Education
Downlink	transmission from a satellite
ECG or EKG	electrocardiogram
FAA	Federal Aviation Administration
FirstNet	First Responder Network Authority (an integrated national network for public safety communications)
GCI	General Communications, Inc., later simply GCI
GED	General Educational Development, a series of tests that, when passed, are recognized as the equivalent of a high school diploma.
GHz	gigahertz: billion cycles per second
GOT	Governor's Office of Telecommunications (also OT)
GSA	(federal) General Services Administration
GSM	Global System for Mobility, a second generation mobile standard (for voice and text)
GTE	General Telephone and Electronics Corporation
HEW	(federal) Department of Health, Education and Welfare
HF	high frequency
IHS	Indian Health Service (branch of PHS)
Intelsat	International Telecommunications Satellite Organization
ISER	Institute of Social and Economic Research
IXC	interexchange carrier
JCET	Joint Council on Educational Telecommunications
KHz	Kilohertz: thousand cycles per second

Ku-band	Portion of the electromagnetic spectrum in the 11to14 GHz range used for satellite communications
LEC	local exchange carrier
LMR	land mobile radio
Lower 48	the 48 contiguous states
LTE	Long Term Evolution, a fourth generation high-speed mobile technology
LTN	Legislative Teleconferencing Network
MHz	Megahertz: millions cycles per second
MTS	Message toll service (long-distance voice service)
NANA	Northwest Arctic Native Association
NASA	National Aeronautics and Space Administration
NEA	National Education Association
NET	Nippon Educational Television
NIE	National Institute of Education
NIH	National Institutes of Health
NSBSD	North Slope Borough School District
NTIA	National Telecommunications and Information Administration (in the U.S. Department of Commerce)
OPASTCO	Organization for the Promotion and Advancement of Small Telecommunication Companies
Open Skies	A U.S. policy authorizing competition in domestic satellite systems
OT	Governor's Office of Telecommunications (also GOT)
PTI	Pacific Telecom, Inc.
PCC	Public Computing Center
PHS	U.S. Public Health Service
PTT	Post, Telegraph and Telephone
RATNET	Rural Alaska Television Network
RCA	Radio Corporation of America; also Regulatory Commission of Alaska (formerly APUC)
REA	Rural Electrification Administration (in the U.S. Department of Agriculture); now known as the Rural Utilities Service
RTFC	Rural Telephone Finance Cooperative
RUS	Rural Utilities Service (in the U.S. Department of Agriculture)
S-band	Portion of the electromagnetic spectrum in the 2.0 to 2.8 GHz range that can be used for satellite communications

SLED	Statewide Library Education Doorway
TAPS	Trans Alaska Pipeline System
Teledensity	telephone lines per 100 population
TWX	Teletypewriter Exchange Service operated by AT&T
UA	University of Alaska
UAA	University of Alaska Anchorage
UACN	University of Alaska Computer Network
UAF	University of Alaska Fairbanks
UAITC	University of Alaska Instructional Telecommunications Consortium
UAS	University of Alaska Southeast
UNESCO	United Nations Educational, Scientific and Cultural Organization
Uplink	Transmission to a satellite
USAC	Universal Service Administrative Company
USTA	United States Telephone Association
VCR	video cassette recorder
VHF	very high frequency
VSAT	very small aperture (satellite) terminal
WACS	White Alice Communications System
WAMCATS	The Washington-Alaska Military Cable and Telegraph System
WATS	Wide Area Telephone Service (toll-free long-distance; "800 service")

Notes

Chapter 1

1. Hudson, Heather E. *How Close They Sound: Applications of Telecommunications for Public Participation in Alaska*. Report to UNESCO, May 1981.
2. Neering, Rosemary. *Continental Dash*. Ganges, BC: Horsdal and Schubart, 1989.
3. For example, "Report of the Governor of the District of Alaska to the Secretary of the Interior, 1906." Washington, DC, Government Printing Office, 1906.
4. Mitchell, William L., US Army Air Corps. "The Opening of Alaska." Missoula, MT: Pictorial Histories Publishing Company and Anchorage: Alaska Historical Society, April 1982.
5. Institute of Social and Economic Research. "Going Private: The 1968 Sale of the Alaska Communication System." Research Summary 59, University of Alaska Anchorage, December 1997.
6. Hudson, H. E., and Parker, E. B. "Medical Communication in Alaska by Satellite." *New England Journal of Medicine*, Vol. 289, Issue 25, December 20, 1973, pp. 1351–1356.
7. Ferguson, S., and Kokesh, J. "What Works: Outcomes Data from AFHCAN and ANTHC Telehealth: An 8 Year Retrospective." Alaska Native Tribal Health Consortium, 2011.
8. *Tobeluk v. Reynolds*, C.A. No. 72-2450 (originally filed *as Hootch v. Alaska State-Operated School System*), Alaska Super. Ct., 3rd Dist. (Anchorage). 1976. (The "Molly Hootch" case.)
9. Hudson, 1981, ibid.
10. Hudson, Heather E. et al. "Universal Access: What have we learned from the E-Rate?" *Telecommunications Policy*, vol. 28, issues 3–4, April–May 2004, pp. 309–321.
11. Derived from annual data available at www.usac.org.
12. Hudson, Heather E. et al. *Toward Universal Broadband in Rural Alaska: A Report to the State Broadband Task Force*. Institute of Social and Economic Research, University of Alaska Anchorage, November 2012.
13. Ibid.
14. State of Alaska Telecommunications and Information Technology Plan Passed by the Telecommunications Information Council, Lt. Governor Fran Ulmer, Chair. December 18, 1996.
15. Chlupach, Robin Ann. *Airwaves over Alaska: The Story of Broadcaster Augie Hiebert*. Issaquah, WA: Sammamish Press, 1992.

Chapter 2

1. Standage, Tom. *The Victorian Internet*. New York: Walker and Company, 1998, p. xiv.
2. *Scientific American*, 1852, quoted in (1) Standage, Tom. *The Victorian Internet*. New York: Walker and Company, 1998, p.57.
3. Standage, *The Victorian Internet*, 80.
4. Standage, *The Victorian Internet*, 87.
5. Quoted in Green, Dianne. *In Direct Touch with the Wide World: Telecommunications in the North 1865-1992*. Whitehorse: Northwestel, 1992, p. 5.
6. Neering, Rosemary. *Continental Dash*. Ganges, BC: Horsdal and Schubart, 1989, p. 31.
7. Green, *In Direct Touch with the Wide World*, 5
8. Ibid., 6.
9. Whymper, Frederick. "A Journey from Norton Sound, Bering Sea, to Fort Youkon (Junction of Porcupine and Youkon Rivers)." London, W. Clowes and Sons. Read before the Royal Geographical Society of London, April 27, 1868, p. 1.
10. Quoted in Duhse, R.J. "Grand Failure." *Alaska Rural Life*, November 1986, p. 21.
11. Kennan, George. *Tent Life in Siberia: And Adventures Among the Koraks and Other Tribes in Kamtchatka and Northern Asia*. New York: G.P. Putnam and Sons, 1870, p. iii.
12. Whymper, "A Journey from Norton Sound, 1–2.
13. Green, *In Direct Touch with the Wide World*, 7.
14. Whymper, "A Journey from Norton Sound," 17.
15. Whymper, "A Journey from Norton Sound, 16–17
16. Standage, *The Victorian Internet*, 88.
17. Whymper, "A Journey from Norton Sound.
18. Ibid., 1–2.
19. Green, *In Direct Touch with the Wide World*, 8–9.
20. Whymper, "A Journey from Norton Sound, 17.
21. Duhse, R.J. "Grand Failure." *Alaska Rural Life*, November 1986, p. 21.
22. Neering, Rosemary. *Continental Dash*. Ganges, BC: Horsdal and Schubart, 1989, p. 210.
23. Woodman, Lyman L. Preface to Mitchell, William L., US Army Air Corps. "The Opening of Alaska." Edited by Lyman L. Woodman. Missoula, MT: Pictorial Histories Publishing Company and Anchorage: Alaska Historical Society, April 1982, p. ix.
24. Jenne, Theron L., and Harry R. Mitchell. "Military Long Lines Communications in Alaska 1900–1976." In *Telecommunications in Rural Alaska*, Robert Walp, ed. Honolulu: Pacific Telecommunications Council, 1982, p. 13.
25. Office of Information, Air Force Communications Service. "The Alaska Communication System Story." Richards-Gebaur Air Force Base, Missouri, p. 2.
26. Dawson City Daily News, Sept 28, 1899 quoted in Green, *In Direct Touch with the Wide World*, 13.
27. Dawson City Daily News, Sept 28, 1899 quoted in Green, *In Direct Touch with the Wide World, Telecommunications in the North 1865-1992*, 14.
28. Ibid.
29. "ACS Observes 49th Anniversary in Alaska." *Alaska Communication System Bulletin*, 1949, p. 1.
30. Report of the Chief Signal Officer, USA, 1902, p. 7
31. Report of the Signal Officer at Ft. St. Michael, Aug 12, 1901, Appendix 3, Report of the Chief Signal Officer, USA, War Department, Washington, 1901, p. 37.
32. Report of the Chief Signal Officer, USA, 1902, p. 6.

33. Jenne and Mitchell, "Military Long Lines Communications, 14.

34. Report of Chief Signal Officer, 1904, p. 12.

35. Report of the Chief Signal Officer, USA, 1901, p. 8

36. Report of the Chief Signal Officer, USA, 1901, p. 10.

37. Mitchell, William. "Building the Alaska Telegraph System." *National Geographic*, Vol. XV, No. 9, September 1904, p. 360.

38. Report of the Chief Signal Officer, USA, 1902, p. 11.

39. Report of the Chief Signal Officer, USA, 1902, p. 11.

40. Mitchell, William. "Billy Mitchell in Alaska." *American Heritage Magazine*, Vol. 12, Issue 2, February 1961.

41. Report of the Chief Signal Officer, USA, 1903, p. 4.

42. Report of the Chief Signal Officer, USA, War Department, Washington, October 3, 1903, p. 4.

43. Mitchell, William. "Building the Alaska Telegraph System." *National Geographic*, Vol. XV, No. 9, September 1904, p. 361.

44. Mitchell, William. "Building the Alaska Telegraph System." *National Geographic*, Vol. XV, No. 9, September 1904, p. 361.

45. Mitchell, William. "Billy Mitchell in Alaska." *American Heritage Magazine*, Vol. 12, Issue 2, February 1961.

46. Office of Information, Air Force Communications Service. "The Alaska Communication System Story." Richards-Gebaur Air Force Base, Missouri, p. 3.

47. Report of the Chief Signal Officer of the United States Army, War Department, Washington, DC, October 4, 1904, p. 3.

48. Ibid., 4.

49. Ibid., 3.

50. Quirk, William A. III. "Historical Aspects of Building of the Washington, D.C.-Alaska Military Cable and Telegraph System, with Special Emphasis on the Eagle-Valdez and Goodpaster Telegraph Lines 1902–1903." U.S. Department of the Interior, Bureau of Land Management, May 1974, p. 9.

51. Quoted in Reid, Sean. "Alaska Calling." *Alaska Magazine*, May 1982, p. 37.

52. Report of Chief Signal Officer, 1904, p. 13.

53. Ibid., 5–6.

54. Ibid., p. 8.

55. Office of Information, Air Force Communications Service. "The Alaska Communication System Story." Richards-Gebaur Air Force Base, Missouri, p. 3

56. Report of Chief Signal Officer, 1904, p. 10.

57. Office of Information, Air Force Communications Service. "The Alaska Communication System Story." Richards-Gebaur Air Force Base, Missouri, p. 2.

58. Report of Chief Signal Officer, 1904, pp. 12–13.

59. Report of the Governor of Alaska, 1906, p. 17.

60. Report of the Governor of Alaska 1908.

61. Report of the Governor of Alaska: 1910, p. 20.

62. Blanchard, Morgan R. 2010. "Wires, Wireless and Wilderness: A Sociotechnical Interpretation of Three Military Communication Stations on the Washington Alaska Military Cable and Telegraph System (WAMCATS)." PhD dissertation, University of Nevada, Reno, pp. 470–471.

63. Report of the Governor of Alaska: 1914, p. 38.

64. Report of the Governor of Alaska: 1915, p. 42.

65. Report of the Governor of Alaska, 1918, p. 77.

66. Ibid.

67. Report of the Governor of Alaska: 1919, p. 19.

68. Ibid., 48.

69. Ibid., p. 49.

70. Blanchard, Morgan R. 2010. "Wires, Wireless and Wilderness: A Sociotechnical Interpretation of Three Military Communication Stations on the Washington Alaska Military Cable and Telegraph System (WAMCATS)." PhD dissertation, University of Nevada, Reno, p. 460.

71. Blanchard, Morgan R. 2010. "Wires, Wireless and Wilderness: A Sociotechnical Interpretation of Three Military Communication Stations on the Washington Alaska Military Cable and Telegraph System (WAMCATS)." PhD dissertation, University of Nevada, Reno, p. 472.

72. Report of the Governor of Alaska: 1932, p. 50.

73. Report of the Governor of Alaska: 1932, p. 49.

74. Jenne and Mitchell, "Military Long Lines Communications," 15.

75. Office of Information, Air Force Communications Service. "The Alaska Communication System Story." Richards-Gebaur Air Force Base, Missouri, p. 4

76. Jenne and Mitchell, "Military Long Lines Communications," 15.

77. Office of Information, Air Force Communications Service. "The Alaska Communication System Story." Richards-Gebaur Air Force Base, Missouri, p. 7

78. Jenne and Mitchell, "Military Long Lines Communications," 15.

79. Ibid., 16.

80. "ACS Observes 49th Anniversary in Alaska." *Alaska Communication System Bulletin*, 1949, p. 2.

81. Commendations of General William O. Butler, General Orders Number 133, Field Headquarters, Eleventh Air Force, Seattle, Washington, 26 November, 1943, p. 3.

82. Jenne and Mitchell, "Military Long Lines," 16.

83. Green, *In Direct Touch with the Wide World*, 28.

84. Jenne and Mitchell, "Military Long Lines," 17.

Chapter 3

1. Gemmill, Henry. "White Alice: She Links Outposts in Alaska, Eventually May Girdle the Globe." *Wall Street Journal*, April 22, 1957.

2. "ACS Observes 49th Anniversary in Alaska." *Alaska Communication System Bulletin*, 1949, p. 3.

3. Bob Bartlett, Alaskan Delegate to Congress in House of Representatives on March 21, 1948, quoted in "ACS Observes 49th Anniversary in Alaska." *Alaska Communication System Bulletin*, 1949, p. 4.

4. "ACS Observes 49th Anniversary in Alaska." *Alaska Communication System Bulletin*, 1949, p. 2.

5. Ibid.

6. Annual Report of the Governor of Alaska, 1955, p. 19.

7. Ibid., 20.

8. Ibid.

9. Schindler, J.F. and D. A. Underwood. "Communications on the National Petroleum Reserve—Alaska (NPRA)." In *Telecommunications in Rural Alaska*, Robert Walp, ed. Honolulu: Pacific Telecommunications Council, 1982, p. 53.

10. Ibid., 49.

11. Ibid.

12. An audio recording of the telephone conversation between Prime Minister Diefenbaker and President Kennedy on July 22, 1961, is available at: http://www.usask.ca/diefenbaker/galleries/virtual_exhibit/cuban_missile_crisis/finlayson_to_dief.php.

13. Green, Dianne. *In Direct Touch with the Wide World: Telecommunications in the North 1865–1992*. Whitehorse: Northwestel, 1992, p. 52.

14. Huber, Louis R. "Urgent Call from Ugashik." *The Alaska Sportsman*, July 1944, p. 32.

15. Ibid.

16. Ibid., 33.

17. Ibid.

18. Ransom, Jay Ellis. "Party Line Phone." *The Alaska Sportsman*, March 1946, p. 10

19. Huber, "Urgent Call from Ugashik," 11.

20. Ransom, "Party Line Phone," 47.

21. Ibid., 11.

22. Ibid., 47.

23. Ibid., 48.

24. Chlupach, Robin Ann. *Airwaves over Alaska: The Story of Broadcaster Augie Hiebert*. Issaquah, WA: Sammamish Press, 1992, p. 46.

25. Hiebert, Augie. "From Tubes to Chips." Presentation at Alaska Broadcasters Association Alaska Day at the FCC, September 26, 1996.

26. Ibid.

27. Gemmill, Henry. "White Alice: She Links Outposts in Alaska, Eventually May Girdle the Globe." *Wall Street Journal*, April 22, 1957.

28. Office of Information, Air Force Communications Service. "The Alaska Communication System Story." Richards-Gebaur Air Force Base, Missouri, p. 4

29. Jenne, Theron L., and Harry R. Mitchell. "Military Long Lines Communications in Alaska 1900–1976." In *Telecommunications in Rural Alaska*, Robert Walp, ed. Honolulu: Pacific Telecommunications Council, 1982, p. 17.

30. Jack Johnson quoted in Zahniser, Jim. "Jack Johnson '30: Alaska radio pioneer." *The Maine Alumnus*, Vol. 60, no. 4, Fall 1979, p. 11.

31. Western Electric Company. "White Alice." Pamphlet, undated. No page numbers.

32. Ibid.

33. Gemmill, "White Alice," 3.

34. Western Electric Company. "The DEW Line Story." Pamphlet undated, p. 6.

35. Ibid., p. 8.

36. "Contractor Chosen for Radio in Arctic." New York Times, February 22, 1955, quoted in Western Electric Company. "The DEW Line Story." Pamphlet undated, p. 13.

37. Western Electric Company, "White Alice."

38. Office of Information Services, Headquarters Alaska Air Command. "White Alice Fact Sheet," March 26, 1958, p. 1.

39. Western Electric Company. "White Alice."

40. Gemmill, "White Alice," 5.

41. Western Electric Company. "White Alice."

42. Ibid.

43. Gemmill, "White Alice," 9.

44. Western Electric Company. "White Alice."

45. Necrason, Brigadier General C.F., USAF. "With the Lady Known as White Alice. . ." *ITT World Wide Service Reporter* 1(2) 1959, p. 3.

46. Ibid.

47. Jenne and Mitchell, "Military Long Lines," 18.

48. Rigert, Joe. "Communicating was Problem in Hours after Earthquake." *Anchorage Daily Times*, April 9, 1964, p. 13.

49. Ibid.

50. Ibid.

51. Office of Information, Air Force Communications Service. "The Alaska Communication System Story." Richards-Gebaur Air Force Base, Missouri, "ACS Sidebars" p. 2

Chapter 4

1. Ransom, Jay Ellis. "Party Line Phone." *The Alaska Sportsman*, March 1946, p. 48.

2. Duncan, Tom. News and Entertainment Radio Broadcasting in Alaska, 1922–1974. Presentation at 10th Annual Alaska Broadcasters' Association Convention, Fairbanks, June 8, 1974, p. 3.

3. Ibid., 5.

4. Ibid., 8.

5. Ibid.

6. Ibid., 11.

7. Ibid., 13.

8. Ibid.

9. Duncan, John Thomas. "Alaska Broadcasting 1922–77: An Examination of Government Influence." Ph.D. Dissertation, University of Oregon, 1982, p. 25.

10. Juneau Sunday-Capitol, October 8, 1922, quoted in Duncan, Tom. News and Entertainment Radio Broadcasting in Alaska, 10.

11. Duncan, Alaska Broadcasting, 33.

12. Quoted in Ibid., 34.

13. Ibid., 61.

14. Duncan, News and Entertainment Radio, 14.

15. Chlupach, Robin Ann. *Airwaves over Alaska: The Story of Broadcaster Augie Hiebert*. Issaquah, WA: Sammamish Press, 1992, p. 34.

16. August Hebert [sic]: Alaska TV Pioneer prepares for the satellite age." *Alaska Business*, October 1982, p. 20.

17. Ibid., 19.

18. Ibid.

19. Ibid, 20.

20. Chlupach, "Airwaves over Alaska," 46–48.

21. Hiebert, "From Tubes to Chips." Presentation at Alaska Broadcasters Association Alaska Day at the FCC, September 26, 1996.

22. *New York Times*, May 23, 1942, Quoted in Chlupach, *Airwaves over Alaska*, 48–49.

23. Chlupach, *Airwaves over Alaska*, 42–43.

24. Ibid., 63.

25. Duncan, "Alaska Broadcasting 1922–77," 164.

26. "History of AFRTS: The First 50 Years." Armed Forces Radio and Television Services, 1992, p. 20. Available at http://afrts.dodmedia.osd.mil/heritage/page.asp?pg=50-years.

27. Stevens, Joe. "KODK Kodiak AFRS Radio 1941." Available at http://www.radio heritage.net/story65.asp

28. "History of AFRTS: The First 50 Years," 19.

29. Ibid.

30. Duncan, "Alaska Broadcasting 1922–77," 3.

31. Ibid., 96.

32. Armed Forces Radio and Television Services Timeline. Available at http://afrts .dodmedia.osd.mil/heritage/heritage.asp

33. Duncan, "Alaska Broadcasting 1922–77," 126.

34. Ibid., 127.

35. Chlupach, *Airwaves over Alaska*, 90.

36. Maxwell, Lauren. "New broadcast center dedicated to Augie Hiebert." See http://www .ktva.com/ktva-celebrates-60-years-on-the-air/.

37. Chlupach, *Airwaves over Alaska*, 89.

38. "About KTUU-TV." See http://articles.ktuu.com/2010-07-14/broadcast-license_ 24129016

39. August Hebert [sic]: Alaska TV Pioneer prepares for the satellite age." *Alaska Business*, October 1982, p. 32.

40. Chlupach, *Airwaves over Alaska*, 216–217.

41. August Hebert [sic]: Alaska TV Pioneer prepares for the satellite age," 17.

42. Duncan, "Alaska Broadcasting 1922–77," 142.

43. Hiebert, "From Tubes to Chips."

44. Bramstedt, Alvin, Jr. interviewed by Hilary Hilscher, July 24, 2001.

45. Chlupach, *Airwaves over Alaska*, 95.

46. Duncan, "Alaska Broadcasting 1922–77," 3.

47. See Duncan, John Thomas. "Alaska Broadcasting 1922–77: An Examination of Government Influence." Ph.D. Dissertation, University of Oregon, 1982, and Duncan, Tom. "News and Entertainment Radio Broadcasting in Alaska, 1922–1974." Presentation at th Annual Alaska Broadcasters' Association Convention, Fairbanks, June 8, 1974.

Chapter 5

1. Institute of Social and Economic Research. "Going Private: The 1968 Sale of the Alaska Communication System." Research Summary 59, University of Alaska Anchorage, December 1997, p.4.

2. Jenne, Theron L. and Harry R. Mitchell. "Military Long Lines Communications in Alaska 1900–1976." In *Telecommunications in Rural Alaska*, Robert Walp, ed. Honolulu: Pacific Telecommunications Council, 1982, p. 18.

3. Ibid., 1–4.

4. Quoted in Ibid., 1–1.

5. Study of the Alaska Communication System—Its Past, Present and Future prepared by staff members for Senator Bartlett; included in Congressional Record, Vol. 111, no. 156, August 24, 1965.

6. Isenson, Ed. Two Major Choices on Disposal of ACS Faces [sic] Congress." *Anchorage Daily News*, May 31, 1966.

7. "Action at Last on Long Distance Rates." *Anchorage Daily News*, March 15, 1965.

8. Act of May 26, 1900 (48 U.S.C. 310).

9. Jones, Douglas N. and Bradford H. Tuck. "Privatization of State-Owned Utility Enterprises: The United States Has Done It Too." *Critical Issues in Cross-National Public Administration*, ed. Stuart S. Nagel. Westport, CN: Quorum Books, 2000.

10. *Middle West Study*, prepared for the Air Force at Air Force request by the Middle West Service Co., September 1, 1963, cited in Ibid.

11. *Perry Report*, submitted by the Deputy of Transportation and Communications as an attachment to a document sent by the Assistant Secretary of the Air Force to the Assistant Secretary of Defense (Installations and Logistics), March 9, 1965, Ibid.

12. *Economic Development in Alaska*, A report to the President prepared by the Federal Field Committee for Development Planning in Alaska, August 1966, p. 36, quoted in Ibid.

13. Senator E. L. Bartlett, "Disposal of Government-owned long-lines communication facilities in Alaska." *Congressional Record*, Vol. 111, no. 156, August 24, 1965.

14. "Study of the Alaska Communication System—Its Past, Present and Future prepared by staff members for Senator Bartlett; included in Congressional Record, Vol. 111, no. 156, August 24, 1965.

15. William Lawrence, ACS senior civilian in Seattle, quoted in Ibid.

16. Cyrus H. McLean, chairman of BC Tel, quoted in Ibid.

17. Ibid.

18. Ibid.

19. Ibid.

20. Ibid.

21. Ibid.

22. "Rate, Service Controls Favored in Sale of ACS." *Anchorage Daily News*, June 1, 1966.

23. Elmer Rasmuson quoted in Isenson, Ed. "Phone, Telegraph Service here below Level in other States." *Anchorage Daily News*, June 1, 1966.

24. Humphries, Harrison. "Lower Rates urged as Key to ACS Sale." *Anchorage Daily Times*, May 31, 1966.

25. Isenson, Ed. "Experts See Better Service in ACS Sale." *Anchorage Daily News*, June 2, 1966.

26. Rate, Service Controls Favored in Sale of ACS." *Anchorage Daily News*, June 1, 1966.

27. Isenson, Ed. "Phone, Telegraph Service Here Below Level in Other States." *Anchorage Daily News*, June 1, 1966.

28. "The Sale of ACS: Let's Get on with It." Editorial in *Anchorage Daily Times*, June 1, 1966.

29. Humphries, Harrison. "ACS Sale Chances Good." *Anchorage Daily Times*, June 1, 1966.

30. Archibald, Janet. "ACS Commander says 'Offer Me $35 Million.'" *Anchorage Daily News*, June 10, 1966.

31. "Lower Rates Said Possible." *Anchorage Daily Times*, June 13, 1966.

32. "New Cable Pact reduces Rate on Alaska Calls." *Anchorage Daily Times*, August 22, 1966.

33. "The Sale of ACS." Editorial, *Anchorage Daily Times*, October 3, 1967.

34. Alaska Communications Disposal Act. Public Law 90-135, 90th Congress, S. 223, November 14, 1967.

35. Ibid.

36. Jones, Douglas N. "What We Thought We Were Doing in Alaska, 1965–1972." *The Journal of Policy History*, Vol. 22, No. 2, 2010, pp. 4–5.

37. Study of the Alaska Communication System.

38. Study of the Alaska Communication System.

39. "Ridiculous Delay." Editorial, *Anchorage Daily Times*, December 7, 1967

40. Brent, Stephen. "Why It's a Pain to Call Long-Distance." *Anchorage Daily News*, August 3, 1968.

41. Senator Ted Stevens, "Modern Communications for Alaska, "*Congressional Record*, Vol. 115, no. 36, Feb 28, 1969

42. Senate Committee [on Armed Services] quoted Ibid.

43. Department of the Air Force, Office of the Secretary. "Notice of Acceptance of Offer." Washington, DC, July 1, 1969. Attachment to letter to Howard R. Hawkins, President, RCA Global Communications, Inc.

44. Ibid.

45. Ibid.

46. Federal Communications Commission. RCA Alaska Communications Inc.: Memorandum Opinion and Order Instituting a Hearing. Docket No. 18823, FCC 70-304, Released March 25, 1970

47. Ibid.

48. Ibid.

49. Ibid.

50. Goldschmidt, Douglas. 1979. Joint Goods, Public Goods and Telecommunications: A Case Study of the Alaskan Telephone System." PhD Dissertation, University of Pennsylvania, p. 131.

51. Regulatory Commission of Alaska. "Commission." Available at http://rca.alaska.gov/RCAWeb/AboutRCA/Commission.aspx.

52. Jones, Douglas N. and Bradford H. Tuck. "Privatization of State-Owned Utility Enterprises: The Alaska Case Revisited Thirty Years Later." Occasional Paper 21, National Regulatory Research Institute, The Ohio State University, November 1997.

53. Alaska Public Utilities Commission. Order Granting Certificate of Public Convenience and Necessity. Order No. Gabel, Richard, Edward C. Hayden, and William H. Melody. "Planning for Telecommunication System Development in Alaska." Appendix B in Staff Paper, The Public Interest Case in Alaska Public Interest Utilities Commission Hearing on Docket U-69-24. Office of Telecommunications, U.S. Department of Commerce, October 2, 1970, p. B-29.

54. Ibid., p. B-34.

55. Jones and Tuck, Ibid.

56. Isenson, Ed. "Many Consider ACS Service Inadequate; Solutions Sought." *Anchorage Daily News*, May 29, 1966.

57. Federal Communications Commission. "In the Matter of RCA ALASKA COMMUNICATIONS, INC." Application for Authority to operate channels of communication. . . . Application of RCA Communications, Inc. No docket number. September 25, 1969.

58. Ibid.

Chapter 6

1. Augie Hiebert, quoted in Chlupach, Robin Ann. *Airwaves over Alaska: The Story of Broadcaster Augie Hiebert.* Issaquah, WA: Sammamish Press, 1992, p. 148

2. For more details on the early days of satellite communications, see Hudson, Heather E. *Communication Satellites: Their Development and Impact.* New York, Free Press, 1990.

3. Chlupach, Robin Ann. *Airwaves over Alaska: The Story of Broadcaster Augie Hiebert.* Issaquah, WA: Sammamish Press, 1992, p. 128.

4. Ibid.

5. Ibid., 127.

6. Ibid., 130.

7. "Alaska's Satellite Station." *Anchorage Daily News*, May 10, 1968.

8. Bartlett, Senator E.L. "Satellite Ground Station for Alaska." Press conference, Anchorage, March 30, 1968.

9. "State Support for Satellite?" *Anchorage Daily News*, March 15, 1968.

10. "A Milestone Decision for Alaska." Editorial, *Anchorage Daily News*, April 25, 1968.

11. Ibid.

12. Bartlett, "Satellite Ground Station."

13. "Comsat to tell Bidders on ACS Satellite Plans." *Anchorage Daily Times*, August 28, 1968

14. Chlupach, *Airwaves over Alaska*, 131.

15. Joint Resolution No. A/F-1-69 A RESOLUTION OF THE CITIES OF ANCHORAGE AND FAIRBANKS ENDORSING A REVIEW OF THE COMMUNICATIONS SERVICES FOR ALASKA. Fairbanks, February 21, 1969.

16. Stevens, Senator Ted. Letter to George M. Sullivan, Mayor of Anchorage, March 6, 1969.

17. Federal Communications Commission. "Comsat Okayed to build Alaska Satellite Earth Station; would link Alaska, Continental U.S., Hawaii, Japan, Pacific area." Nonbroadcast and General Action Report No. 3560, Public Notice, May 14, 1969.

18. Chlupach, *Airwaves over Alaska*, 132.

19. Ibid., 133.

20. Statement of Senator Ted Stevens. In the Matter of and in re Applications of Communications Satellite Corporation For construction permits for three new stations in the Domestic Public Point-to-Point Microwave Radio Service at Talkeetna, Scotty Lake, and Twelvemile, Alaska et al To the Review Board, Before the Federal Communications Commission, December 19, 1969.

21. Alaska Conference on Satellite Telecommunications, Anchorage, Alaska, August 28 and 29, 1969. Agenda and attendees.

22. Sharrock, George O. Chairman, Federal Field Committee for Development Planning in Alaska. Letter to Honorable Myron Tribus, Assistant Secretary of Commerce for Science and Technology, September 4, 1969.

23. Chlupach, *Airwaves over Alaska*, 135–136

24. Communications Satellite Corporation. "Comsat offers plan for early start on U.S. satellite demonstrations. Press release, Washington, DC, June 13, 1969.

25. Charyk, Joseph V., President, Communications Satellite Corporation. Letter to Honorable Keith H. Miller, Governor of Alaska, October 6, 1969.
26. McCormack, James, Chairman, Communications Satellite Corporation. Letter to Senator Ted Stevens, July 18, 1969.
27. McCormack, James. Letter to Robert W. Sarnoff, President, Radio Corporation of America, June 26, 1969.
28. Charyk, Joseph V., President, Communications Satellite Corporation. Letter to Howard A. Hawkins, President, RCA Alaska Communications, November 4, 1969.
29. Charyk, Joseph V., President, Communications Satellite Corporation Letter to Honorable Dean Burch, Chairman, Federal Communications Commission, October 19, 1970.
30. "Dedication is Tuesday for Bartlett Earth Station." *Anchorage Daily Times*, June 25, 1970.
31. "Satellite Era Begun by Alaska." *Anchorage Daily Times*, June 30, 1970.
32. Chlupach, *Airwaves Over Alaska*, 152.
33. Ibid., 222–223
34. Chlupach, pp. 149–50. FULL TRANSCRIPT: A.G. Hiebert, "Statement Prepared for Public Service Commission RCA Alascom Hearings, June 10, 1970. Chlupach, pp. 218–221.
35. Quoted in Williams, Andy. "Anchorage Television Beamed Live to Juneau." *Anchorage Daily News*, May 8, 1972.
36. "Gravel Requesting Satellite 'Favors.'" *Anchorage Daily Times*, April 28, 1971.
37. Chlupach, *Airwaves over Alaska*, 162–163
38. Comsat. "Alaska/Comsat cooperative demonstration to provide TV and voice communications via satellite to remote Alaskan areas." Press release, Washington, DC, April 16, 1972.
39. Chlupach, *Airwaves over Alaska*, 152–153
40. *Southeast Alaska Empire*, May 4, 1972 quoted in Chlupach, p. 163.
41. "Tests to Prove Small Receiver," *Anchorage Daily News*, May 8, 1972.
42. "Satellite Television Demonstration." *The Nome Nugget*, June 2, 1972.
43. Chlupach, *Airwaves over Alaska*, 164.
44. Comsat. Preliminary Results of Alaskan Test/Demonstration Program. Attachment No. 1 to ICSC-62-33E W/12/72.
45. Ibid.
46. "More Communities for ETV System?" *Anchorage Daily News*, December 1969, p. 3.
47. Burch, Dean, Chairman, FCC. Letter to Senator Ted Stevens re Comsat, August 13, 1973.
48. Ibid.

Chapter 7

1. Quoted in Satellite House Call." Film produced by Department of Communication, Stanford University, 1974. Available at http://www.youtube.com/watch?v=GzclgBfn_yY.
2. Merritt, Robert P. "Alaska Telecommunications." In *Telecommunications in Rural Alaska*, Robert Walp, ed. Honolulu: Pacific Telecommunications Council, 1982, p. 6.

3. "ATS." NASA Goddard Space Flight Center Mission Archives. http://www.nasa.gov/centers/goddard/missions/ats.html

4. Stanley, Glenn M. "Radio Communications in Alaska and other Remote Areas with Special Reference to Satellite Telecommunications." University of Alaska Geophysical Institute Annual Report, 1971–72. Fairbanks, 1972, pp. 123–124.

5. Rostow, Eugene S. President's Task Force on Communications Policy. Final Report. Washington, DC, December 1968, p. 14.

6. Senator Mike Gravel, speaking at Presentation at Institute of Social and Economic Research (ISER) symposium "The Evolution of Telecommunications in Alaska: From Bush Telephones to Broadband," University of Alaska Anchorage, June 13, 2011.

7. "The Lister Hill Center's Experimental Satellite Communications Project." December 1970. (copy from office of Senator Mike Gravel).

8. Ibid.

9. Northrip, Charles M. "Developments in Alaska." Alaska Educational Broadcasting Newsletter, Volume III, Number 5, Fairbanks, May–June, 1970, p. 1.

10. Ibid., p. 3.

11. Ibid., p. 5.

12. "The Lister Hill Center's Experimental Satellite Communications Project." December 1970 (copy from office of Senator Mike Gravel).

13. Stanley, Glenn M. "Radio Communications in Alaska and Other Remote Areas with Special Reference to Satellite Telecommunications." University of Alaska Geophysical Institute Annual Report, 1971–72. Fairbanks, 1972, p. 124.

14. Hudson, Heather E., and Parker, E. B. "Medical Communication in Alaska by Satellite." *New England Journal of Medicine*, Vol. 289, Issue 25, December 20, 1973, pp. 1351–1356.

15. Hudson, Heather E. "Medical Communication in Rural Alaska." In *Telecommunications in Rural Alaska*, Robert Walp, ed. Honolulu: Pacific Telecommunications Council, 1982, p. 59.

16. Kreimer, Oswaldo, et al. "Health Care and Satellite Radio Communication in Village Alaska: Final Report of the ATS-1 Satellite Biomedical Experiment Evaluation." Stanford University: Institute for Communication Research, June 1974, p. 53.

17. *Satellite House Call.* Film produced by Department of Communication, Stanford University, 1974. The film was the Master's thesis of Judy Irving for her degree in documentary film at Stanford. Available at http://www.youtube.com/watch?v=GzclgBfn_yY

18. Hudson and Parker, "Medical Communication in Alaska by Satellite."

19. National Aeronautics and Space Administration. "ATS aides in Alaskan Emergencies." NASA News. Washington, DC, April 5, 1972.

20. Cassirer, Henry R., and Wigren, Harold E. Implications of Satellite Communication for Education: August–September 1970. Paris: UNESCO, 1970.

21. Alaska Educational Broadcasting Commission. "Alaska Satellite Project." August 1970. Request for Unesco consultant to help with planning of ATS-1 education project.

22. Northrip, "Developments in Alaska," p. 1.

23. National Education Association. "Satellite Seminar Launched for Teachers in Remote Alaskan Villages." Washington, DC, January 24, 1973.

24. Karen Michel, personal communication, August 2014. Karen Michel and Roger McPherson coproduced the ATS-1 audio programs.

25. Parker, Walter. "Village Satellite III: The Third Evaluation of the Action Study of Educational Uses of Satellite Communications in Remote Alaskan Communities." undated, p. 68.

26. Ibid., p. 50.

27. "Report on Alaskan Use of the ATS-1 Satellite through September 30, 1972," quoted in Goldschmidt, Douglas. "The Benefits of Satellite Telecommunications in Alaska." In *Telecommunications in Rural Alaska*, Robert Walp, ed. Honolulu: Pacific Telecommunications Council, 1982, p. 55.

28. Chlupach, Robin Ann. *Airwaves over Alaska: The Story of Broadcaster Augie Hiebert.* Issaquah, WA: Sammamish Press, 1992, p. 167

29. "ATS." NASA Goddard Space Flight Center Mission Archives. http://www.nasa.gov/centers/goddard/missions/ats.html

30. Nurse Hilda Silva, personal interview with the author during the Stanford ATS-6 evaluation, Fort Yukon, 1975.

31. Foote, Dennis, Parker, Edwin, and Hudson, Heather. "Telemedicine in Alaska: The ATS-6 Satellite Biomedical Demonstration." Institute for Communication Research, Stanford University, February 1976, p. x.

32. Ibid., p. xi.

33. Ibid., p. xii.

34. Ibid.

35. Walp, Robert M. "First-Hand: S-Band Story." IEEE Global History Network. Available at http://www.ieeeghn.org/wiki/index.php/First-Hand:S-Band_Story.

36. Ibid.

37. Robert D. Arnold interview with Hilary Hilscher, June 30, 2000.

38. Pittman, Theda Sue, and Orvik, J. M. "ATS-6 and State Communications Policy for Rural Alaska: An Analysis of Recommendations." Center for Northern Educational Research, University of Alaska Fairbanks, 1976, p. 2.

39. Ibid., p. 1.

40. Mark O. Badger quoted in James, Beverly, and Daly, P. "Origination of State-Supported Entertainment Television in Rural Alaska." *Journal of Broadcasting and Electronic Media*, Vol. 31, No. 2, Spring 1987, p. 175.

41. Pittman, and Orvik, ATS-6 and State Communications Policy for Rural Alaska, p. vi.

42. Ibid., p.

43. Ibid., p. 22.

44. Ibid., p. 8.

45. Cotton, Stephen E. "Alaska's 'Molly Hootch Case': High Schools and The Village Voice." *Educational Research Quarterly*, Volume Number 4, 1984.

46. Pittman, and Orvik, ATS-6 and State Communications Policy for Rural Alaska, p. 6.

47. Marvin Weatherly interviewed by Hilary Hilscher, December 26, 2000.

48. Chlupach, *Airwaves over Alaska*, p. 169.

49. Quoted in Goldschmidt, Douglas. 1979. "Joint Goods, Public Goods and Telecommunications: A Case Study of the Alaskan Telephone System." PhD Dissertation, University of Pennsylvania, p. 150.

Chapter 8

1. Senator Ted Stevens in *Satellite House Call*. Film produced by Department of Communication, Stanford University. Available at http://www.youtube.com/watch?v=GzclgBfn_yY

2. The author of this book took photographs of the participants to document the demonstration while a doctoral student in communication at Stanford.

3. Recollection of Edwin Parker.

4. Letter from Senator Ted Stevens to Edwin Parker, March 13, 1974. Edwin Parker Alaska correspondence files.

5. Ibid.

6. Albert Horley received a Howard Hughes doctoral fellowship for graduate study at Stanford and was mentored by Robert Walp while Walp was at Hughes. He had completed all of his studies for a Stanford Ph.D. before being appointed to the position at HEW. He later completed his dissertation on broadband electronic communications for Brazil after leaving HEW, with Professor Edwin Parker as a dissertation advisor.

7. RCA Task Force Report from Edwin Parker correspondence files.

8. Letter from Edwin Parker to Governor William Egan, April 17, 1974. Edwin Parker Alaska correspondence files.

9. Letter from Governor William Egan to Howard Hawkins, April 16, 1974. Edwin Parker correspondence files.

10. Letter from Governor William Egan to Howard Hawkins, May 7, 1974. Edwin Parker Alaska correspondence files.

11. Letter from Professors Edwin Parker and Bruce Lusignan to Governor William Egan on May 13, 1974. Edwin Parker Alaska correspondence files.

12. Ibid.

13. Letter from Governor William Egan to Howard Hawkins, June 22, 1974. Edwin Parker Alaska correspondence files.

14. Howard Hawkins letter to Governor William Egan, June 28, 1974. Edwin Parker Alaska correspondence files.

15. Letter from Governor William Egan to Walter Hinchman, June 24, 1974. Edwin Parker Alaska correspondence files.

16. Letters from Senator Ted Stevens to Senator John C. Stennis, chair of the Senate Armed Services Committee, and Representative F. Edward Hebert, chair of the House Armed Services Committee, March 13, 1974. Edwin Parker correspondence files.

17. Cloe, John Haile, with Monaghan, M. F. *Top Cover for America: The Air Force in Alaska 1920–1983.* Anchorage Chapter-Air Force Association; Missoula, Mont.: Pictorial Histories Publishing Company, 1984.

18. Reid, S. "Earth Stations Bring an End to White Alice." *Communications News,* June 1, 1985.

19. Millsap, Pam. "Hickel Raps Egan and RCA." *Anchorage Daily News,* June 28, 1974.

20. "Hickel Urges Satellite Halt." *Anchorage Daily Times,* June 28, 1974, p. 3.

21. Albert Feiner memo to Alaska Native Health Service, May 14, 1974. Edwin Parker Alaska correspondence files.

22. Letter from Professors Edwin Parker and Bruce Lusignan to Gordon Zerbetz, June 18, 1974. Edwin Parker Alaska correspondence files.

23. Hupe, Howard. "Alaska's Communication Needs and Potential Solutions Utilizing Satellites" DHEW internal staff report, undated. Edwin Parker Alaska correspondence files.

24. Letter from Walter Hinchman to Howard Hawkins, July 23, 1974. Edwin Parker Alaska correspondence files.

25. Walp, Robert M. "First-Hand: C-Band Story" IEEE Global History Network. Available at http://www.ieeeghn.org/wiki/index.php/First-Hand:C-Band_Story.

26. Ibid.

27. State of Alaska. "Comments on the September 20 RCA-Alascom Communications Plan." Submitted to the FCC on October 21, 1974. Edwin Parker Alaska correspondence files.

28. Ibid.

29. Ibid.

Chapter 9

1. Robert Walp, interview with Hilary Hilscher, October 2000.

2. Letter from Governor Jay Hammond to Walter Hinchman, February 14, 1975. Edwin Parker Alaska correspondence files.

3. Letter from Edwin Parker to Governor William Egan, July 16, 1974. Edwin Parker correspondence files.

4. Walp, Robert M. "First-Hand: C-Band Story." IEEE Global History Network. Available at http://www.ieeeghn.org/wiki/index.php/First-Hand:C-Band_Story.

5. Ibid.

6. State of Alaska press release, February 21, 1975. Edwin Parker Alaska correspondence files.

7. Walp, "First-Hand: C-Band Story."

8. Marvin Weatherly, interview with Hilary Hilscher, December 26, 2000.

9. Ibid.

10. Walp, "First-Hand: C-Band Story."

11. Edwin Parker Alaska correspondence files.

12. Memo from Glenn Stanley to Governor Jay Hammond, April 9, 1975. Edwin Parker Alaska correspondence files.

13. Memo from Walter B. Parker to Governor Jay Hammond, April 9, 1975. Edwin Parker Alaska correspondence files.

14. Memo from Robert M. Walp to Marvin Weatherly, April 10, 1975. Edwin Parker Alaska correspondence files.

15. Letter from Governor Jay Hammond to Richard Wiley, April 11, 1975. Edwin Parker Alaska correspondence files.

16. Submission of Philip Schneider of RCA to the FCC, April 21, 1975. Edwin Parker Alaska correspondence files.

17. Edwin Parker Alaska correspondence files.

18. *Telecommunication Reports*, May 5, 1975.

19. Letter from Edwin Parker to Marvin Weatherly, May 7, 1975. Edwin Parker Alaska correspondence files.

20. Letter from John W. Pettit, of the firm Hamel, Park, McCabe, and Saunders, counsel for the State of Alaska, to FCC Chairman Richard Wiley, May 30, 1975. Edwin Parker Alaska correspondence files.

21. Walp, "First-Hand: C-Band Story."

22. Richard Edge, also an attorney at the firm Hamel, Park, McCabe, and Saunders.

23. Walp, "First-Hand: C-Band Story."

24. Quoted in Walp, "First-Hand: C-Band Story."

25. Robert Walp, interview with Hilary Hilscher, October 2000.

Chapter 10

1. Robert M. Walp, interview with Hilary Hilscher, October, 2000.

2. Duncan, John Thomas. "Alaska Broadcasting 1922–77: An Examination of Government Influence." Ph.D. Dissertation, University of Oregon, 1982, p. 244.

3. Ibid., p. 268.

4. Ibid., pp. 286–287.

5. See www.knom.org.

6. Duncan, "Alaska Broadcasting 1922–77," p. 4.

7. Ibid., p. 273.

8. Bob Arnold interviewed by Hilary Hilscher, June 30, 2000.

9. Ibid.

10. Alex Hills interviewed by Hilary Hilscher, August 23, 2000.

11. Madigan, Robert J., and Peterson, Jack W. "Television and Social Change on the Bering Strait." University of Alaska Anchorage, April 1974.

12. Duncan, "Alaska Broadcasting 1922–77," p. 352.

13. Marvin Weatherly, interview with Hilary Hilscher, December 26, 2000.

14. Alaska Broadcasting Association. "2009 Special Awards: Hall of Fame: Michael 'Mike' Porcaro." http://www.alaskabroadcasters.org/aba/porcaro.html

15. Hiebert, Augie. "From Tubes to Chips." Presentation at Alaska Broadcasters Association Alaska Day at the FCC, September 26, 1996.

16. Chlupach, Robin Ann. *Airwaves over Alaska: The Story of Broadcaster Augie Hiebert.* Issaquah, WA: Sammamish Press, 1992, pp. 169–173.

17. Hiebert, "From Tubes to Chips."

18. Chlupach, Robin Ann. *Airwaves over Alaska: The Story of Broadcaster Augie Hiebert.* Issaquah, WA: Sammamish Press, 1992, p. 175.

19. Office of Telecommunications, Office of the Governor. "Satellite Television Demonstration Project: Final Report." Juneau, State of Alaska, February 1, 1978, p. 41.

20. Walp, Robert M. "First-Hand: C-Band Story" IEEE Global History Network. Available at http://www.ieeeghn.org/wiki/index.php/First-Hand:C-Band_Story.

21. Office of Telecommunications, Office of the Governor, 1.

22. Walp, "First-Hand: C-Band Story."

23. Office of Telecommunications, Office of the Governor, 7.

24. Quoted in Ibid., 54, italics added in report.

25. Ibid., p. 57.

26. Chlupach, *Airwaves over Alaska*, pp. 175–176.

27. Walp, "First-Hand: C-Band Story."

28. Hiebert, Augie, Chlupach, *Airwaves over Alaska*, p. 176.

29. Hiebert, "From Tubes to Chips."

30. McIntyre, Theodore W. "The Alaska Television Demonstration Project: Past, Present and . . . Future?" In *Telecommunications in Rural Alaska*, Robert Walp, ed. Honolulu: Pacific Telecommunications Council, 1982, p. 89.

31. Wilke, Jennifer. "Everything I Need to Know I Learned from LearnAlaska," Presentation at Institute of Social and Economic Research (ISER) symposium "The Evolution of Telecommunications in Alaska: From Bush Telephones to Broadband," University of Alaska Anchorage, June 13, 2011.

32. Office of Telecommunications, Office of the Governor. "Satellite Television Demonstration Project: Final Report." Juneau, State of Alaska, February 1, 1978, p. 30.

33. Ibid., p. 28.

34. Ibid., p. 23.

35. Ibid., p. 36.

36. McIntyre, "The Alaska Television Demonstration Project," p. 90.

37. Hiebert, Augie, quoted in Chlupach, *Airwaves over Alaska*, p. 178.

38. Hills, Alexander H. "Television and Telephone Service in Alaskan Villages: A Technological, Economic, and Organizational Analysis." Ph.D. Dissertation, Carnegie-Mellon University, 1979, p. 15.

39. Pittman, Theda Sue, Heather E. Hudson and Edwin B. Parker. "Planning Assistance for the Development of Rural Broadcasting." Prepared for the Alaska Public Broadcasting Commission, Anchorage, July 1977, p. iv.

40. Ibid., p. 3.

41. Ibid., pp. v–vi.

42. Ibid., p. vii.

43. Ibid., p. 2.

44. Wilke, Jennifer. "Everything I Need to Know I Learned from LearnAlaska."

45. Bramble, William J. "Technology and Education in Public Schools." In Telecommunications Symposium. 33rd Alaska Science Conference, American Association for the Advancement of Science, Arctic Division, Fairbanks, September 18, 1982, p. 95.

46. Wilke, Jennifer. "Everything I Need to Know I Learned from LearnAlaska."

47. Bramble, "Technology and Education in Public Schools," p. 98.

48. Bennett, F. Lawrence. "Engineering Education via Telecommunications in Alaska." In Telecommunications Symposium. 33rd Alaska Science Conference, American Association for the Advancement of Science, Arctic Division, Fairbanks, September 18, 1982, pp. 102–103.

49. Ibid., pp. 105–106.

50. Senate Concurrent Resolution No. 35 in the Legislature of the State of Alaska, Eleventh Legislature—First Session, "Directing the Legislative Council to conduct a feasibility study relating to educational television," April 23, 1979.

51. Greely, John. "School officials seek $23.5 million for instructional TV Satellite." *Anchorage Daily News*, February 25, 1980.

52. Wilke, Jennifer. "Everything I Need to Know I Learned from LearnAlaska."

53. Demmert, Jane P., and Wilke, Jennifer L. "The LearnAlaska Networks: Instructional Telecommunications in Alaska." In *Telecommunications in Rural Alaska*, Robert Walp, ed. Honolulu: Pacific Telecommunications Council, 1982, p. 70.

54. Bramble, William J. "Technology and Education in Public Schools," p. 94.

55. Ibid., p. 99.

56. Demmert and Wilke. "The LearnAlaska Networks," p. 72.

57. Cotton, Stephen E. "Alaska's 'Molly Hootch Case': High Schools and The Village Voice." *Educational Research Quarterly*, Vol. No. 4, 1984.

58. Demmert and Wilke. "The LearnAlaska Networks," p. 72.

59. Metty, Michael P. "Policy Implications and Constraints of Educational Telecommunications in a SubArctic Region." In Telecommunications Symposium. 33rd Alaska Science Conference, American Association for the Advancement of Science, Arctic Division, Fairbanks, September 18, 1982, p. 80.

60. Fitzgerald, Doreen. "Telecommunications: Where Do We Go from Here?" *University of Alaska Magazine*, Vol. 1, No. 2, Spring 1983.

61. Demmert and Wilke. "The LearnAlaska Networks," p. 73.

62. Fitzgerald, Doreen. "Telecommunications: Where Do We Go from Here?" *University of Alaska Magazine*, Vol. 1, No. 2, Spring 1983, p. 22.

63. Baltes, Kathleen (Teleconference Coordinator, Legislative Affairs Agency, State of Alaska). "Telecommunications Systems Enhance Participatory Government or It Doesn't Have to End When You Leave the Voting Booth." In *Telecommunications in Rural Alaska*, Robert Walp, ed. Honolulu: Pacific Telecommunications Council, 1982, p. 75.

64. Ibid., p. 77.

65. Ibid., p. 78.

66. Ibid., p. 79.

67. Metty, "Policy Implications and Constraints," p. 80.

68. Ibid., p. 82.

69. Office of Telecommunications, Office of the Governor. "Satellite Television Demonstration Project: Final Report." Juneau, State of Alaska, February 1, 1978, pp. 70–71.

70. Bramble, "Technology and Education in Public Schools," p. 100.

71. Chlupach, *Airwaves over Alaska*, p. 181.

Chapter 11

1. Valaskakis, Gail. "The Issue is Control: Northern Native Communications in Canada." Proceedings of the Chugach Conference: Communication Issues of the '90s. Anchorage, October 5–6, 1990, p. 44.

2. Boucher, H.A. "Red." An Interim Report: House Special Committee on Telecommunications. January 13, 1986, pp. 17–18.

3. Ibid., p. 19.

4. Schaible, Grace Berg, Alaska Attorney General. "Alaska request that state pay space and power charges for LPTV minitransmitters." Memo to John Andrews, Deputy Commissioner, Department of Administration, May 12, 1987.

5. Taylor, Richard. "A Crisis in Alaska Television Programming." Unpublished manuscript, January 1987. Cited in Pearson, Larry and Doug Barry. "Talking to Each Other, Talking to Machines: Alaska's Telecommunications Future." Report to the Joint Committee on Telecommunications of the Alaska State Legislature, January 1987, p. 13.

6. Department of Administration, Department of Telecommunications, The House Special Committee on Telecommunications, The Alaska Public Broadcasting Commission, The Office of Management and Budget Division of Policy. "A Report to the Alaska Legislature in Response to Intent Language Regarding Telecommunications in the FY 88 Operating Budget." Juneau, January 1988, p. 10.

7. Ibid., p. 19.

8. Ibid., pp. 14–15.

9. "A Report to the Alaska Legislature in Response to Intent Language Regarding Telecommunications in the FY 88 Operating Budget." Juneau, January 1988, p. i.

10. Department of Administration, "A Report to the Alaska Legislature," p. 6.

11. Andrews, John M. Commissioner, Department of Administration. "RATNET equipment funding." Memo to Garrey Peska, Chief of Staff, Office of the Governor, June 13, 1989.

12. Ibid.

13. Ibid.

14. "Visions of Alaska's Future." Statewide Telecommunications Forum. Juneau, March 29–30, 1993, pp. 16–17.

15. Quoted in Forbes, Norma, and Lonner, W. J. "The Effects of Television Viewing on Cognitive Skills: Implications for Schooling." In Telecommunications Symposium. 33rd Alaska Science Conference, American Association for the Advancement of Science, Arctic Division, Fairbanks, September 18, 1982, p. 44.

16. Pittman, Theda Sue, Hudson, Heather E., and Parker, Edwin B. "Planning Assistance for the Development of Rural Broadcasting." Prepared for the Alaska Public Broadcasting Commission, Anchorage, July 1977, p. 1.

17. Forbes, and Lonner, "The Effects of Television Viewing," p. 44.

18. McConnell, Douglas G. "The Cultural Impact of Television in Alaska." In Telecommunications in Rural Alaska, Robert Walp, ed. Honolulu: Pacific Telecommunications Council, 1982, p. 94.

19. Ibid., p. 95.

20. Forbes, and Lonner, "The Effects of Television Viewing," p. 45.

21. Ibid., p. 49.

22. Forbes, Norma. "Television's Effects on Rural Alaska. Social and Cognitive Effects of the Introduction of Television on Rural Alaskan Native Children." College of Human and Rural Development, University of Alaska at Fairbanks, 1984.

23. Marvin Weatherly interviewed by Hilary Hilscher, December 26, 2000.

24. Alexander, Rosemary, in "Proceedings of the Chugach Conference: Finding Our Way in the Communication Age." Anchorage, October 3–5, 1991, p. 61.

25. "Proceedings of the Chugach Conference: Finding Our Way in the Communication Age." Anchorage, October 3–5, 1991, p. 62.

26. Alexander, Rosemary in "Proceedings of the Chugach Conference," p. 62.

27. "Proceedings of the Chugach Conference: Finding Our Way in the Communication Age." Anchorage, October 3–5, 1991, p. 56.

28. Rasmus-Dede, Paula, in "Proceedings of the Chugach Conference: Communication Issues of the '90s." Anchorage, October 5–6, 1990, p. 41.

29. Ibid., p. 41.

30. Alexander, Rosemary in "Proceedings of the Chugach Conference," pp. 48–50.

31. Ibid., p. 61.

32. Pearson, Larry, in "Proceedings of the Chugach Conference: Communication Issues of the '90s." Anchorage, October 5–6, 1990, p. 37.

33. Charles Northrip, Charles, in "Proceedings of the Chugach Conference: Communication Issues of the '90s." Anchorage, October 5–6, 1990, p. 37.

34. Hiebert, Augie, in "Proceedings of the Chugach Conference: Communication Issues of the '90s." Anchorage, October 5–6, 1990, p. 39.

35. "Proceedings of the Chugach Conference: Communication Issues of the '90s." Anchorage, October 5–6, 1990.

36. Hiebert, Augie, in "Proceedings of the Chugach Conference: Communication Issues of the '90s." Anchorage, October 5–6, 1990, p. 39.

37. Vandor, Marjorie L., Assistant Attorney General. Memorandum to The Honorable Bob Poe, Commissioner, Department of Administration. "Alaska Rural Communications Service (ARCS) Council: Programming Decisions." Juneau, February 8, 1999.

38. Ibid.

39. Source: http://www.nac.nu.ca/history_e.htm

40. Valaskakis, Gail. "The Issue Is Control: Northern Native Communications in Canada." "Proceedings of the Chugach Conference: Communication Issues of the '90s." Anchorage, October 5–6, 1990, p. 44.

41. Metty, Michael P. "Policy Implications and Constraints of Educational Telecommunications in a SubArctic Region." In Telecommunications Symposium. 33rd Alaska Science Conference, American Association for the Advancement of Science, Arctic Division, Fairbanks, September 18, 1982, p. 83.

42. Ibid., p. 83.

Chapter 12

1. Ron Duncan, interview with Hilary Hilscher, December 18, 2000.

2. Robert M. Walp, interview with Hilary Hilscher, 2000.

3. Ibid.

4. Ronald Duncan, interview with Hilary Hilscher, December 18, 2000.

5. Walp, interview with Hilary Hilscher, 2000.

6. Federal Communications Commission. Notice of Inquiry. "In the Matter of Rates and Services for the Provision of Communications by Authorized Common Carriers between the Contiguous States and Alaska, Hawaii, Puerto Rico, and the Virgin Islands." Washington, DC, released January 5, 1984, p. 5.

7. Walp, interview with Hilary Hilscher, 2000.

8. Federal Communications Commission. Notice of Inquiry, 4.

9. Melody, William H. "Telecommunications Networks: Internal Subsidies and System Extension in Alaska." In Telecommunications in Rural Alaska, Robert Walp, ed. Honolulu: Pacific Telecommunications Council, 1982, p. 123.

10. Melody, William H. "Telecommunications in Alaska: Economics and Public Policy." Report to the Alaska State Legislature, April 1978, p. 137.

11. Alascom, Inc. "Policies in Conflict: A Briefing Paper on Alaska Telecommunications Issues." Undated, p. 4.

12. Melody, "Telecommunications in Alaska: Economics and Public Policy," 130.

13. Derived from data in Goldschmidt, Douglas. 1979. Joint Goods, Public Goods and Telecommunications: A Case Study of the Alaskan Telephone System." PhD Dissertation, University of Pennsylvania, p. p. 4.

14. Melody, "Telecommunications in Alaska: Economics and Public Policy," 137.

15. Goldschmidt, Douglas. 1979. Joint Goods, Public Goods and Telecommunications: A Case Study of the Alaskan Telephone System." PhD Dissertation, University of Pennsylvania, pp. 305–6.

16. Office of Telecommunications. "Jurisdictional Cost Separations and Toll Rate Disparities in Alaska." Office of the Governor, Juneau, October 1978, p. 14.

17. Ibid., 4.

18. Melody, "Telecommunications in Alaska: Economics and Public Policy," 140.

19. Office of Telecommunications. "Jurisdictional Cost Separations and Toll Rate Disparities in Alaska," 13.

20. Melody, "Telecommunications in Alaska: Economics and Public Policy," 141.

21. Ibid., 142.

22. Melody, William H. "Telecommunications Networks: Internal Subsidies and System Extension in Alaska." In *Telecommunications in Rural Alaska*, Robert Walp, ed. Honolulu: Pacific Telecommunications Council, 1982, p. 124.

23. Melody, "Telecommunications in Alaska: Economics and Public Policy," Report to the 143.

24. Office of Telecommunications, "Jurisdictional Cost Separations and Toll Rate Disparitics in Alaska," 19–20.

25. Goldschmidt, Douglas. 1979. Joint Goods, Public Goods and Telecommunications: A Case Study of the Alaskan Telephone System." PhD Dissertation, University of Pennsylvania, p. 187.

26. Allan, Daniel S. "Economic Feasibility of Upgrading Telephone Service in Alaskan Villages." In *Telecommunications in Rural Alaska*, Robert Walp, ed. Honolulu: Pacific Telecommunications Council, 1982.

27. Melody, "Telecommunications Networks: Internal Subsidies and System Extension in Alaska," 126.

28. Ronald Duncan, interview with Hilary Hilscher, December 18, 2000.

29. Federal Communications Commission. Notice of Inquiry. 3.

30. Ibid.

31. Ibid., 2.

32. Ibid., 6.

33. Alascom, Inc. "Policies in Conflict: A Briefing Paper on Alaska Telecommunications Issues." Undated, p. 5.

34. Boucher, H.A. "Red." "Alascom and GCI Controversy." Memorandum to Governor Steve Cowper, November 14, 1986.

35. Ibid.

36. Kleeshulte, Chuck. Telephone Rate Hikes on Line?" Juneau Empire, March 29, 1985, p. 15.

37. Ibid.

38. Ibid.

39. Kleeshulte, Chuck. "GCI Calling: Firm Seeks to Tap Phone Market." *Juneau Empire*, March 29, 1985, p. 15.

40. Ibid.

41. Ibid.

42. Ibid.

43. Ronald Duncan, interview with Hilary Hilscher, December 18, 2000.

44. Walp, interview with Hilary Hilscher, 2000.

45. Ronald Duncan, interview with Hilary Hilscher, December 18, 2000.

46. General Communication, Inc. "Comments of General Communication, Inc." Before the Federal Communications Commission, In the Matter of Decreased Regulation of Certain Basic Telecommunications Services, Mary 6, 1987, p. 4.

47. Ibid., 5–6.

48. Pearson, Larry and Doug Barry. "Talking to Each Other, Talking to Machines: Alaska's Telecommunications Future." Report to the Joint Committee on Telecommunications of the Alaska State Legislature, January 1987, p. 44.

49. Ibid., 57.

50. Dunham, Vernon K. Direct Testimony before the Alaska Public Utilities Commission, In the Matter of the Investigation of the Complaint by Alascom, Inc., against General Communication, Inc., for providing Unauthorized Intrastate Telecommunications Services, U-86-99, p. 2.

51. General Communication, Inc. Comments of GCI in Response to APUC Notice of Inquiry In the Matter of the Consideration of Conditions of Interstate Equal Access Involving Intrastate Telephone Services and Prevention of Arbitrage, August 25, 1986, p. 7.

52. Hughes, J. Letter from GCI to GCI customers, Anchorage, November 25, 1986.

53. Alaska Consumer Advocacy Program. "Telecommunications Policy in Alaska." April 23, 1985, p. 25.

54. Alaska Public Utilities Commission and Public Utilities Reports. *Alaska Public Utilities Commission Digest: For the Years 1980–1990*. Arlington, VA: Public Utilities Reports, 1992, p. 401.

Chapter 13

1. State Representative H.A. "Red" Boucher in "Proceedings of the Chugach Conference: Discussing the Future of Communications in Alaska." University of Alaska Anchorage, August 18–19, 1989, p. 29.

2. Hammond, Jay. S., Governor. "Executive Order No. 050." July 1, 1981.

3. Pearson, Larry and Doug Barry. "Talking to Each Other, Talking to Machines: Alaska's Telecommunications Future." Report to the Joint Committee on Telecommunications of the Alaska State Legislature, January 1987, p. 7.

4. Melody, William H. "Telecommunications in Alaska: Economics and Public Policy." Report to the Alaska State Legislature, April 1978.

5. Pearson and Barry, p. 9.

6. Pearson and Barry, p. 11.

7. Melody, "Telecommunications in Alaska: Economics and Public Policy."

8. Pearson and Barry, p. 10.

9. Alaska 2001 Advisory Committee, Lt. Governor Fran Ulmer, Chair. Draft Report to the Alaska Public Utilities Commission. August 22, 1995, p. 9.

10. Boucher, H.A. "Red." An Interim Report: House Special Committee on Telecommunications. January 13, 1986, pp. 2–3.

11. Pearson and Barry, pp. 7–8.

12. Ibid., p. 6.

13. Ibid., p. 5.

14. Boucher, H.A. "Red." An Interim Report, p. 4.

15. State Representative H.A. "Red" Boucher in "Proceedings of the Chugach Conference: Discussing the Future of Communications in Alaska," p. 29.

16. Hills, Alex. "The Changing Structure of Alaskan Telecommunications." In *Telecommunications in Rural Alaska*, Robert Walp, ed. Honolulu: Pacific Telecommunications Council, 1982, p. 1.

17. Pearson and Barry, p. 13.

18. Cowper, Steve, governor. "Executive Order No. 066." January 19, 1987.

19. Alaska State Legislature House of Representatives Special Committee on Telecommunications. "Telecommunications: Critical Issues." March 15, 1989, p. 1.

20. Boucher, H.A. "Red." An Interim Report, p. 7.

21. Ibid., p. 8.

22. Ibid., pp. 9–10.

23. Alaska State Legislature House of Representatives Special Committee on Telecommunications, p. 2.

24. Pearson and Barry, pp. 87–88.

25. Boucher, H.A. "Red." An Interim Report, p. 1.

26. Pearson, Larry. "The Telecommunications Information Council: Legislative History and Implementation Plan." Alaska House Special Committee on Telecommunications. Juneau, September 1987, pp. 4–5.

27. Telecommunications Information Council. "First Annual Report." Juneau, March 23, 1988.

28. Pearson, "The Telecommunications Information Council, p. 2.

29. Ibid., p. 3.

30. Ibid., p. 23.

31. Ibid., p. 10.

32. Ibid., p. 8.

33. Ibid., p. 7.

34. Hills, Alex. "Melting the Ice Curtain between Russia and Alaska." *Business Communications Review*, December 1993, p. 26.

35. Alascom Video: Provideniya Friendship Flight: 1988.

36. Alascom Video: Provideniya Friendship Flight: 1988.

37. Personal communication, Lee Wareham, October 2014.

38. Personal communication, Lee Wareham, October 2014.

39. William Harris in Proceedings of the Chugach Conference: Discussing the Future of Communications in Alaska." University of Alaska Anchorage, August 18–19, 1989, p. 42.

40. Hills, Alex. "Melting the Ice Curtain," p. 27.

41. Personal communication, Lee Wareham, October 2014.

42. Hills, A., "Beaming E-Mail to the Russian Far East," chapter in M. Stone-Martin and L. Breeden, *51 Reasons: How We Use the Internet and What It Says about the Information Superhighway*, Farnet, Inc., Lexington, MA: 1994.

43. Personal communication, Lee Wareham, October 2014.

44. Chuck Schumann, Microcom, personal communication, October 2014.

45. Chuck Schumann, Microcom, personal communication, October 2014.

46. William Harris in Proceedings of the Chugach Conference, p. 42.

47. Pearson, "The Telecommunications Information Council," p. 7.

48. Hudson, Heather E. "Electronic Trails across the North: New Directions for Alaska." Keynote speech, "Visions of Alaska's Future." Statewide Telecommunications Forum. Juneau, March 29–30, 1993.

49. State of Alaska Telecommunications and Information Technology Plan Passed by the Telecommunications Information Council, Lt. Governor Fran Ulmer, Chair. December 18, 1996, p.2.

50. Ulmer, Fran, Lieutenant Governor. Letter introducing "State of Alaska Draft Telecommunications and Information Technology Plan." Juneau: September 11, 1996.

51. State of Alaska, Office of the Lieutenant Governor. "Lieutenant Governor Eyes the Future with Telecommunications Plan." Juneau, September 11, 1996.

52. State of Alaska Telecommunications and Information Technology Plan Passed by the Telecommunications Information Council, Lt. Governor Fran Ulmer, Chair. December 18, 1996, p. 2.

53. Alaska 2001 Advisory Committee, Lt. Governor Fran Ulmer, Chair. Report to the Alaska Public Utilities Commission. Juneau, March 1996, p. 5.

54. Ibid., pp. 3–4.

55. Telecommunications Act of 1996, Public Law 104-104, 110 Stat. 56 (1996). Passed February 8, 1996.

56. Brown, Ronald H., Irving, Larry, Prabhakar, Arati, and Katzen, Sally. *The National Information Infrastructure: Agenda for Action*. Washington, DC: U.S. Department of Commerce, September 15, 1993, p. 5.

57. Gore, Albert. "Remarks at the National Press Club." Washington, DC, White House, Office of the Vice President, December 21, 1993.

58. *Telecommunications Act of 1996*, Public Law 104-104, 110 Stat. 56 (1996).

59. Ibid.

60. "AUSF Overview." http://rca.alaska.gov/RCAWeb/Documents/Telecomm/ausf_overview.pdf

61. Ibid.

62. http://rca.alaska.gov/RCAWeb/Documents/Telecomm/ausf_overview.pdf

63. Alaska Universal Service Administrative Company. "AUSF Annual Summary: Period ending December 31, 2013."

Chapter 14

1. Vandelaar, Janis. "ATA's President Eller: A Personal Profile." *Tip'n'Ring*. Publication of the Alaska Telephone Association, Vol. 5, No. 1, 1984.

2. Alaska 2001 Advisory Committee, Lt. Governor Fran Ulmer, Chair. Draft Report to the Alaska Public Utilities Commission. August 22, 1995, p. 7.

3. Ibid., p. 8.

4. Ibid., p. 12.

5. Vandelaar, "ATA's President Eller: A Personal Profile."

6. Eller, Don. Presentation at symposium on "The Evolution of Telecommunications in Alaska," Institute of Social and Economic Research, University of Alaska Anchorage, June 13, 2011.

7. Ibid.

8. Vandelaar, "ATA's President Eller."

9. www.yukontel.com

10. Tribute to Jack Rhyner by Senator Ted Stevens. http://capitolwords.org/date/2008/04/28/S3440-3_tribute-to-jack-h-rhyner/

11. www.telalaska.com

12. Ibid.

13. http://newscenter2.verizon.com/press-releases/verizon/2000/page-29760214.html

14. www.aptalaska.com

15. Ibid.

16. "GCI purchases United Companies." *Juneau Empire*, Tuesday, June 10, 2008.
17. http://www.unicom-alaska.com/
18. Carson, Wesley E. "New PTI owners ready to deliver quality service." *Juneau Empire*, April 16, 1999.
19. *PR Newswire*, New York, May 17, 1999.
20. Holst, Svend. "PTI sale approved." *The Juneau Empire*, April 26, 1999.
21. http://adaktu.net/
22. Joyce, Stephanie. "Adak's Phone Company Calls on Feds to Restore Subsidies." Alaska Public Radio Network (APRN) broadcast July 30, 2013. http://www.alaskapublic .org/2013/07/30/adaks-phone-company-calls-on-feds-to-restore-subsidies/
23. Federal Communications Commission "In the Matter of Adak Eagle Enterprises, LLC and Windy City Cellular, LLC Petitions for Waiver of Certain High-Cost Universal Service Rules." WC Docket No. 10-90 and WT Docket No. 10-208: ORDER. Adopted July 15, 2013, para. 15.
24. Federal Communications Commission. In the Matters of Connect America Fund WC Docket 10-90 and Universal Service Reform—Mobility Fund WT Docket No. 10-208, Petitions for Waiver of Windy City Cellular, LLC and Adak Eagle Enterprises, LLC. Order. Adopted January 10, 2014.
25. http://www.city.ketchikan.ak.us/public_utilities/kputel/index.html
26. Gordon, Bill "Selling a Utility: The Fairbanks, Alaska Story." *Water Engineering & Management*, 146.4 (April 1999), p. 20.
27. Ibid., p. 20.
28. Ibid., p. 22.
29. *Communications Daily*, July 30, 1991.
30. Rinehart, Steve. "High Bid Spurned; ATU Goes to Citizens." *Anchorage Daily News*, July 24, 1991, p. A-1.
31. Stange, Jay. "PTI plan infuriates Citizens." *Anchorage Daily News*, August 2, 1991.
32. Ibid.
33. Rinehart, Steve, and Wohlforth, Charles. "PTI stalls Petition Decision." *Anchorage Daily News*, August 2, 1991.
34. Boucher, H. A. "Red," Anchorage Assembly. Letter to Larry and Alice Pearson, September 25, on Pacific Telecom, Inc. letterhead, 1991.
35. *Communications Daily*, January 8, 1996.
36. *Communications Daily*, January 29, 1996.
37. *Communications Daily*, April 2, 1997.
38. "Anchorage OKs phone utility sale." *Juneau Empire*, April 22, 1998.
39. "Anchorage trust fund shrinks in downturn." *Juneau Empire*, December 28, 2008.
40. https://www.mtasolutions.com/
41. Ibid.
42. "MTA takeover bid fails." *Peninsula Clarion*, September 20, 2000.
43. http://www.cvinternet.net/
44. U.S. Department of Agriculture. (2011). "Advancing Broadband: A Foundation for Strong Rural Communities." Broadband Initiatives Program: Awards Report. Washington, DC, p. 11.
45. http://www2.ctcak.net/
46. Ibid.
47. Ibid.

48. Bouker, David. "30th Year Recognition of Telephone Operations Background Material of NTC Benefactors." Nushagak Telephone Cooperative website, http://www.nushtel .com/home-about-us.htm

49. www.bristolbay.com

50. www.otz.net

51. http://www.astac.net/servlet/content/about.html

Chapter 15

1. Reid, Sean. "Can Anyone Figure out Alaska's Phone Wars?" *Alaska Business Monthly*, August 1993, p. 17.

2. APUC Commissioner Carolyn Guess in "Proceedings of the Chugach Conference: Discussing the Future of Communications in Alaska." University of Alaska Anchorage, August 18–19, 1989, p.34.

3. Alaska 2001 Advisory Committee, Lt. Governor Fran Ulmer, Chair. Report to the Alaska Public Utilities Commission. Juneau, March 1996, pp. 10–11.

4. Ronald Duncan, interview with Hilary Hilscher, December 18, 2000.

5. Ibid.

6. Susan Knowles interviewed by Hilary Hilscher, August 29, 2000.

7. Ibid.

8. Eller, Don. Presentation at symposium on "The Evolution of Telecommunications in Alaska," Institute of Social and Economic Research, University of Alaska Anchorage, June 13, 2011.

9. William Harris in Proceedings of the Chugach Conference: Discussing the Future of Communications in Alaska." University of Alaska Anchorage, August 18–19, 1989, p. 42.

10. Alascom. "Intrastate Competitive Entry Presentation Draft." Draft testimony before the Legislature, undated.

11. Dr. Ben Johnson quoted in Alascom. "Intrastate Competitive Entry Presentation Draft." Draft testimony before the Legislature, undated.

12. AT&T. "AT&T's Petition to Terminate the Joint Service Arrangement. Summary," Before the Federal Communications Commission. Washington, DC: FCC CC Docket No. 83-1376, March 31, 1993.

13. Ibid.

14. Katz, John A. "Outline of Testimony for State Telecommunications Hearing." Washington, DC, Office of the Governor, August 24, 1993, p. 3.

15. Reid, "Can Anyone Figure out Alaska's Phone Wars?" p. 18.

16. General Communication, Inc. United States Securities and Exchange Commission Form 10-K, Annual Report for fiscal year ended December 31, 1994, pp. 10–11.

17. Susan Knowles interviewed by Hilary Hilscher, August 29, 2000.

18. "AT&T to Acquire Alascom for $365 Million." *New York Times*, October 18, 1994.

19. Telecommunications Act of 1996, Public Law 104-104, 110 Stat. 56 (1996), Section 251(c).

20. McAllister, Bill. "A Hold on Local Phone Competition: Court Action Allows ACS to Fend off GCI's Entry." *Juneau Empire*, February 4, 2001.

21. Supreme Court of Alaska. ACS of Alaska, Inc. v. Regulatory Commission of Alaska (12/12/2003) sp-5762.

22. Alaska Supreme Court Remands Local Telephone Competition Decision to RCA. http://www.sec.gov/Archives/edgar/containers/fix031/808461/000080846103000034/gci8k121203exhibit.txt

23. GCI Announces Settlement of Rural Exemption Proceeding for Fairbanks and Juneau. http://www.prnewswire.com/news-releases/gci-announces-settlement-of-rural-exemption-proceeding-for-fairbanks-and-juneau-72568497.html

24. Dart, Greg. "GCI Plans to End Local Telephone Service Monopoly." *Alaska Star*, October 21, 2005. http://classic.alaskastar.com/stories/102005/hom_20051020018.shtml

25. "GCI Rolls Out Local Phone Service in Eagle River." http://www.prnewswire.com/news-releases/gci-rolls-out-local-phone-service-in-eagle-river-54152202.html

Chapter 16

1. Barry, Douglas K. "Distance Learning by Satellite (and other technologies): A Catalog of Available Services." Juneau: Alaska Department of Education, Office of Instructional Support, April 1998, p. 2.

2. Stapler, Dale C. "The Economic Development Potential of Satellite Delivery Education." *Economic Development Review*, Winter 1990, p. 33.

3. Ibid., p. 31.

4. Barry, Douglas K. "Distance Learning by Satellite (and other technologies): A Catalog of Available Services." Juneau: Alaska Department of Education, Office of Instructional Support, April 1998, pp. 1–2.

5. Source: https://www2.ed.gov/programs/starschools/index.html

6. Barry, "Distance Learning by Satellite," p. 33.

7. Ibid., p. 36.

8. Ibid., p. 3.

9. Ibid., p. 33.

10. Ibid., p. 29.

11. Visions of Alaska's Future." Statewide Telecommunications Forum. Juneau, March 29–30, 1993, pp. 3–4

12. Raab, Diane. "Earning College Credit by Satellite." *Juneau Empire*, February 8, 1993, p. 1.

13. Visions of Alaska's Future.," pp. 1–2.

14. "New Technology: Two-way video classes help village schools."

15. Kane, Karen. Preserving the Past while preparing for the Future." *Offices That Work*, Wang Laboratories, Lowell, MA, January 1993.

16. Bower, David. "The Most Northern CODEC." *Pixel-to-Pixel*. Austin, TX, VTEL User Group Association, Summer 1993, p. 5.

17. "Students Master New Craft: TV Production." "School Happenings: North Slope School District." Quoted in *Alaska Daily News*, January 4, 1993.

18. Smith, Carlton. "Model Alaskan School District Relies on PC/VS WAN." WANG Dialogue, Vol. 1, No. 21, December 15, 1992, p. 1.

19. Ibid., p. 2.

20. Kane, Preserving the Past."

21. Smith, "Model Alaskan School District Relies on PC/VS WAN," p. 1.

22. Baesler, Jeff. "On Top of the World." *Communicating" The Northwest Information Community's Bimonthly Forum*, Vol. 2, No.2, Oct/Nov 1992, p. 3.

23. Ahmaogak, George, Sr. "From the desk of the Mayor. . ." *Quasagniq: A New Dawn. The North Slope Borough Newsletter.* Vol. VII, No. 9, October 1995, p. 2.

24. Ringland, Natalie. AuroraNet: The Future has Arrived!" *Quasagniq: A New Dawn. The North Slope Borough Newsletter.* Vol. VII, No. 9, October 1995, pp. 4–5.

25. Finkler, Earl. "And It's a Lot Quicker than Mushing Dog Teams." *Anchorage Daily News*, February 21, 1994, p. C3.

26. Ibid.

27. Quoted in Gore, Albert. "Remarks at the National Press Club." Washington, DC, White House, Office of the Vice President, December 21, 1993.

28. Senator Jay Rockefeller (D-WVA) quoted in Dickard, Norris, ed. "Great Expectations: Leveraging America's Investment in Educational Technology." Benton Foundation and Education Development Center, 2002.

29. Cary, Martin. "Universal Services in Alaska: The Impact of the E-Rate on Rural Alaska." Presentation to IEEE International Conference on Communications, Anchorage, May 2003.

30. . Interview with Della Mathis, Anchorage, August 1999.

31. Matthis, Della. "E-Rate in Alaska: Telecommunications—Expanding Education and Library Service," June 2006. Available at http://library.state.ak.us/usf/Bandwidthreport5-06.doc.

32. Democratic Leadership Council's Progressive Policy Institute "The State New Economy Index: Benchmarking Economic Transformation in the States." See http://neweconomyindex.org/states.

33. . Karen R. Crane, quoted in Kelley, Tina. "Internet Showing Its Value in Remote Alaskan Villages." *The New York Times*, August 5, 1999.

34. Derived from Universal Service Administrative Company. "2013 Annual Report." Washington, DC: 2014.

35. Personal interviews with Della Matthis, May 1999 and March 2002.

36. See www.schoolaccess.net.

37. Ibid.

38. . See, for example, www.mta-telco.com (Matanuska Telephone Cooperative).

39. Alaska Waiver Order : CC Docket NO. 96-45: Order FCC 01-350, adopted November 29, 2001; released December 3, 2001.

40. See http://www.library.state.ak.us/usf/waiver.cfm#P102.

41. See http://kc3.cilc.org/.

42. See http://www.nsbsd.org/Page/2840.

43. Hopkins, Kyle. "Rosetta Stone produces Inupiaq language software." *Juneau Empire*, January 23, 2011.

44. http://www.schoolaccess.net/public-general/case-studies/kashunamuit

45. Presentation by Gloria O'Neill, President and Chief Executive Officer, Cook Inlet Tribal Council, Arctic Federal Listening Session, Anchorage, August 14, 2014.

46. State of Alaska Department of Education and Early Development. "State to Award Digital Teaching Initiative Grants." Juneau, July 2, 2014.

47. http://www.schoolaccess.net/public-general/success-stories/alaska-owl

48. See http://library.alaska.gov/dev/owl.html.

Chapter 17

1. Quoted in Nice, Philip and Walter Johnson. *The Alaska Health Aide Program: A Tradition of Helping Ourselves*. Anchorage: University of Alaska Anchorage, Center for Circumpolar Health Studies, 1999.

2. American Telemedicine Association: http://www.americantelemed.org/about-telemedicine/what-is-telemedicine#.U0R3XpVOVpo

3. Ferguson, Stewart. "What Works: Outcomes Data from AFHCAN and ANTHC Telehealth: An 8 Year Retrospective." Anchorage. Alaska Native Tribal Health Consortium, May 2011.

4. Perdue, Karen, et al. "Evolution and Summative Evaluation of the Alaska Federal Health Care Access Network Telemedicine Project." University of Alaska Statewide Health Programs and University of Alaska Anchorage Center for Human Development, November 2004, pp. 15–16.

5. Quoted in Nice, Philip, and Johnson, Walter. *The Alaska Health Aide Program: A Tradition of Helping Ourselves*. Anchorage: University of Alaska Anchorage, Center for Circumpolar Health Studies, 1999.

6. Ibid.

7. Perdue, Karen, et al. "Evolution and Summative Evaluation of the Alaska Federal Health Care Access Network Telemedicine Project." University of Alaska Statewide Health Programs and University of Alaska Anchorage Center for Human Development, November 2004, p. 16.

8. Quoted in Nice and Johnson, *The Alaska Health Aide Program*.

9. For example, see the comment by George Winer, health aide for the village of Beaver "Satellite House Call" Film produced by Department of Communication, Stanford University. Available at http://www.youtube.com/watch?v=GzclgBfn_yY

10. See Hudson, Heather E. and Edwin B. Parker. "Medical Communication in Alaska by Satellite." *New England Journal of Medicine*, December 20, 1973.

11. Perdue, "Evolution and Summative Evaluation," 25.

12. Taft, Destyn E. "Telemedicine and Telehealth Services Going Online for Providence Health System Alaska." *Telemedicine in Alaska News*, June 26, 1998.

13. Perdue, "Evolution and Summative Evaluation," 29.

14. Telecommunications Act of 1996, Public Law 104-104, 110 Stat. 56 (1996).

15. Perdue, "Evolution and Summative Evaluation," 30.

16. Pearce, Frederick W. "Presentation to the 17th Annual Conference of the Caribbean Association of National Telecommunications Organizations (CANTO)." Puerto Rico, May 2001.

17. Perdue, "Evolution and Summative Evaluation," p. 30.

18. Carlsson, Sylvia. "Telemedicine in the Bush: A Glimpse into the New Century." Undated.

19. See www.afhcan.org.

20. Ferguson, Stewart. "What Works: Outcomes Data from AFHCAN and ANTHC Telehealth: An 8 Year Retrospective." Anchorage. Alaska Native Tribal Health Consortium, May 2011.

21. Sivitz, H., Hanna, B., and Wregglesworth, B. "AFHCAN Deployment: Completing the Installation of 235 Sites in Alaska." *Telemedicine Journal and E-Health* 9 (Supplement 1): S 64.

22. Personal interview, Kotzebue, May 2000.

23. See www.afhcan.org/training.

24. Ferguson, Stewart. "Next Steps: The AFHCAN Telehealth Program." Unpublished Presentation. Anchorage: Alaska Native Medical Center, 2012.

25. Kokesh, John, A. Stewart Ferguson, Chris Patricoski. "Telehealth In Alaska: Delivery of Health Care Services from a Specialist's Perspective." *International Journal of Circumpolar Health* 63:4 2004, pp. 387–400.

26. Ferguson, "What Works."

27. See Hudson, Heather E. and Edwin B. Parker. "Medical Communication in Alaska by Satellite." *New England Journal of Medicine*, December 20, 1973.

28. Ferguson, "What Works."

29. Ibid.

30. John Kokesh, MD, A. Stewart Ferguson, PhD, and Chris Patricoski, MD. "Preoperative Planning for Ear Surgery Using Store-and-Forward Telemedicine." *Otolaryngology–Head and Neck Surgery* (2010) 143, p. 257.

31. Hofstetter, Philip J., John Kokesh, A. Stewart Ferguson, and Linda J. Hood. "The Impact of Telehealth on Wait Time for ENT Specialty Care." (2010) *Telemedicine and E-Health*. Vol. 16 No. 5, June 2010, p. 551.

32. Ibid., 555.

33. Kokesh, Ferguson, and Patricoski. "Preoperative Planning for Ear Surgery," 253.

34. Public Law 104-104, the Telecommunications Act of 1996. See 47 U.S.C. (h)(2)(A)

35. For details, see www.rhc.universalservice.org/eligibility/services.asp.

36. The Federal Communications Commission (FCC) defines "most cost effective method" as "the method of least cost after consideration of the features, quality of transmission, reliability, and other factors relevant to choosing a method of providing the required services." See www. rhc.universalservice.org.

37. Derived from: Universal Service Administrative Company. "2013 Annual Report." Washington, DC: 2014.

38. Federal Communications Commission. "In the Matter of Rural Health Care Support Mechanism," WC Docket No. 02-60, adopted September 26, 2006, pp. 1–2.

39. "FCC Launches Initiative To Increase Access To Health Care In Rural America Through Broadband Telehealth Services." November 19, 2007. See www.fcc.gov/cgb/rural/rhcp.html.

40. www.usac.org

41. Alaska eHealth Network. "March 2013 Quarterly Data Report." Available at http://apps.fcc.gov/ecfs/document/view?id=7022263579

42. Based on Hudson, H. E. (2007). "Rural Telemedicine: Lessons from Alaska for Developing Regions." *Journal of eHealth Technology and Application*, Vol. 5, No. 3, September 2007, pp. 460–467.

43. Ferguson, Stewart. "What Works."

44. Personal interview with project director Fred Pearce, Anchorage, August 1999.

Chapter 18

1. Willie Hensley in "Proceedings of the Chugach Conference: Finding Our Way in the Communication Age." Anchorage, October 3–5, 1991, p. 85.

2. Milroy, Cam. "The Tundra Telephone." Arctic Life and Traditions. http://arcticlifeandtraditions.blogspot.com/2010/02/tundra-telephone.html

3. Hundt, Reed E. Statement of the Honorable Reed E. Hundt, Chairman. Washington, DC, Federal Communications Commission, November 1, 1994.

4. Dizard, John W. "Gold Rush at the FCC." *Fortune*, July 12, 1982, p. 104.

5. Ibid., p. 102.

6. Alaska Consumer Advocacy Program. "Telecommunications Policy in Alaska." April 23, 1985, p. 10.

7. Treadwell, Mead. "Proposal: Cellular Telephone Expansion Project." Tundra Telephone, unpublished memo, Anchorage, August 15, 1982.

8. Alaska 2001 Advisory Committee, Lt. Governor Fran Ulmer, Chair. Report to the Alaska Public Utilities Commission. Juneau, March 1996, pp. 10–11.

9. Quoted in General Communication, Inc. United States Securities and Exchange Commission Form 10-K, Annual Report for fiscal year ended December 31, 1994, p. 4.

10. GCI staff, personal communication, July 2014.

11. "Alaska Communications." May 2011. www.alsk.com

12. GCI & Alaska Communications form The Alaska Wireless Network, LLC, June 5, 2012. http://www.gci.com/awn/joint-media-release.

13. "AT&T Invests More than $650 Million in Alaska from 2009 Through 2011 to Improve Local Networks." Anchorage, Alaska, Feb. 6, 2012. http://www.corp.att.com/alaska/press/pr_20120213.html

14. Telegeography. "AlaskaComm, GCI announce wireless pact in a bid to fend off Verizon." June 8, 2012. http://www.telegeography.com/products/commsupdate/articles/2012/06/08/alaskacomm-gci-announce-wireless-pact-in-a-bid-to-fend-off-verizon/

15. Edge, Megan. "As Its Focus Shifts, Alaska Communications Will Sell Wireless Business to GCI." *Anchorage Dispatch News*, December 4, 2014.

16. Augie Hiebert, quoted in Chlupach, Robin Ann. *Airwaves over Alaska: The Story of Broadcaster Augie Hiebert*. Issaquah, WA: Sammamish Press, 1992, p. 148.

17. B.C. "Chuck" Russell in "Proceedings of the Chugach Conference: Discussing the Future of Communications in Alaska." University of Alaska Anchorage, August 18–19, 1989, pp. 18–20.

18. "GCI Milestones," available at www.gci.com.

19. Bauman, Margaret. "Cable Company links Kodiak, Kenai to Fiber Optic System." *Alaska Journal of Commerce*, January 28, 2007.

20. Alaska Communications Systems. "ACS Launches Commercial Traffic on AKORN, Brings Competition and Route Diversity to Alaska." Press Release, April 7, 2009.

21. *American Recovery and Reinvestment Act* of 2009, Pub. L. No. 111-5, 123 Stat, 115, 516 (Feb. 19, 2009).

22. See http://terra.gci.com.

23. U.S. Department of Agriculture. (2011). "Advancing Broadband: A Foundation for Strong Rural Communities." Broadband Initiatives Program: Awards Report. Washington, DC, p. 11.

24. See www.rurdev.usda.gov/UTP_BIPResources.html.

25. See www2.ntia.doc. gov/Alaska.

26. http://www2.ntia.doc.gov/grantee/university-of-alaska-fairbanks

27. See www.recovery.gov. The maps are available at www.connectak.org.

28. See www2.ntia.doc.gov/Alaska and www.recovery.gov

29. Alaska Statewide Broadband Task Force. "A Blueprint for Alaska's Broadband Future." Anchorage, August 2013. Available at www.alaska.edu/oit/bbtaskforce/homepage.html

30. Derived from Universal Service Administrative Company. "2013 Annual Report." Washington, DC: 2014.

31. Federal Communications Commission. Universal Service Fund and Intercarrier Compensation Transformation Order, November 2011. See https://apps.fcc.gov/edocs_public/attachmatch/FCC-11-161A1.pdf.

32. Office of Native Affairs and Policy. "2012 Annual Report." Federal Communications Commission. Washington, DC: 2013." Available at www.fcc.gov/native.

33. See http://wireless.fcc.gov/auctions/901/reports/901winning_bids_by_state_county.pdf

34. Federal Communications Commission. "Tribal Mobility Fund Phase I Auction—Winning Bids Sorted by State and Tribal Land." March 2014. Available at http://wireless.fcc.gov/auctions/default.htm?job=auction_summary&id=902

35. Office of Native Affairs and Policy. "2012 Annual Report."

36. Federal Communications Commission. "Public Notice: Wireline Competition Bureau Seeks Further Comment on Issues Regarding The Design of The Remote Areas Fund." Wc Docket No. 10-90, January 17, 2013.

37. Office of Native Affairs and Policy. "2012 Annual Report."

38. Ibid.

39. Ibid.

40. See www.firstnet.gov

41. Regulatory Commission of Alaska. "Rural Alaska Broadband Internet Access Grant Program Round Five Grant Application Guide(2010)." See www.rca.alaska.gov.

42. Hudson, Heather E. *From Rural Village to Global Village: Telecommunications for Development in the Information Age.* New York: Routledge, 2006.

43. Hudson, Heather E., et al. "Toward Universal Broadband in Rural Alaska: A Report to the State Broadband Task Force." Institute of Social and Economic Research, University of Alaska Anchorage, November 2012.

44. www.gci.com/internet/plans, February 2015.

45. Hardin, Kyle. "The Digital Divide: How Poor Internet Access Pulls the Plug on Rural Alaska." Bristol Bay Native Association, August 22, 2014.

46. Hudson, Heather E. "After Broadband" study in Southwest Alaska. Institute of Social and Economic Research (ISER), University of Alaska Anchorage, forthcoming 2015.

Chapter 19

1. FCC Chairman Tom Wheeler, Presentation at FCC Roundtable, Anchorage, Alaska, August 27, 2014.

2. Ron Duncan interviewed by Hilary Hilscher, December 18, 2000.

3. "CITC Fab Lab." http://citci.org/wp-content/uploads/2014/06/FABLAB_1-pager1406251.pdf.

4. See http://www.uaf.edu/anlc/

5. See www.ancsaat40.com.

6. See www.alaskanative.net.

7. See www.inuit.org and www.iccalaska.org.

8. Data derived from Universal Service Administrative Company. "2013 Annual Report." Washington, DC: 2014.

9. William Harris in Proceedings of the Chugach Conference: Discussing the Future of Communications in Alaska." University of Alaska Anchorage, August 18–19, 1989, p. 42.

10. Pearson, Larry, and Barry, Doug. "Talking to Each Other, Talking to Machines: Alaska's Telecommunications Future." Report to the Joint Committee on Telecommunications of the Alaska State Legislature, January 1987, p. 20.

11. Ibid, p. 5.

12. FCC Chairman Tom Wheeler, Presentation at FCC Roundtable, Anchorage, Alaska, August 27, 2014.

13. Pearson and Barry. "Talking to Each Other, Talking to Machines: Alaska's Telecommunications Future," p. 11.

14. See http://www.koahnicbroadcast.org/.

15. Augie Hiebert in Proceedings of the Chugach Conference: Communication Issues of the '90s." Anchorage, October 5–6, 1990, p. 39.

16. Metty, Michael P. "Policy Implications and Constraints of Educational Telecommunications in a Subarctic Region." In Telecommunications Symposium. Thirty-third Alaska Science Conference, American Association for the Advancement of Science, Arctic Division, Fairbanks, September 18, 1982, p. 83.

17. Points a, g, h, and i were contributed by Professor William Melody at the symposium on "The Evolution of Telecommunications in Alaska," Institute of Social and Economic Research, University of Alaska Anchorage, June 13, 2011.

18. The latter two points were made in Pearsonand Barry. "Talking to Each Other, Talking to Machines: Alaska's Telecommunications Future," p. 20.

19. Alaska State Commission on Research (SCoR). "To Build a Fire: A Plan for the Future of Science and Technology in Alaska." Final report, January 2014.

20. Edwin Parker, personal communication, 2014.

21. U.S. Census, 2010 data. Available at http://live.laborstats.alaska.gov/cen/dparea.cfm.

22. Chlupach, pp. 149–50. FULL TRANSCRIPT: A.G. Hiebert, "Statement Prepared for Public Service Commission RCA Alascom Hearings, June 10, 1970. Chlupach, pp. 218–221.

Bibliography

"A Milestone Decision for Alaska." Editorial, *Anchorage Daily News*, April 25, 1968.

"About KTUU-TV." See http://articles.ktuu.com/2010-07-14/broadcast-license_24129016

ACS draft press release. "Telephone and telegraph rate reductions proposed today by the U.S. Air Force will save Alaska Communications System customers an estimated $7 million from December 1, 1969 to July 1, 1970." October 17, 1969.

"ACS Observes 49th Anniversary in Alaska." *Alaska Communication System Bulletin*, 1949.

"Action at Last on Long Distance Rates." *Anchorage Daily News*, March 15, 1965.

"Agreement Paves the Way . . . to Start of New Era." *Anchorage Sunday Times*, August 10, 1975.

Ahmaogak, George, Sr. "From the desk of the Mayor. . ." *Quasagniq: A New Dawn. The North Slope Borough Newsletter*. Vol. VII, No. 9, October 1995.

Alascom. "Intrastate Competitive Entry Presentation Draft." Draft testimony before the Legislature, undated.

Alascom, Inc. "Policies in Conflict: A Briefing Paper on Alaska Telecommunications Issues." Undated.

Alascom. Provideniya Friendship Flight. Video. Anchorage: Alascom, 1988.

"Alascom is looking up." *Alaska Industry*, January 1982.

Alaska 2001 Advisory Committee, Lt. Governor Fran Ulmer, Chair. Draft Report to the Alaska Public Utilities Commission. August 22, 1995.

Alaska 2001 Advisory Committee, Lt. Governor Fran Ulmer, Chair. Report to the Alaska Public Utilities Commission. Juneau, March 1996.

Alaska Broadcasters Association. Alaska Day at the FCC. Presentations. September 26, 1996.

"Alaska Communications." See http://www.alaskacommunications.com/About-ACS.aspx.

Alaska Communications Systems. "ACS Launches Commercial Traffic on AKORN, Brings Competition and Route Diversity to Alaska." Press Release, April 7, 2009.

Alaska Communications Systems Group, Inc. U.S. Securities and Exchange Commission Form 10-K, for the fiscal year ended December 31, 2000.

Alaska Conference on Satellite Telecommunications, Anchorage, Alaska, August 28 and 29, 1969. Agenda and attendees; telegram from Senator Ted Stevens listing VIPs who would attend.

Alaska Consumer Advocacy Program. "Telecommunications Policy in Alaska." April 23, 1985.

Alaska Educational Broadcasting Commission. "Alaska Satellite Project." August 1970.

Alaska Educational Broadcasting Commission. ATS-F/Alaska: Program Plans. September 22, 1972.

Alaska eHealth Network. "March 2013 Quarterly Data Report." Available at http://apps
.fcc.gov/ecfs/document/view?id=7022263579

Alaska Federal Health Care Access Network (AFHCAN). Alaska Native Tribal Health
Consortium. See: www.afhcan.org.

Alaska House Special Committee on Telecommunications. "The Telecommunications
Information Council: Legislative History and Implementation Plan." Juneau, September
1987.

"Alaska Is One Step into Satellite Age." *Anchorage Daily Times*, February 26, 1973, p. 6.

Alaska Office of Telecommunications. "Report on Alaskan Use of the ATS-1 Satellite
through September 30, 1972." Juneau, 1973.

Alaska Office of Telecommunications. "FCC Rules on RCA Satellite Ownership." November
3, 1975.

"Alaska Population Overview: 2012 Estimates." Juneau, Alaska Department of Labor,
November 2013.

Alaska Public Utilities Commission. Order Granting Certificate of Public Convenience
and Necessity. Order No. 12, Docket U-69-24, August 31, 1970.

Alaska Public Utilities Commission and Public Utilities Reports. *Alaska Public Utilities
Commission Digest: For the Years 1980–1990.* Arlington, VA: Public Utilities Reports,
1992.

Alaska State Commission on Research (SCoR). "To Build a Fire: A Plan for the Future of
Science and Technology in Alaska." Final report, January 2014.

Alaska State Legislature House of Representatives Research Agency. "Public Broadcasting
in Alaska: A Long-Range Plan." Juneau, December 1984.

Alaska State Legislature House of Representatives Special Committee on
Telecommunications. "Telecommunications: Critical Issues." March 15, 1989.

Alaska State Legislature, House. "Speaker Names Telecommunications Task Force." Juneau:
April 20, 1993.

Alaska Statewide Broadband Task Force. "A Blueprint for Alaska's Broadband Future."
Anchorage, August 2013. Available at www.alaska.edu/oit/bbtaskforce/homepage.html.

Alaska Supreme Court Remands Local Telephone Competition Decision to RCA. http://
www.sec.gov/Archives/edgar/containers/fix031/808461/000080846103000034/
gci8k121203exhibit.txt.

Alaska Universal Service Administrative Company. "AUSF Annual Summary: Period
ending December 31, 2013.

"Alaska's Satellite Station: Summer Construction Eyed." *Anchorage Daily News*, May 10,
1968.

Alexander, Paul "A Tale of St. Lazaria Island." (wartime communications for Sitka region).
September 27, 2000.

Allan, Daniel S. "Economic Feasibility of Upgrading Telephone Service in Alaskan
Villages." In *Telecommunications in Rural Alaska*, Robert Walp, ed. Honolulu: Pacific
Telecommunications Council, 1982, pp. 107-116.

American Recovery and Reinvestment Act of 2009 (ARRA). Pub. L. No. 111-5,123 Stat.
115, 516, Feb. 19, 2009.

"Anchorage OKs Phone Utility Sale." *Juneau Empire*, April 22, 1998.

Anchorage Telephone Utility. "Strategic Management Plan." Anchorage: July 30, 1992.

"Anchorage Trust Fund Shrinks in Downturn." *Juneau Empire*, December 28, 2008.

Andrews, John M. Commissioner, Department of Administration. "RATNET equipment funding." Memo to Garrey Peska, Chief of Staff, Office of the Governor, June 13, 1989.

Andrews, John M. Commissioner, Department of Administration. "RATNET" Memo, August 29, 1989.

Annual Report. Geophysical Institute, University of Alaska Fairbanks, 1971-72.

Annual Report. Geophysical Institute, University of Alaska Fairbanks, 1972-73.

Annual Report. Geophysical Institute, University of Alaska Fairbanks, 1973-74.

Anthropos. "Evaluation of the Impact of Mini TV Stations upon Three Remote Communities in Alaska." Report to the Corporation for Public Broadcasting, Anchorage, 1974.

Archibald, Janet. "ACS Commander says 'Offer me $35 Million.'" *Anchorage Daily News*, June 10, 1966.

Armed Forces Radio and Television Services Timeline. Available at http://afrts.dodmedia .osd.mil/heritage/heritage.asp

AT&T. "AT&T's Petition to Terminate the Joint Service Arrangement. Summary" Before the Federal Communications Commission. Washington, DC: FCC CC Docket No. 83-1376, March 31, 1993.

"AT&T Invests More than $650 Million in Alaska From 2009 Through 2011 to Improve Local Networks."Anchorage, Alaska, Feb. 6, 2012. See http://www.corp.att.com/alaska/ press/pr_20120213.html

"AT&T to Acquire Alascom for $365 Million." *The New York Times*, October 18, 1994.

Atlantic Monthly. "At Last Count: No Phone Homes." June 1993, p. 77.

"ATS." NASA Goddard Space Flight Center Mission Archives. http://www.nasa.gov/ centers/goddard/missions/ats.html

"August Hebert [sic]: Alaska TV Pioneer Prepares for the Satellite Age." *Alaska Business*, October 1982.

Baesler, Jeff. "On Top of the World." *Communicating The Northwest Information Community's Bimonthly Forum*, Vol. 2, No.2, Oct/Nov 1992, p. 3.

Baltes, Kathleen (Teleconference Coordinator, Legislative Affairs Agency, State of Alaska). "Telecommunications Systems Enhance Participatory Government or It Doesn't Have to End When You Leave the Voting Booth." In *Telecommunications in Rural Alaska*, Robert Walp, ed. Honolulu: Pacific Telecommunications Council, 1982, pp. 74-84.

Barry, Douglas K. "Distance Learning by Satellite (and other technologies): A Catalog of Available Services." Juneau: Alaska Department of Education, Office of Instructional Support, April 1998.

Bartlett, Senator E.L. Proposal to Dispose of Alaska Communication System. Congressional Record, Washington, DC, July 8, 1965.

Bartlett, Senator E.L. Disposal of Government-Owned Long-Lines Communication Facilities in Alaska. *Congressional Record*, Washington, DC, August 24, 1965.

Bartlett, Senator E.L. "Satellite Ground Station for Alaska." Press conference, Anchorage, March 30, 1968.

Batra, Vinod. "Modern Phone Service Comes to Alaska Bush." *Telephony*, March 3, 1980, pp. 20–23.

Batra, Vinod. "Local Exchange Growth in Bush Alaska." In *Telecommunications in Rural Alaska*, Robert Walp, ed. Honolulu: Pacific Telecommunications Council, 1982, pp. 36–40.

Battle, Lucius D., vice president for corporate relations, COMSAT. Letter to Senator Ted Stevens re COMSAT proposal to provide aeronautical communications by satellite, May 18, 1970.

Bauman, Margaret. "Cable Company Links Kodiak, Kenai to Fiber Optic System." *Alaska Journal of Commerce*, January 28, 2007.

Bennett, F. Lawrence. "Engineering Education via Telecommunications in Alaska." In Telecommunications Symposium. 33rd Alaska Science Conference, American Association for the Advancement of Science, Arctic Division, Fairbanks, September 18, 1982, pp. 101–107.

Blume, Allen D., and Stuart P. Browne. "Telematics: Alaska and the Information Frontier." In *Telecommunications in Rural Alaska*, Robert Walp, ed. Honolulu: Pacific Telecommunications Council, 1982, pp. 128–131.

Bookey, Dennis. "Making Waves: Radio Meets the Challenge." In Alaska Broadcasters Association. Alaska Day at the FCC. Presentations. September 26, 1996.

Boucher, H. A. "Red." "Alascom and GCI Controversy." Memorandum to Governor Steve Cowper, November 14, 1986.

Boucher, H. A. "Red." An Interim Report: House Special Committee on Telecommunications. January 13, 1986.

Boucher, H. A. "Red," Anchorage Assembly. Letter to Larry and Alice Pearson, September 25, on Pacific Telecom, Inc. letterhead, 1991.

Boucher, H. A "Red," and Larry Pearson. "Connecting to a Competitive Future: A Report on the Status and Prospects of ATU to Tom Fink, Mayor of Anchorage, February 1994.

Bouker, David. "30th Year Recognition of Telephone Operations Background Material of NTC Benefactors." Nushagak Telephone Cooperative website, http://www.nushtel.com/home-about-us.htm.

Bower, David. "The Most Northern CODEC." *Pixel-to-Pixel*. Austin, TX, VTEL User Group Association, Summer 1993.

Bramble, William J. "Technology and Education in Public Schools." In Telecommunications Symposium. 33rd Alaska Science Conference, American Association for the Advancement of Science, Arctic Division, Fairbanks, September 18, 1982, pp. 94–100.

Bramstedt, Al. " 'Strange Things Done. . .': Television through Teamwork." In Alaska Broadcasters Association. Alaska Day at the FCC. Presentations. September 26, 1996.

Brent, Stephen. "Why It's a Pain to Call Long-Distance." *Anchorage Daily News*, August 3, 1968.

Brown, Ronald H., Larry Irving, Arati Prabhakar, and Sally Katzen. *The National Information Infrastructure: Agenda for Action*. Washington, DC: U.S. Department of Commerce, September 15, 1993.

Burch, Dean, Chairman, FCC. Letter to Senator Ted Stevens re COMSAT, August 13, 1973.

Burke, Jill. "Free Television in Rural Alaska Will Go Dark Without $5.3 Million Upgrade." *Alaska Dispatch*, April 4, 2013.

"Bush Telephone Service." *Alaska Industry*, October 1977, pp. 27–28, 50.

Butler, Major General William O. Commendation for Robert J. Gleason, major, Air Corps, United States Army. Field Headquarters, Eleventh Air Force, Seattle, November 26, 1943.

"Canadian Satellite to Aid RCA Here." *Anchorage Daily Times*, April 20, 1973.

Carlsson, Sylvia. "Telemedicine in the Bush: A Glimpse into the New Century." Undated.

Carson, Wesley E. "New PTI owners ready to deliver quality service." *Juneau Empire*, April 16, 1999.

Cassirer, Henry R., and Harold E. Wigren. Implications of Satellite Communication for Education: August-September 1970. Paris: UNESCO, 1970.

Charyk, Joseph V., President, Communications Satellite Corporation. Letter to Howard A. Hawkins, President, RCA Alaska Communications, November 4, 1969.

Charyk, Joseph V., President, Communications Satellite Corporation. Letter to Keith H. Miller, Governor of Alaska, October 6, 1969.

Charyk, Joseph V., President, Communications Satellite Corporation. Letter to Honorable Dean Burch, Chairman, Federal Communications Commission, October 19, 1970.

Chassman, Mark. "Shrinking Space: Linking Alaskans via Satellite." In Alaska Broadcasters Association. Alaska Day at the FCC. Presentations. September 26, 1996.

Chesbrough, Geof. Memo re Alaska Communications to Walt Sutter. Office of Telecommunications Policy, Executive Office of the President, May 9, 1974.

Chlupach, Robin Ann. *Airwaves over Alaska: The Story of Broadcaster Augie Hiebert.* Issaquah, WA: Sammamish Press, 1992.

"CITC Fab Lab." http://citci.org/wp-content/uploads/2014/06/FABLAB_1-pager1406251 .pdf

Clarke, Jim. "Ulmer Offers Plan to Move Alaska into the Wired World." *Anchorage Daily News*, September 12, 1996.

Cloe, John Haile with Michael F. Monaghan (1984). *Top Cover for America the Air Force in Alaska 1920–1983.* Anchorage Chapter-Air Force Association; Missoula, MN: Pictorial Histories Pub Co.

Communications Conference, September 25, 1968. Transcript.

Communications Satellite Corporation. "COMSAT offers plan for early start on U.S. satellite demonstrations. Press release, Washington, DC, June 13, 1969.

Communications Working Group, Federal Field Committee for Development Planning in Alaska. "The Need for a Long-Range Communications Development Plan for Alaska." 1969.

"Complete Phase 1 for Bush Phones." *Anchorage Daily Times*, February 7, 1973, p. 2.

COMSAT. "In the matter of the Application of Communications Satellite Corporation for authority to construct four high capacity satellites to be used as part of a domestic communications satellite system to provide the use of satellites and associated services to AT&T." Before the Federal Communications Commission, October 19, 1970.

COMSAT. "Alaska/COMSAT cooperative demonstration to provide TV and voice communications via satellite to remote Alaskan areas." Press release, Washington, DC, April 16, 1972.

COMSAT. Preliminary Results of Alaskan Test/Demonstration Program. Attachment No. 1 to ICSC-62-33E W/12/72.

"Comsat to Tell Bidders on ACS Satellite Plans." *Anchorage Daily Times*, August 28, 1968.

Connect Alaska. (2011). "Broadband and Business: Leveraging Technology in Alaska to Stimulate Economic Growth." An Alaska Business White Paper. Washington, DC: Connected Nation. Accessible at: http://www.connectak.org/_documents/AK_BizWhitePaper_FINAL.pdf.

Connecting Our Communities. Video. Palmer, Alaska: Matanuska Telephone Association (MTA), 2003.

Cotton, Stephen E. "Alaska's 'Molly Hootch Case': High Schools and The Village Voice." *Educational Research Quarterly*, Volume Number 4, 1984.

Cowper, Steve, Governor. "Executive Order No. 066." January 19, 1987.

Dailey, D., A. Bryne, A. Powell, J. Karaganis, and J. Chung. (2010) *Broadband Adoption in Low-Income Communities*. Social Science Research Council. Report commissioned by the Federal Communications Commission as part of the National Broadband Plan.

Dart, Greg. "GCI Plans to end local telephone service monopoly. *Alaska Star*, October 21, 2005. http://classic.alaskastar.com/stories/102005/hom_20051020018.shtml.

"Dedication Is Tuesday for Bartlett Earth Station." *Anchorage Daily Times*, June 25, 1970.

Demmert, Jane P., and Jennifer L. Wilke. "The Learn/ALASKA Networks: Instructional Telecommunications in Alaska." In *Telecommunications in Rural Alaska*, Robert Walp, ed. Honolulu: Pacific Telecommunications Council, 1982, pp. 69–73.

Department of Administration. "Annual Report on Teleconferencing to Governor Bill Sheffield." Juneau, May 1985 (Quoted in Pearson and Barry.)

Department of Administration, Department of Telecommunications, The House Special Committee on Telecommunications, The Alaska Public Broadcasting Commission, The Office of Management and Budget Division of Policy. "A Report to the Alaska Legislature in Response to Intent Language Regarding Telecommunications in the FY 88 Operating Budget." Juneau, January 1988.

DeVries, Anne. "Description of Alaska Telecommunications Systems: Research Request." Juneau: Alaska State Legislature House of Representatives Research Agency, January 9, 1981.

Dinero, S. C., E. Mariotz, and P. Bhagat. (2007) "Bridging the Technology Gap: Building a Development Model for Rural Alaska." *Journal of Cultural Geography*, Vol. 24, Issue 2. Accessible at: http://www.tandfonline.com/doi/abs/10.1080/08873630709478211#preview.

Dischner, Molly. "Telecoms combine networks; GCI vs. KTUU; Verizon enters." *Alaska Journal of Commerce*, December 24, 2013.

Dizard, John W. "Gold Rush at the FCC." *Fortune*, July 12, 1982, pp. 102–108.

Dowling, Richard P. "The Alaska Small Earth Station Program: Mitigating Isolation through Technology." In *Telecommunications in Rural Alaska*, Robert Walp, ed. Honolulu: Pacific Telecommunications Council, 1982, pp. 31–35.

Duhse, R. J. "Grand Failure." *Alaska Rural Life*, November 1986, pp. 20–21.

Duncan, John Thomas. "Alaska Broadcasting 1922-77: An Examination of Government Influence." Ph.D. Dissertation, University of Oregon, 1982.

Duncan, Ronald A., Executive Vice President, GCI. Letter to Representative H.A. "Red" Boucher, Alaska State Legislature, April 11, 1985.

Duncan, Tom. News and Entertainment Radio Broadcasting in Alaska, 1922–1974. Presentation at 10th Annual Alaska Broadcasters' Association Convention, Fairbanks, June 8, 1974.

Dunham, Vernon K. Direct Testimony before the Alaska Public Utilities Commission. In the Matter of the Investitgation of the Complaint by Alascom, Inc., against General Communication, Inc, for providing Unauthorized Intrastate Telecommunications Services, U-86-99.

Economics and Statistics Administration, and National Telecommunications and Information Administration (NTIA). *Digital Nation: 21st Century America's Progress toward Universal Broadband Internet Access*. US Department of Commerce, February 2010.

Economics and Statistics Administration, and National Telecommunications and Information Administration. *Exploring the Digital Nation: Home Broadband Internet Adoption in the United States.* US Department of Commerce, November 2010.

Economics and Statistics Administration, and National Telecommunications and Information Administration. *Exploring the Digital Nation: Computer and Internet Use at Home.* US Department of Commerce, November 2011.

Eller, Don. Presentation at "The Evolution of Telecommunications in Alaska: From Bush Telephones to Broadband." ISER Symposium, University of Alaska Anchorage, June 13, 2011.

Federal Communications Commission. "COMSAT Okayed to build Alaska Satellite Earth Station; would link Alaska, Continental U.S., Hawaii, Japan, Pacific area." Nonbroadcast and General Action Report No. 3560, Public Notice, May 14, 1969.

Federal Communications Commission. "In the Matter of Adak Eagle Enterprises, LLC and WC Docket No. 10-90 and WT Docket No. 10-208. Order. Adopted July 15, 2013. Windy City Cellular, LLC Petitions for Waiver of Certain High-Cost Universal Service Rules."

Federal Communications Commission. "In the Matter of RCA ALASKA COMMUNICA-TIONS, INC." Application for Authority to operate channels of communication. . . . Application of RCA Communications, Inc. No docket number. September 25, 1969.

Federal Communications Commission. RCA Alaska Communications Inc: Memorandum Opinion and Order Instituting a Hearing. Docket No. 18823, FCC 70-304, Released March 25, 1970.

Federal Communications Commission. "RCA Ordered to Transfer Domestic Satellite Operation in Lower 48 States." Washington, DC, December 27, 1974.

Federal Communications Commission. Notice of Inquiry. "In the Matter of Rates and Services for the Provision of Communications by Authorized Common Carriers between the Contiguous States and Alaska, Hawaii, Puerto Rico, and the Virgin Islands." Washington, DC, released January 5, 1984.

Federal Communications Commission. "Action in Docket Case: Federal-State Joint Board completes Study of NTS Cost Issues and Adopts Recommendations concerning Subscriber Line Charges, Federal Lifeline Assistance, High Cost Assistance, and Pooling of Common Line Costs." March 12, 1987.

Federal Communications Commission. "Final Recommended Decision: Executive Summary, CC Docket No. 8-1376, October 29, 1993.

Federal Communications Commission. National Broadband Plan Workshop: Diversity and Civil Rights Issues in Broadband Deployment and Adoption, 2009.

Federal Communications Commission. *Connecting America: The National Broadband Plan.* Washington, DC, 2010.

Federal Communications Commission. Universal Service Fund and Intercarrier Compensation Transformation Order, November 2011. See https://apps.fcc.gov/edocs_public/attachmatch/FCC-11-161A1.pdf

Federal Communications Commission, Office of Native Affairs and Policy. "Annual Report 2012." Available at www.fcc.gov/native.

Federal Communications Commission. "In the Matters of Connect America Fund" WC Docket 10-90 and Universal Service Reform—Mobility Fund WT Docket No. 10-208, Petitions for Waiver of Windy City Cellular, LLC and Adak Eagle Enterprises, LLC. Order. Adopted January 10, 2014.

Federal Communications Commission. "Tribal Mobility Fund Phase I Auction—Winning Bids Sorted by State and Tribal Land." March 2014. Available at http://wireless.fcc.gov/auctions/default.htm?job=auction_summary&id=902.

Ferguson, S., J. Kokesh, C. Patricoski, P. Hofstetter, and N. Hogge. (2009) *Impact of Store-and-Forward Telehealth in Alaska: A Seven Year Retrospective; Impact and Experiences from Seven Years of Utilization within the AFHCAN System.* Alaska Native Tribal Health Consortium.

Ferguson, Stewart. "What Works: Outcomes Data from AFHCAN and ANTHC Telehealth: An 8 Year Retrospective." Anchorage: American Telemedicine Association, 2011 Mid-Year Meeting, September 19–21, 2011.

Ferguson, Stewart. "Next Steps: The AFHCAN Telehealth Program." Unpublished Presentation. Anchorage: Alaska Native Medical Center, 2012.

Finkler, Earl. "And It's a Lot Quicker Than Mushing Dog Teams." *Anchorage Daily News*, February 21, 1994, p. C3.

Fitzgerald, Doreen. "Telecommunications: Where Do We Go from Here?" *University of Alaska Magazine*, Vol. 1, No. 2, Spring 1983.

Foote, Dennis, Edwin Parker, and Heather Hudson. "Telemedicine in Alaska: The ATS-6 Satellite Biomedical Demonstration." Institute for Communication Research, Stanford University, February 1976.

Forbes, Norma. "Television's Effects on Rural Alaska. Social and Cognitive Effects of the Introduction of Television on Rural Alaskan Native Children." College of Human and Rural Development, University of Alaska Fairbanks, 1984.

Forbes, Norma, and Walter J. Lonner. "The Effects of Television Viewing on Cognitive Skills: Implications for Schooling." In Telecommunications Symposium. 33rd Alaska Science Conference, American Association for the Advancement of Science, Arctic Division, Fairbanks, September 18, 1982, pp. 43–52.

Fortuine, Robert. "Medicine by Satellite Telephone—A New Breakthrough." *Alaska Medicine* 18(5), 72–73, 1976.

Foster, David. "Alaska-Siberia 'Friendship Flight' Helps Thaw Icy Relations with Soviets" Associated Press, Jun. 13, 1988.

"From Tundra to Technology: Alaska Native Claims Settlement Act @ 40." Panelist comments and copies of *Tundra Times* articles compiled for ANCSA@40, Anchorage, 2011.

Gabel, Richard, Edward C. Hayden, and William H. Melody. "Planning for Telecommunication System Development in Alaska." Staff Paper, The Public Interest Case in Alaska Public Interest Utilities Commission Hearing on Docket U-69-24. Office of Telecommunications, U.S. Department of Commerce, October 2, 1970.

Garrick, R. "Elder examines Aboriginal Artifacts via Videoconference." Keewaytinook Okimakanak Research Institute (KORI) research article, first published online April 14, 2008. Accessible at: http://research.knet.ca/node/194

Gazaway, Deborah. "Alaska Cable Television." Division of Telecommunications, Department of Administration, Juneau, January 1987.

GCI & Alaska Communications form The Alaska Wireless Network, LLC, June 5, 2012. http://www.gci.com/awn/joint-media-release.

"GCI Milestones." Available at www.gci.com.

"GCI purchases United Companies." *Juneau Empire*, June 10, 2008.

"GCI Rolls Out Local Phone Service in Eagle River." http://www.prnewswire.com/news-releases/gci-rolls-out-local-phone-service-in-eagle-river-54152202.html

Gemmill, Henry. "White Alice: She Links Outposts in Alaska, Eventually May Girdle the Globe." *Wall Street Journal*, April 22, 1957.

General Communication, Inc. "Comments of GCI in Response to APUC Notice of Inquiry In the Matter of the Consideration of Conditions of Interstate Equal Access Involving Intrastate Telephone Services and Prevention of Arbitrage," August 25, 1986.

General Communication, Inc. "Comments of General Communication, Inc." Before the Federal Communications Commission, In the Matter of Decreased Regulation of Certain Basic Telecommunications Services, Mary 6, 1987.

General Communication, Inc. United States Securities and Exchange Commission Form 10-K, Annual Report for fiscal year ended December 31, 1994.

Goldschmidt, Douglas. 1979. Joint Goods, Public Goods and Telecommmunications: A Case Study of the Alaskan Telephone System." PhD Dissertation, University of Pennsylvania.

Goldschmidt, Douglas. "The Benefits of Satellite Telecommunications in Alaska." In *Telecommunications in Rural Alaska*, Robert Walp, ed. Honolulu: Pacific Telecommunications Council, 1982, pp. 54-57.

Gordon, Bill. "Selling a Utility: The Fairbanks, Alaska Story." *Water Engineering & Management*146.4 (Apr 1999): 20–23.

Gore, Albert. "Remarks at the National Press Club." Washington, DC, White House, Office of the Vice President, December 21, 1993.

Gravel, Mike. "Special Report on Communications for Alaska." Washington, DC. Office of Senator Mike Gravel, June 1970.

"Gravel Requesting Satellite 'Favors.'" *Anchorage Daily Times*, April 28, 1971.

Greely, John. "School Officials Seek $23.5 Million for Instructional TV Satellite." *Anchorage Daily News*, February 25, 1980.

Green, Dianne. *In Direct Touch with the Wide World: Telecommunications in the North 1865–1992*. Whitehorse: Northwestel, 1992.

Gruening, Clark. Transmittal letter for "Electronic Access to Government: A Report to Lieutenant Governor Fran Ulmer." January 12, 1995.

Gustafson, Pam. "Educators Most Hit by Learn Alaska's Loss." *Anchorage Daily Times*, April 30, 1986.

Guy, Julianna. "Formula for Fairness: Markets-based Fees." Presentation at Alaska Broadcasters Association Alaska Day at the FCC, September 26, 1996.

Hahn, S., and Lehman, L. (2006) "The Half-Million-Square-Miles Campus: University of Alaska Fairbanks Off-Campus Library Services." *Journal of Library and Information Services in Distance Learning* Vol. 2, Issue 3.

Hammond, Jay. S., Governor. "Executive Order No. 050." July 1, 1981.

Hardin, Kyle. "The Digital Divide: How Poor Internet Access pulls the Plug on Rural Alaska." Bristol Bay Native Association, August 22, 2014.

Hawkins, Scott, and Chaterine Palmer. "Anchorage: Future Software Capital?" *Alaska Journal of Commerce*, October 21, 1991, p. 23.

Hayden, Edgar C., Office of Telecommunications, U.S. Department of Commerce. "Planning for Telecommunication System Development for Alaska: Notes for a Meeting with Governor Keith Miller of Alaska, December 8, 1969.

Hayden, Edgar C. "Planning for Telecommunication System Development in Alaska." Office of Telecommunications, U.S. Department of Commerce, April 1971.

"Hickel Urges Satellite Halt." *Anchorage Daily Times*, June 28, 1974, p. 3.

Hiebert, A.G. "Anatomy of Alaska's Telecommunications Requirements." Prepared for Brookings Institution Seminar Four, December 14-17, 1969.

Hiebert, A.G., President, Northern Television Inc. Letter to Senator Ted Stevens re opportunities with NASA, July 12, 1971.

Hiebert, A.G. Alaska Educational Broadcasting Commission Telecommunications Advisory Committee, Anchorage, August 24, 1973.

Hiebert, Augie. "From Tubes to Chips." Presentation at Alaska Broadcasters Association Alaska Day at the FCC, September 26, 1996.

Hills, A. "Beaming E-Mail to the Russian Far East," chapter in M. Stone-Martin and L. Breeden, *51 Reasons: How We Use the Internet and What It Says about the Information Superhighway.* Lexington, MA: Farnet, Inc., 1994.

Hills, Alex. "Rural Telephony and the Economic Challenge Ahead." *IEEE Communications Magazine*, September 1981.

Hills, Alex. "The Changing Structure of Alaskan Telecommunications." In *Telecommunications in Rural Alaska*, Robert Walp, ed. Honolulu: Pacific Telecommunications Council, 1982, pp. 1–3.

Hills, Alex. "Alaska's Giant Satellite Network." *IEEE Spectrum*, July, 1983, pp. 50–55.

Hills, Alex. "Subzero Engineering." *IEEE Spectrum*, December 1986, pp. 52–56.

Hills, Alex. "Melting the Ice Curtain between Russia and Alaska." *Business Communications Review*, December 1993, pp. 26–29.

Hills, Alexander H. "Television and Telephone Service in Alaskan Villages: A Technological, Economic, and Organizational Analysis." Ph.D. Dissertation, Carnegie-Mellon University, 1979.

Hills, Alex and M. Granger Morgan. "Telecommunications In Alaskan Villages." *Science*, Vol. 211, 16 January, 1981, pp. 241–248.

Hilscher, Hilary. Memo to Senator Ted Stevens re trip to Alaska to attend communications meetings, September 4, 1974.

Hilscher, Hilary. Transcripts of interviews with Alaska telecommunications leaders and innovators: Robert Arnold, Alvin Bramstedt, Jr., John Dudley, Alex Hills, Ed Istvan, Tom Jensen, Susan Knowles, William Miller, Walter Parker, Theda Pittman, Glenn Stanley, Robert Walp, Lee Wareham, and Marvin Weatherly. Anchorage: University of Alaska Anchorage Archives. Recorded 2000 and 2001.

"History of AFRTS: The First 50 Years." Armed Forces Radio and Television Services, 1992. Available at http://afrts.dodmedia.osd.mil/heritage/page.asp?pg=50-years.

Hofstetter, Philip J., John Kokesh, A. Stewart Ferguson, and Linda J. Hood. "The Impact of Telehealth on Wait Time for ENT Specialty Care." *Telemedicine and E-Health*. Vol. 16, No. 5. June 2010, pp. 551–556.

Holst, Svend. "PTI sale approved." *The Juneau Empire*, April 26, 1999.

Hopkins, Kyle. "Rosetta Stone Produces Inupiaq Language Software." *Juneau Empire*, January 23, 2011.

Horrigan, J. B. *Broadband Adoption and Use in America.* The Federal Communications Commission Omnibus Broadband Initiative (OBI) Working Paper Series No. 1, 2009.

House Concurrent Resolution No. 60 in the Legislature of the State of Alaska Ninth Legislature, First Session, Relating to the Telecommunications Policy of the State. April 30, 1975.

House Finance Committee. "A Bill for an Act entitled "An Act establishing the Alaska education technology program; and providing for an effective date." Alaska State Legislature, Seventeenth Legislature, Second Session, 1992.

Hove, Henry. "Making Advances: Easing into the Digital World." Presentation at Alaska Broadcasters Association Alaska Day at the FCC, September 26, 1996.

Huber, Louis R. "Urgent Call from Ugashik." *The Alaska Sportsman*, July 1944, pp. 10–11, 32–34.

Hudson, Heather E. "After Broadband in Southwest Alaska." *Institute of Social and Economic Research* (ISER), University of Alaska Anchorage, forthcoming 2015.

Hudson, Heather E. "Alaska Rural Telecommunications Milestones." Prepared for the "The Evolution of Telecommunications in Alaska: From Bush Telephones to Broadband." ISER Symposium, University of Alaska Anchorage, June 13, 2011.

Hudson, Heather E. *Communication Satellites: Their Development and Impact*. New York: Free Press, 1990.

Hudson, Heather E. "Connectivity across the Arctic: A Comparative Analysis of Broadband Access in Alaska, Northern Canada, and Greenland." Paper presented at the International Congress of Arctic Social Sciences, Akureyri, Iceland, June 22nd–26th, 2011.

Hudson, Heather E. "Digital Diversity: Broadband and Indigenous Populations in Alaska" *Journal of Information Policy*, Vol. 1, 2011, pp. 378–393.

Hudson, Heather E. "Electronic Trails across the North: New Directions for Alaska." Keynote speech, "Visions of Alaska's Future." Statewide Telecommunications Forum. Juneau, March 29–30, 1993.

Hudson, Heather E. *From Rural Village to Global Village: Telecommunications for Development in the Information Age*. New York: Routledge, 2006.

Hudson, Heather E. "Medical Communication in Rural Alaska." In *Telecommunications in Rural Alaska*, Robert Walp, ed. Honolulu: Pacific Telecommunications Council, 1982, pp. 58–63.

Hudson, Heather E. "Rural Broadband: Strategies and Lessons from North America." *Intermedia*, Vol. 39, Issue 2, May 2011, pp. 12–19.

Hudson, Heather E. "The Future of the E-Rate: U.S. Universal Service Fund Support for Public Access and Social Services." In . . . *and Communications for All: An Agenda for a New Administration*, A. Schejter (ed.). Lanham, MD: Lexington Books, 2009.

Hudson, Heather E. "Rural Telemedicine: Lessons from Alaska for Developing Regions." *Journal of eHealth Technology and Application*, Vol. 5, No. 3, September 2007, pp. 460–467.

Hudson, Heather E., ed. *New Directions in Satellite Communications: Challenges for North and South*. Dedham, MA: Artech House, 1985.

Hudson, Heather E. "Telemedicine in Alaska: From ATS-1 to AFHCAN." Presentation at "The Evolution of Telecommunications in Alaska: From Bush Telephones to Broadband." ISER Symposium, University of Alaska Anchorage, June 13, 2011.

Hudson, Heather E., et al. "Toward Universal Broadband in Rural Alaska: A Report to the State Broadband Task Force." Institute of Social and Economic Research, University of Alaska Anchorage, November 2012. Available at http://www.iser.uaa.alaska.edu/Publications/2012_11-TERRA.pdf.

Hudson, Heather E., and Edwin B. Parker. "Medical Communication in Alaska by Satellite." *New England Journal of Medicine*, Vol. 289, Issue 25, December 20, 1973, pp. 1351-1356.

Hudson, Heather E., and Edwin B. Parker. "Telecommunication Planning for Rural Development." *IEEE Transactions on Communications*, Fall 1975.

Hudson, Heather E., and Theda S. Pittman. "From Northern Village to Global Village: Rural Communications in Alaska." *Pacific Telecommunications Review*, Vol. 21, No. 2, 1999.

Hughes, J. Letter from GCI to GCI customers, Anchorage, November 25, 1986.

Humphries, Harrison. "ACS Sale Chances Good." *Anchorage Daily Times*, June 1, 1966.

Humphries, Harrison. "Lower Rates Urged as Key to ACS Sale." *Anchorage Daily Times*, May 31, 1966.

Hundt, Reed E. Statement of the Honorable Reed E. Hundt, Chairman. Washington, DC, Federal Communications Commission, November 1, 1994.

Imaituk Inc. (2011) "A Matter of Survival: Arctic Communications Infrastructure in the 21st Century." Arctic Communications Infrastructure Assessment (ACIA) Report. Prepared for the Northern Communications and Information Systems Working Group, April 30, 2011. Canadian Northern Economic Development Agency (CanNor).

Institute of Social and Economic Research. "Going Private: The 1968 Sale of the Alaska Communication System." Research Summary 59, University of Alaska Anchorage, December 1997.

Institute of Social and Economic Research. (2011). "The Evolution of Telecommunications in Alaska: From Bush Telephones to Broadband." Symposium organized by the Institute of Social and Economic Research (ISER), University of Alaska Anchorage, June 13, 2011. PowerPoint presentation slides and a video of the proceedings.

Instructional Television Program Schedule, January 5 to March 13, 1981.

Isenson, Ed. "Experts see Better Service in ACS Sale." *Anchorage Daily News*, June 2, 1966.

Isenson, Ed. "Many Consider ACS Service Inadequate; Solutions Sought." *Anchorage Daily News*, May 29, 1966.

Isenson, Ed. "Phone, Telegraph Service Here Below Level in Other States." *Anchorage Daily News*, June 1, 1966.

Isenson, Ed. "Two Major Choices on Disposal of ACS Faces [sic] Congress." *Anchorage Daily News*, May 31, 1966.

James, Beverly, and Patrick Daly. "Origination of State-Supported Entertainment Television in Rural Alaska." *Journal of Broadcasting and Electronic Media*, Vol. 31, No. 2, Spring 1987, pp. 169–180.

Jenne, Theron L. "Communications in Alaska." Unpublished research paper, University of Oklahoma, October 1977.

Jenne, Theron L., and Harry R. Mitchell. "Military Long Lines Communications in Alaska 1900-1976." In *Telecommunications in Rural Alaska*, Robert Walp, ed. Honolulu: Pacific Telecommunications Council, 1982, pp. 12–19.

Joint Resolution No. A/F-1-69. "A Resolution of the Cities of Anchorage and Faibanks Endorsing a Review of the Communications Services for Alaska." Fairbanks, February 21, 1969.

Jones, Douglas N. "What We Thought We Were Doing in Alaska, 1965–1972." *The Journal of Policy History*, Vol. 22, No. 2, 2010.

Jones, Douglas N., and Bradford H. Tuck. "Privatization Of State-Owned Utility Enterprises: The Alaska Case Revisited Thirty Years Later." Occasional Paper 21, National Regulatory Research Institute, The Ohio State University, November 1997.

Jones, Douglas N., and Bradford H. Tuck. "Privatization of State-Owned Utility Enterprises: The United States has done it too." *Critical Issues in Cross-National Public Administration*, ed. Stuart S. Nagel. Westport, CN: Quorum Books, 2000.

Joyce, Stephanie. "Adak's Phone Company Calls on Feds to Restore Subsidies." Alaska Public Radio Network (APRN) broadcast July 30, 2013. http://www.alaskapublic.org/2013/07/30/adaks-phone-company-calls-on-feds-to-restore-subsidies/

Kane, Karen. "Preserving the Past while Preparing for the Future." *Offices That Work*. Wang Laboratories: Lowell, MA, January 1993.

Katz, John A. "Outline of Testimony for State Telecommunications Hearing." Washington, DC, Office of the Governor, August 24, 1993.

Katz, Raul. *The Impact of Broadband on the Economy: Research to Date and Policy Issues*. An International Telecommunication Union (ITU) Broadband Report, April 2012.

Kennan, George. *Tent Life in Siberia: and adventures among the Koraks and other tribes in Kamtchatka and northern Asia*. New York: G.P. Putnam and Sons, 1870.

Kitka, Julie, President, Anchorage Federation of Natives. "Testimony for State Telecommunications Hearing," August 27, 1993.

Kleeshulte, Chuck. "GCI Calling: Firm Seeks to Tap Phone Market." *Juneau Empire*, March 29, 1985, p. 15.

Kleeshulte, Chuck. "Telephone Rate Hikes on Line?" *Juneau Empire*, March 29, 1985, p. 15.

Kleinfeld, J., and Bloom, J. "A Long Way from Home: Effects of Public High Schools on Village Children Away From Home." Fairbanks: Center of Northern Educational Research and Institute of Social, Economic and Government Research, University of Alaska, 1973.

Knowles, Susan. Testimony for State Telecommunications Hearing. Regulatory timeline, undated.

Kokesh, John, A. Stewart Ferguson, and Chris Patricoski. "Telehealth in Alaska: Delivery of Health Care Services from a Specialist's Perspective." *International Journal of Circumpolar Health* 63:4 2004, pp. 387–400.

Kokesh, John, A. Stewart Ferguson, and Chris Patricoski, "Preoperative planning for ear surgery using store-and-forward telemedicine." *Otolaryngology–Head and Neck Surgery* (2010) 143, pp. 253–257.

Kokesh, John, Chris Patricoski, and A. Stewart Ferguson. "Telemedicine for Otolaryngology." Johnson, Jonas T., and Clark A. Rosen, eds. *Bailey's Head and Neck Surgery: Otolaryngology, Fifth Edition*. Kluwers Health: Alphen aan den Rijn, Netherlands, 2013.

Kreimer, Oswaldo et al. "Health Care and Satellite Radio Communication in Village Alaska: Final Report of the ATS-1 Satellite Biomedical Experiment Evaluation." Stanford University: Institute for Communication Research, June 1974.

Kreinheder, Jack. "Synopsis of Telecommunications Reports." House Research Agency, Alaska State Legislature, February 23, 1981.

"KUAC-TV Power to Full 50,000 Watts." *Fairbanks News-Miner*, December 23, 1972.

Laschever, Eric. "Telecommunications Programs in Alaska." Memorandum to Representative-Elect Marco Pignalberi. Alaska State Legislature, House of Representatives Research Agency, December 31, 1984.

"Learn/Alaska Network." Juneau, Spring 1982.

Leslie, Tim. "'Doctor Call' Satellite Severs its Usefulness." *Anchorage Times*, April 29, 1985.

Liston, Donn. "Native Leader Raps RCA on Bush Satellite Proposal." *Anchorage Daily News*, 1974.

"Living Color in the Bush: Television Debuts in Bethel—and It's a Sellout." *Anchorage Daily News*, September 19, 1972.

Longenbaugh, Betsy. "Budget Cuts Forcing LearnAlaska off TV." *Juneau Empire*, June 18, 1986.

"Lower Rates Said Possible." *Anchorage Daily Times*, June 13, 1966.

Madigan, Robert J., and W. Jack Peterson. "Television and Social Change on the Bering Strait." University of Alaska Anchorage, April 1974.

Mansell, Robin. "The History of Telecommunications Subsidies in Northwest Canada and Alaska: A Comparative Analysis." In *Telecommunications in Rural Alaska*, Robert Walp, ed. Honolulu: Pacific Telecommunications Council, 1982, pp. 117–121.

Maracle, Brian. "Broadcasting as a Tool for Social Change." In "Proceedings of the Chugach Conference: Finding Our Way in the Communication Age." Anchorage, October 3–5, 1991, pp. 44–52.

Maxwell, Lauren. "New broadcast center dedicated to Augie Hiebert." See http://www.ktva .com/ktva-celebrates-60-years-on-the-air/.

McAllister, Bill. "A Hold on Local Phone Competition: Court Action Allows ACS to Fend off GCI's Entry." *Juneau Empire*, February 4, 2001.

McClanahan, A. J. "Satellite beams Native Convention to the Bush." *Anchorage Daily Times*, October 21, 1982.

McConnell, Douglas G. "The Cultural Impact of Television in Alaska." In *Telecommunications in Rural Alaska*, Robert Walp, ed. Honolulu: Pacific Telecommunications Council, 1982, pp. 93–95.

McCormack, James. Letter to Robert W. Sarnoff, President, Radio Corporation of America, June 26, 1969.

McCormack, James. Letter to Senator Ted Stevens re Governor's Task Force on Satellite Communications for Alaska, July 18, 1969.

McDowell Group. (2009). *Alaska Geographic Differential Study*. Prepared for the Alaska State Department of Administration.

McIntyre, Theodore W. "The Alaska Television Demonstration Project: Past, Present and . . . Future?" In *Telecommunications in Rural Alaska*, Robert Walp, ed. Honolulu: Pacific Telecommunications Council, 1982, pp. 89–92.

McKinney, Virginia. Interior Telephone leading in Race to serve Bush Communities." *Alaska Industry*, October 1977, pp 28-30.

McMahon, Rob. "The Institutional Development of Indigenous Broadband Infrastructure in Canada and the United States: Two Paths to 'Digital Self-Determination.'" *Canadian Journal of Communication*. Vol. 36, Issue 1, January 2011, pp. 115–140.

Melody, William H. "Telecommunications in Alaska: Economics and Public Policy." Report to the Alaska State Legislature, April 1978.

Melody, William H. "Telecommunications Networks: Internal Subsidies and System Extension in Alaska." In *Telecommunications in Rural Alaska*, Robert Walp, ed. Honolulu: Pacific Telecommunications Council, 1982, pp. 122–127.

Melody, William H. "Economic and Policy Issues in Establishing a Telecommunications Network in Alaska." Presentation at "The Evolution of Telecommunications in Alaska: From Bush Telephones to Broadband." ISER Symposium, University of Alaska Anchorage, June 13, 2011.

Merritt, Robert P. "Alaska Telecommunications." In *Telecommunications in Rural Alaska*, Robert Walp, ed. Honolulu: Pacific Telecommunications Council, 1982, pp. 5–10.

Metty, Michael P. "Policy Implications and Constraints of Educational Telecommunications in a SubArctic Region." In Telecommunications Symposium. 33rd Alaska Science Conference, American Association for the Advancement of Science, Arctic Division, Fairbanks, September 18, 1982, pp. 77–85.

Millsap, Pam. "Hickel Raps Egan and RCA." *Anchorage Daily News*, June 28, 1974.

Milroy, Cam. "The Tundra Telephone." Arctic Life and Traditions. http://arcticlifeand traditions.blogspot.com/2010/02/tundra-telephone.html

Mitchell, William L., U.S. Army Air Corps. "The Opening of Alaska." Missoula, MT: Pictorial Histories Publishing Company and Anchorage: Alaska Historical Society, April 1982.

Mittauer, Richard. "Alaska Rescue via ATS-1." Press release. National Aeronautics and Space Administration, October 7, 1971.

"More Communities for ETV System?" *Anchorage Daily News*, December 1969, p. 3.

Moyers, Al. "Sale of ACS celebrates 25th anniversary." January 13, 1996. http://public.afca .scott.af.mil/96jan/96jan13.htm

"MTA takeover bid fails." *Peninsula Clarion*, September 20, 2000.

National Aeronautics and Space Administration. "Alaska—NASA Radio/TV." NASA News, February 3, 1970.

National Aeronautics and Space Administration. "ATS aides in Alaskan Emergencies." NASA News. Washington, DC, April 5, 1972.

National Education Association. "Satellite Seminar Launched for Teachers in Remote Alaskan Villages." Washington, DC, January 24, 1973.

Necrason, Brigadier General C.F., USAF. "With the Lady Known as White Alice. . ." *ITT World Wide Service Reporter* 1(2): 3,14, 1959.

Neering, Rosemary. *Continental Dash*. Ganges, BC: Horsdal and Schubart, 1989.

"New Cable Pact reduces Rate on Alaska Calls." *Anchorage Daily Times*, August 22, 1966.

"New Long Distance Rates." *Anchorage Daily Times*, August 15, 1968.

"News Release: State of Alaska: Office of the Governor: William A. Egan, Governor." Juneau, October 11, 1972.

"Night Phone Tolls Drop Oct. 1." *Anchorage Daily Times*, August 8, 1968.

"No Asterisks, Please." Editorial, *Anchorage Daily Times*, February 6, 1973, p. 6.

Norris, Frank. "Keeping Time in Alaska: National Directives, Local Responses." *Alaska History*. Volume 16, Numbers 1 & 2, Spring/Fall 2001.

Northern Television Symposium. Report of Proceedings. Yellowknife, January 15-16, 1987.

Northrip, Charles M. "Developments in Alaska." Alaska Educational Broadcasting Newsletter, Volume III, Number 2, Fairbanks, February 1970.

Northrip, Charles M. "Developments in Alaska." Alaska Educational Broadcasting Newsletter, Volume III, Number 5, Fairbanks, May-June, 1970.

Nussbaum, Paul. "State Drops Efforts to own Earth Stations." *Anchorage Daily News*, February 29, 1980, p. A-1.

Obazuaye, Sunday P. "Using Telecommunications for Distance Education: The LearnAlaska Case." Unpublished MA Thesis, University of Alaska Anchorage, 1989.

Office of Information, Air Force Communications Service. "The Alaska Communication System Story." Richards-Gebaur Air Force Base, Missouri, undated.

Office of Telecommunications, Office of the Governor. "Satellite Television Demonstration Project: Final Report." Juneau: State of Alaska, February 1, 1978.

Office of Telecommunications. "Jurisdictional Cost Separations and Toll Rate Disparities in Alaska." Office of the Governor, Juneau, October 1978.

Office of the Governor. Press Release. May 1, 1975. Re agreement with RCA for installation of satellite stations in Alaska.

Page, G. A., and M. Hill. (2008) "Information, Communication, and Educational Technologies in Rural Alaska." *New Directions for Adult and Continuing Education*, Vol. 2008, Issue 117, Spring 2008, pp. 59–70.

Park, Bernard. Letter to Governor William A. Egan re Educational Television Committee, January 21, 1963.

Parker, Edwin B. "From NASA Experiment to Operational Service: Alaska Telecom Transitions 1970-1975." Presentation at "The Evolution of Telecommunications in Alaska: From Bush Telephones to Broadband." ISER Symposium, University of Alaska Anchorage, June 13, 2011.

Parker, Edwin B., and Heather E. Hudson. *Electronic Byways: State Policies for Rural Development Through Telecommunications*. Boulder, CO: Westview, 1992. Second edition: Aspen Institute, 1995.

Parker, Edwin B., Heather E. Hudson, Don A. Dillman, and Andrew D. Roscoe. *Rural America in the Information Age: Telecommunications Policy for Rural Development*. Washington, D.C.: University Press of America, 1989.

Parker, Edwin B., and Bruce Lusignan. Letter to Gordon Zerbetz, Chairman, Alaska Public Utilities Commission, re RCA's plans for serving Alaska, June 18, 1974.

Parker, Walter B. "An Inventory of Alaska's Communication Systems." State of Alaska, Office of the Governor, Division of Planning and Research, January 1974.

Parker, Walter B. "Village Satellite III: The Third Evaluation of the Action Study of Educational Uses of Satellite Communications in Remote Alaskan Communities." 1974.

Parker, Walter B. "The Evolution of the Present Alaska Satellite System." In *Telecommunications in Rural Alaska*, Robert Walp, ed. Honolulu: Pacific Telecommunications Council, 1982, pp. 20–21.

Parker, Walter B. "The Future of Universal Telephone Service in Alaska." A Policy Paper for the Alaska Senate Telecommunications Project. Anchorage, February 5, 1985.

Parker, Walter B. "Telecommunications and Information System History of Alaska." Institute of the North, February 2008.

Patricoski, Chris. "Alaska Telemedicine: Growth through Collaboration." *International Journal of Circumpolar Health*, 63:4 2004.

Pearce, Frederick W. "Presentation to the 17th Annual Conference of the Caribbean Association of National Telecommunications Organizations (CANTO)." Puerto Rico, May 2001.

Pearson, Larry. "The Telecommunications Information Council: Legislative History and Implementation Plan." House Special Committee on Telecommunications, September 1987.

Pearson, Larry, and Doug Barry. "Talking to Each Other, Talking to Machines: Alaska's Telecommunications Future." Report to the Joint Committee on Telecommunications of the Alaska State Legislature, January 1987.

Peck, Major Brad. "Cooperative Communications: The Military Mission." In Alaska Broadcasters Association. Alaska Day at the FCC. Presentations. September 26, 1996.

Perdue, Karen, et al. "Evolution and Summative Evaluation of the Alaska Federal Health Care Access Network Telemedicine Project." University of Alaska Statewide Health

Programs and University of Alaska Anchorage Center for Human Development, November 2004.

Pittman, Theda Sue, Heather E. Hudson, and Edwin B. Parker. "Planning Assistance for the Development of Rural Broadcasting." Prepared for the Alaska Public Broadcasting Commission, Anchorage, July 1977.

Pittman, Theda Sue, and James M. Orvik. "ATS-6 and State Communications Policy for Rural Alaska: An Analysis of Recommendations." Center for Northern Educational Research, University of Alaska Fairbanks, 1976.

Probst, Cheryl. "Says RCA is Pressed in Sale: Private Opinions accuse Air Force in White Alice Deal." *Anchorage Daily Times*, March 9, 1973.

"Proceedings of the Chugach Conference: Communication Issues of the '90s." Anchorage, October 5–6, 1990.

"Proceedings of the Chugach Conference: Discussing the Future of Communications in Alaska." University of Alaska Anchorage, August 18–19, 1989.

"Proceedings of the Chugach Conference: Finding our way in the Communication Age." Anchorage, October 3–5, 1991.

Public Law 90-13. An Act to authorize the disposal of the government-owned long-lines communications facilities in the State of Alaska, and for other purposes. 90th Congress, S.223, November 14, 1967.

Quirk, William A. III. "Historical Aspects of Building of the Washington, D.C.-Alaska Military Cable and Telegraph System, with Special Emphasis on the Eagle-Valdez and Goodpaster Telegraph Lines 1902-1903." U.S. Department of the Interior, Bureau of Land Management, May 1974.

Raab, Diane. "Earning College Credit by Satellite." *Juneau Empire*, February 8, 1993, p. 1.

Rainery, Richard. "State Sponsored Television in Alaska: Alternatives for Delivery and Distribution." Rural Research Agency, Juneau, August 1984.

Ransom, Jay Ellis. "Party Line Phone." *The Alaska Sportsman*, March 1946, pp. 10–11, 44–48.

Rate, Service Controls Favored in Sale of ACS." *Anchorage Daily News*, June 1, 1966.

RCA Alascom. Memorandum of Understanding re Special Satellite Facilities. n.d.

RCA Alaska Communications. "Executive Summary: Alaska Communications Plan 1972-1980." Anchorage: RCA Alaska Communications, Inc. Sept. 20, 1974.

RCA Alaska Communications. "RCA Alascom begins large-scale program to extend modern communications to isolated villages throughout Alaska." Press release, Anchorage, July 9, 1974.

RCA Alaska Communications. RCA Alascom President expresses concern over legislative action on bush communications. Press release, May 6, 1975.

"RCA brings the Telephone to Alaska's Bush." n.d. RCA NEWSLETTER

"RCA Could Lose Phone Authority." *Anchorage Daily Times*, April 18, 1972.

"RCA Plans New Satellite System." *Anchorage Daily News*, March 13, 1973.

"RCA Says Phone Bill Detrimental to State." *Anchorage Daily Times*, February 23, 1973.

"RCA Shaves Estimate for Bush Phone Job." *Anchorage Daily Times*, June 5, 1972.

Regulatory Commission of Alaska. Annual Report Fiscal Year 2000. Anchorage, January 2001.

Regulatory Commission of Alaska. "Commission." Available at http://rca.alaska.gov/ RCAWeb/AboutRCA/Commission.aspx

Reid, S. "Earth Stations Bring an End to White Alice." *Communications News*, June 1, 1985.

Reid, Sean. "Alaska Calling." *Alaska Magazine*, May 1982, pp. 36–41, 77.

Reid, Sean. "Can Anyone Figure out Alaska's Phone Wars?" *Alaska Business Monthly*, August 1993, pp. 14–19.

Report of the Alaskan Task Force: Seventh Report of the preparedness subcommittee of the Committee on Armed Services. United States Senate, Washington, DC, 1951.

Report of the Chief Signal Officer, War Department, Office of the Chief Signal Officer, Washington, DC, October 4, 1904.

Report of the Governor of the District of Alaska to the Secretary of the Interior, 1906. Washington, DC, Government Printing Office, 1906.

Rigert, Joe. "Communicating Was Problem in Hours after Earthquake." *Anchorage Daily Times*, April 9, 1964, p. 13.

Rinehart, Steve. "High Bid Spurned; ATU Goes to Citizens." *Anchorage Daily News*, July 24, 1991, p. A-1.

Rinehart, Steve. "Voters Spurn ATU Sale." *Anchorage Daily News*, October 2, 1991.

Rinehart, Steve, and Charles Wohlforth. "PTI Stalls Petition Decision." *Anchorage Daily News*, August 2, 1991.

Ringland, Natalie. AuroraNet: The Future Has Arrived!" *Quasagniq: A New Dawn. The North Slope Borough Newsletter*. Vol. VII, No. 9, October 1995.

Rostow, Eugene S. President's Task Force on Communications Policy. Final Report. Washington, DC, December 1968.

Samini-Moore, Douglas. "Unique Connections: Public Broadcasting in Alaska." In Alaska Broadcasters Association. Alaska Day at the FCC. Presentations. September 26, 1996.

"Satellite Era Begun by Alaska." *Anchorage Daily Times*, June 30, 1970.

"Satellite House Call." Film produced by Department of Communication, Stanford University. Available at http://www.youtube.com/watch?v=GzclgBfn_yY

"Satellite Television Demonstration." *The Nome Nugget*, June 2, 1972.

"Satellite Will See Use Here." *Anchorage Daily Times*, December 21, 1971.

Schaible, Grace Berg, Alaska Attorney General. "Alaska request that state pay space and power charges for LPTV minitransmitters." Memo to John Andrews, Deputy Commissioner, Department of Administration, May 12, 1987.

Schatz, Carol. "Moving Mountains: Television for Alaskans." In Alaska Broadcasters Association. Alaska Day at the FCC. Presentations. September 26, 1996.

Schindler, J. F. and D. A. Underwood. "Communications on the National Petroleum Reserve—Alaska (NPRA)." In *Telecommunications in Rural Alaska*, Robert Walp, ed. Honolulu: Pacific Telecommunications Council, 1982, pp. 49–53.

Schultz, Caroline. "The Growth of Telecommunications." *Alaska Economic Trends*, April 2014, pp. 14–19.

Secretary of Defense, Department of the Air Force. Notice of Acceptance (of sale of ACS to RCA), July 1, 1969.

Senate Concurrent Resolution no. 35 in the Legislature of the State of Alaska, Eleventh Legislature—First Session, "Directing the Legislative Council to conduct a feasibility study relating to educational television," April 23, 1979.

Shaginaw, George. "Alaska and Alascom: A Decade of Telecommunications Development." In *Telecommunications in Rural Alaska*, Robert Walp, ed. Honolulu: Pacific Telecommunications Council, 1982, pp. 26–30.

Sharrock, George O. Letter from George O. Sharrock, Federal Field Committee for Development Planning in Alaska to Honorable Myron Tribus, Assistant Secretary of Commerce for Science and Technology, September 4, 1969.

Shillito, Barry J., Assistant Secretary of Defense. Letter to Senator Ted Stevens re sale of ACS, April 17, 1969.

Smith, Carlton. "Model Alaskan School District Relies on PC/VS WAN." *WANG Dialogue*, Vol. 1, No. 21, December 15, 1992.

Southwest Alaska Municipal Conference (SWAMC). (2010) Southwest *Alaska Broadband Strategy: "Planning for Opportunity."* Draft published October 6, 2010.

Standage, Tom. *The Victorian Internet.* New York: Walker and Company, 1998.

Stanford University Medical Center. Press release on ATS-1 telemedicine evaluation, February 19, 1974.

Stange, Jay. "PTI Plan Infuriates Citizens." *Anchorage Daily News*, August 2, 1991.

Stanley, Glenn M. "Radio Communications in Alaska and other Remote Areas with Special Reference to Satellite Telecommunications." University of Alaska Geophysical Institute Annual Report, 1971-72. Fairbanks, 1972, pp. 119–127.

Stanley, Glenn M. Interim Report to Representative Fred Brown, Chairman, House Commerce Committee, January 13, 1980.

Stapler, Dale C. "The Economic Development Potential of Satellite Delivery Education." *Economic Development Review*, Winter 1990, pp. 30–37.

State of Alaska Department of Education and Early Development. "State to Award Digital Teaching Initiative Grants." Juneau, July 2, 2014.

State of Alaska Telecommunications and Information Technology Plan Passed by the Telecommunications Information Council, Lt. Governor Fran Ulmer, Chair. December 18, 1996.

State of Alaska, House of Representatives, Telecommunications Task Force. Public Hearing "Agenda and State of Alaska Concerns." Anchorage Legislative Information Office, August 27, 1993.

State of Alaska, Office of Management and Budget. (2009) *Response to NTIA, USDA RUS Joint Request for Information*. April 9, 2009.

State of Alaska, Office of the Lieutenant Governor. "Lieutenant Governor Eyes the Future with Telecommunications Plan." Juneau, September 11, 1996.

"State Plans Earth Station Bids." *Alaska Business News Letter*, Vol. II, No. 12, February 28, 1975.

"State Support for Satellite?" *Anchorage Daily News*, March 15, 1968.

Stevens, Senator Ted. Letter to George M. Sullivan, Mayor of Anchorage, March 6, 1969.

Stevens, Ted, Senator. Letter to George Sullivan, mayor of Anchorage, re resolution by cities of Anchorage and Fairbanks endorsing proposed sale of ACS and COMSAT license request for an earth station at Talkeetna, March 6, 1969.

Stevens, Senator Ted. "Modern Communications for Alaska." Congressional Record, February 28, 1969.

Stevens, Ted. Statement of Senator Ted Stevens. In the Matter of and in re Applications of Communications Satellite Corporation For construction permits for three new stations in the Domestic Public Point-to-Point Microwave Radio Service at Talkeetna, Scotty Lake, and Twelvemile, Alaska et al. To the Review Board, Before the Federal Communications Commission, December 19, 1969.

Stevens, Ted. United States Senator for Alaska. Press release re satellite frequencies for educational television, November 30, 1970.

Stevens, Ted. Testimony of Ted Stevens, United States Senator, before the Interim Committee on the Problems of the Unorganized Borough, State Senator George Hohman, Chairman, October 28, 1974. Presented by Hilary Hilscher.

"Storytellers reach listeners through UA satellite project." Available at http://sourdough jim.com/KUAC/Scrapbook/1972/Pages/1972Pages14_jpg.html

Strait, Steve. "Public Broadcasting in Alaska: A Long-Range Plan," 1991 edition. Anchorage, Alaska Public Broadcasting Commission. Available at http://030c78c.netsolhost.com/ APBI/apbc/archives/Public%20Broadcasting%20in%20Alaska%20LRP%201991%20 Edition.pdf

"Students Master New Craft: TV Production." "School Happenings: North Slope School District." Quoted in *Alaska Daily News*, January 4, 1993.

"Summary of Events in "Domestic Satellite Service Proceeding." October 23, 1969. Re FCC NOI of March 2, 1966 on legal authority of FCC to authorize construction and operation of communication satellite facilities by private entities. Author unknown.

Supreme Court of Alaska. ACS of Alaska, Inc. v. Regulatory Commission of Alaska (12/12/2003) sp-5762.

Sutter, Walter, Assistant Director, Office of Telecommunications Policy, Executive Office of the President. Memorandum for the Record: Communication Services for Alaska, December 27, 1973.

Taft, Destyn E. "Telemedicine and Telehealth Services Going Online for Providence Health System Alaska." *Telemedicine in Alaska News*, June 26, 1998.

Telecommunications Act of 1996, Public Law 104-104, 110 Stat. 56 (1996).

Telecommunications Information Council, Fran Ulmer, Lieutenant Governor, chair. "State of Alaska Draft Telecommunications and Information Technology Plan." Juneau: September 11, 1996.

Telecommunications Information Council. "First Annual Report." Juneau, March 23, 1988.

Telegeography. "AlaskaComm, GCI announce wireless pact in a bid to fend off Verizon." June 8, 2012. http://www.telegeography.com/products/commsupdate/articles/2012/06/08/ alaskacomm-gci-announce-wireless-pact-in-a-bid-to-fend-off-verizon/

Telegeography. "Submarine Cable Landing Directory." See http://www.telegeography.com/ telecom-resources/submarine-cable-landing-directory/

"Television, Telephone Service Coming for State's Bush Areas." *Anchorage Daily News*, May 13, 1971.

"Tests to Prove Small Receiver," *Anchorage Daily News*, May 8, 1972.

Thadathil, George. " AFRTS: the First Sixty Years." Available at http://afrts.dodmedia.osd .mil/heritage/page.asp?pg=60-years

The Lieutenant Governor's Working Group on Electronic Access. "Electronic Access to Government: A Report to Lieutenant Governor Fran Ulmer." January 12, 1995.

"The Lister Hill Center's Experimental Satellite Communications Project." December 1970. (copy from office of Senator Mike Gravel)

"The Sale of ACS." Editorial, *Anchorage Daily Times*, October 3, 1967.

Tobeluk v. Reynolds, C.A. No. 72-2450 (formerly captioned Tobeluk v. Lind, originally filed as Hootch v. Alaska State-Operated School System), Alaska Super. Ct., 3rd Dist. (Anchorage).

Treadwell, Mead. "Proposal: Cellular Telephone Expansion Project." Tundra Telephone, unpublished memo, Anchorage, August 15, 1982.

Triplett, Duane L. "Commercial Television in Alaska." In *Telecommunications in Rural Alaska*, Robert Walp, ed. Honolulu: Pacific Telecommunications Council, 1982, pp. 101–104.

Tuckerman, J. G. "Long-Distance Communications in Alaska." Unpublished research paper, University of Alaska Anchorage, July 2, 1980.

U.S. Department of Agriculture. *Alaska Forum on Jobs and Economic Development*: Final Report. Palmer, Alaska: USDA Service Center, 2010.

U.S. Department of Agriculture. "Advancing Broadband: A Foundation for Strong Rural Communities." Broadband Initiatives Program: Awards Report. Washington, DC, 2011.

Uchitel, Robert N. "Cable Television in Alaska." In *Telecommunications in Rural Alaska*, Robert Walp, ed. Honolulu: Pacific Telecommunications Council, 1982, pp. 105–106.

Ulbrich, Jeffrey. "Real Northern Exposure: All-Native TV Network Unites Canada's Arctic." *Anchorage Daily News*, January 22, 1992, p. A8.

Universal Service Administrative Company. "2013 Annual Report." Washington, DC: 2014.

University of Alaska Geophysical Institute Annual Report, 1973-74. Fairbanks, pp. 82–83.

Valaskakis, Gail. "The Issue is Control: Northern Native Communications in Canada." "Proceedings of the Chugach Conference: Communication Issues of the '90s." Anchorage, October 5–6, 1990, pp. 15–20.

Vandelaar, Janis. "ATA's President Eller: A Personal Profile." *Tip'n'Ring*. Publication of the AlaskaTelephone Association, Vol. 5, No. 1, 1984.

Vandor, Marjorie L., Assistant Attorney General. Memorandum to The Honorable Bob Poe, Commissioner, Department of Administration. "Alaska Rural Communications Service (ARCS) Council: Programming Decisions." Juneau, February 8, 1999

"Visions of Alaska's Future." Statewide Telecommunications Forum. Juneau, March 29–30, 1993.

Von Braun, Wernher. "Lighting up the Earth's Dark Corners with—TV from the Sky." *Popular Science*, March 1975, pp. 71–73, 144, 145.

Walp, Robert M. "First-Hand: C-Band Story" IEEE Global History Network. Available at http://www.ieeeghn.org/wiki/index.php/First-Hand:C-Band_Story.

Walp, Robert M. "First-Hand: Ku-Band Story" IEEE Global History Network. Available at http://www.ieeeghn.org/wiki/index.php/First-Hand:Ku_Band_Story.

Walp, Robert M. "First-Hand: S-Band Story" IEEE Global History Network. Available at http://www.ieeeghn.org/wiki/index.php/First-Hand:S-Band_Story.

Walp, Robert M. "Several Paths leading to the use of small earth stations for satellite communications systems." Abstract. IEEE Conference on the History of Telecommunications, Memorial University, St. John's, Newfoundland, July 25-27, 2001.

Warf, B. (2011) "Contours, Contrasts, and Contradictions of the Arctic Internet." *Polar Geography*, Vol. 34, Issue 3, September 2011, pp. 193–208.

Western Electric Company. "The DEW Line Story," undated.

Western Electric Company. "White Alice," undated.

Whalen, David J. "Satellite Communications in Alaska: Commercial Satellite Options." Presentation at Institute of Social and Economic Research (ISER) symposium "The Evolution of Telecommunications in Alaska: From Bush Telephones to Broadband," University of Alaska Anchorage, June 13, 2011.

White Alice Fact Sheet, March 26, 1958.

Whitehead, Clay T. Letter to Ted Stevens re federal expenditures for communications in Alaska. Washington, DC. Office of Telecommunications Policy, Executive Office of the President, March 22, 1974.

Whymper, Frederick. Whymper's Journey in Alaska, or Russian America. "A Journey from Norton Sound, Bering Sea, to Fort Youkon (Junction of Porcupine and Youkon Rivers)." Paper read before the Royal Geographical Society of London, April 27, 1868.

Wilke, Jennifer. "Everything I Need to Know I Learned from Learn/Alaska." Presentation at Institute of Social and Economic Research (ISER) symposium "The Evolution of Telecommunications in Alaska: From Bush Telephones to Broadband," University of Alaska Anchorage, June 13, 2011.

Williams, Andy. "Anchorage Television Beamed Live to Juneau." *Anchorage Daily News*, May 8, 1972.

Wilson, Martha R., M.D. (Chief, Office of Program Development, Alaska Area Native Health Service). Memorandum re Discussion with RCA Officials, August 30, 1974.

Young, Patricia. "Rural Alaska Television Network—Programming and Cost." Alaska State Legislature, Legislative Research Agency, November 28, 1989.

Zahniser, Jim. "Jack Johnson '30: Alaska Radio Pioneer." *The Maine Alumnus*, Vol. 60, no. 4, Fall 1979.

Index

Dr. Heather E. Hudson is Professor of Communications Policy at the Institute of Social and Economic Research (ISER), University of Alaska Anchorage. She was previously Director of ISER and founding Director of the Telecommunications Management and Policy Program at the University of San Francisco. Her research focuses on applications of information and communication technologies for socio-economic development, regulatory issues, and policies and strategies to extend affordable access to communications, particularly in rural areas. She has planned and evaluated communication projects in Alaska and northern Canada and more than 50 developing countries and emerging economies. Dr. Hudson is the author of numerous articles and several books and is the recipient of two Fulbright awards and a Sloan Industry Fellowship. She has consulted for international organizations, government agencies, the private sector, and consumer and indigenous organizations, and has testified before regulators in the United States and Canada. She received an Honours B.A. from the University of British Columbia, M.A. and Ph.D. from Stanford University, and J.D. from the University of Texas at Austin.